ASTROPHYSICS OF THE SUN

Astrophysics of the Sun

HAROLD ZIRIN

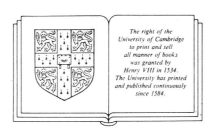

The right of the
University of Cambridge
to print and sell
all manner of books
was granted by
Henry VIII in 1534.
The University has printed
and published continuously
since 1584.

CAMBRIDGE UNIVERSITY PRESS

Cambridge

New York   New Rochelle

Melbourne   Sydney

Published by the Press Syndicate of the University of Cambridge
The Pitt Building, Trumpington Street, Cambridge CB2 1RP
32 East 57th Street, New York, NY 10022, USA
10 Stamford Road, Oakleigh, Melbourne 3166, Australia

First published 1988

Printed and bound in Great Britain at The Bath Press, Avon

British Library cataloguing in publication data

Zirin, Harold
    Astrophysics of the sun.
    1. Sun
    I. Title
    523.7   QB521

Library of Congress cataloging in publication data

Zirin, Harold.
    Astrophysics of the sun.
    Bibliography: p.
    Includes index.
    1. Sun.   2. Astrophysics.   I. Title.
QB521.Z57   1987   523.7   86-26367

ISBN 0 521 302684 hard covers
ISBN 0 521 316073 paperback

Contents

Preface

There is a star, a real star, that is only 140 million km from us, that we can study with small telescopes, even, and see the most wondrous things. It is a key to understanding the universe around us, to learning the meaning of the few photons or particles that we may pick up from more distant bodies. In this book we will talk about that star and see if we can understand how it works.

Readers of this book will quickly see that it is, or at least started out to be, a revision of my earlier text on the Sun, *The Solar Atmosphere*, published in 1966. Originally there was to be a simple revision for a reprint house; when that project fell through, Cambridge University Press agreed to publish a completely new version, and the second half of the book has been completely rewritten to reflect twenty years of progress in the field. Chapters 3 through 5, where the basic physical theory is outlined, have not changed much.

Looking back, I can see that it was a bit daring to write a textbook about the Sun in 1966 when my knowledge of it was somewhat limited. Writing the book helped me learn more. Building an observatory and observing the Sun taught me immensely more, and I have mostly written the book to share with colleagues and students what I have learned. I have had the privilege over these years of observing the Sun on an almost daily basis; I recommend at least a short stint of daily observation of the Sun to anyone who wishes to understand it.

Like the rest of astronomy, the field has seen great progress in recent years. The opening of the ultraviolet, X-ray and microwave regions was exploited early and effectively, revealing the marvelous high-energy aspects of solar activity. Unexpected phenomena such as γ-ray lines, coronal holes and solar oscillations have been revealed. Where we previously worried whether an ion of such high ionization potential as Ca XV (4 million deg) could exist in the corona, we now study line ratios in the spectrum of Fe XXV (30 million deg).

It is often remarked that the trouble with studying the Sun is that we know too much about it. That statement is foolish and untrue. If we knew so much about it how could so many totally unexpected phenomena have been discovered? The fact is that until recent years the extent of our observational knowledge was quite limited. In 1966 high resolution observations of more than a few frames were virtually unknown. The original solar observatories were under the clouds of Europe and the eastern United States. Hale's observatory on Mt. Wilson offered clear skies but inferior seeing, while the national observatories in New Mexico and Arizona were subject to a summer monsoon and the mediocre seeing of sun-baked mountaintops. The work of the non-establishment Lockheed observatory in California demonstrated the possibilities and benefits of continuous cinematography of the evolution of solar activity. The establishment of observatories in California and Hawaii at sites chosen specifically for solar work enabled us to learn the details of the evolution of solar activity and the construction of vacuum reflectors permitted the utilization of these sites. As astronomers learn about solar seeing patterns, similar observatories are beginning to appear elsewhere.

While we certainly don't know enough about the Sun, what we do know is far too much for a single person to cover definitively in a single textbook. Still, I felt it was worthwhile to put together a coherent (I hope) picture of the Sun as I see it, at least as a guide to research and the literature. I have tried to understand everything I was writing about and make it clear to the reader. Where I couldn't, I tried to present the ideas of others, or at least to summarize the observational data.

In addition to the bibliography, I have provided references for more detailed reading after each chapter. There is also an Appendix defining the many abbreviations used. Those familiar with computerized typesetting will see that I have set this book in TeX. I am indebted to Prof. Knuth for inventing this marvelous tool, although I did not always have that feeling as I dug through his manual.

While I have tried to discuss physical explanations, I have not pursued the theory in detail. Few solar problems are well enough understood, and long series of equations will simply bore the reader. If you need theoretical details, consult the references, where the material is set forth by someone who understands it better. Similarly, while the advent of big computers has made possible massive theoretical modeling, little has been added to our understanding by these calculations and I do not pursue them in detail. The input parameters are too poorly known and the fine structure is complex and obscure. In researching the material for the book I have been surprised by how often high-quality theoretical analysis has been applied to primitive and often erroneous concepts of the observational data that were to be explained. Typically the theoretician seems to spend a few hours learning what the problem might be and many months calculating his model for it. A better impedance match between basic assumptions and the efforts spent on modeling is greatly needed. A good example is the Petschek mechanism, which explains a possible way in which magnetic fields might reconnect. It is a fine idea, but describes a configuration which I have never observed in the solar atmosphere. I do not understand why the theorists who followed this model could not examine the observed data to see if it actually happened.

The rules of experimental or observational science are clear: collect data, formulate explanations with verifiable observational consequences and look for them. If you have a good new theory, develop observational tests and use them. To explain a newly observed phenomenon *post hoc* is all too easy; to predict something which is later found is the true challenge. Parker's prediction of the solar wind stands out as one of the few solar phenomenon that was predicted before it was observed; even in that case there was some prior observational evidence. Another remarkable example was Wolff's (1972) prediction of the existence of solar *p*-modes. Phenomena like coronal holes, the high-energy aspects of flares, or the reversal of the polar fields were never predicted and some are still orthogonal to our concepts of the behavior of solar plasma.

I fell into solar physics by accident in 1953 when I went to work for the late Donald Menzel. I had been engrossed in stellar and theoretical astronomy and thought solar astronomers were funny people who observed in the daytime. I have since found it an evergreen subject, full of new challenges. There is a certain presence in the changing aspects of a star whose surface we can view almost every day that does not exist in the point images of distant stars. The excitement of new eruptions of solar activity or of new observations we can carry on with modern techniques is always there.

Just like green fields and virgin forests, the granules, the sunspots, the elegant prominences reflect the pure beauty of nature. They offer aesthetic pleasure, as well as scientific challenge, to those who study them.

I am concerned about the state of solar physics in our country. What was once one of the major fields of endeavor in astronomy is now only a small patch. Aggressive attempts are being made to close solar observatories. There are only a few universities in this country where courses on the Sun are taught and even fewer universities that would consider making faculty appointments in this field. The chairman of a major astronomy department once said to me, "We teach all aspects of astronomy here – except, of course, the Sun." How he expected his students to understand his galaxies without understanding the stars they were composed of, how he expected them to understand those stars without some rudimentary knowledge of the Sun is a mystery to me.

How did this melancholy state of affairs come about? Partly it is because the strong presence of space research that has contributed so much to the field has led to a heavy concentration in government or industrial laboratories, where it loses the vitality that the presence of students brings. Partly it is the lack of good observing facilities in sites suited to solar work, where people would have a better chance of obtaining successful observations. Although the excitement and religious significance of cosmological research has virtually driven solar physics from the campuses, society as a whole will always feel the need to understand the Sun. Obviously it is the best place to begin the study of the physical processes in the universe. In unexpected ways progress in solar physics has been essential for progress in astronomy as a whole. Aside from the fact that until recently most of the big nighttime telescopes were built by solar astronomers like Hale, McMath, Babcock and Goldberg, the development of new observational techniques has usually flowed from initial observations on the Sun. Radio, X-ray and gamma-ray astronomy started out in the study of the nearest star and worked their way outward. Progress in theoretical disciplines like magnetohydrodynamics and plasma physics owed much to initial application to solar problems. Helioseismology techniques are now applied to the search for stellar oscillations. Exciting new work on magnetic activity in the stars has been inspired by our growing knowledge of solar activity. I hope that the unity of astronomy will somehow be preserved and people will continue to study the Sun for all the marvelous knowledge it brings us. What should the solar astronomers do? Set the highest standards for their work and teach the best students they can find.

I wish to thank many people who helped me on the road to writing this book. Bob Leighton and Bob Howard were my early colleagues in solar physics at Caltech and partners in the site survey that led to Big Bear. Walter Orr Roberts and the late Henry J. Smith helped me raise the funds to build the observatory. Lee duBridge supported the building of the Big Bear Observatory with his usual enthusiasm.

I also wish to thank the members of the Big Bear group who have obtained so many of the observations presented in this book. And I thank my colleagues throughout the solar physics community for all the fun of arguments and discussions and competitions in our study of the Sun. Particular thanks are due to Dr. Charles Lawrence, who worked out the TeX macros for placing the figures and figure captions, to Nora Knicker and Susan Grubbs, who carefully proofread and checked and arranged.

Above all I thank my wife Mary, who has put up with the various travails of being associated with my work. She went through every word of the text to translate it into comprehensible English, and in the process taught me a little bit about writing.

Harold Zirin
Pasadena, Calif. Aug 4 1987

1

Looking at the Sun

Astronomers love stars, and we have a fine one right near us. It may be studied in the daytime, a great advantage for those who, like me, have trouble adjusting to all-night observing. And the more we study it the more wondrous facts we uncover. Contrary to popular belief, it is a fairly large star: eighth brightest among the 100 nearest stars. Although it falls in the middle of the sequence of spectral classes, most stars are dwarfs of later and smaller types. The Sun is not too old to have sunspot activity, nor has it so much as to be dangerous to life (about twice as much, had we a choice, would give us more sunspots to study and do little harm). Careful measures have shown it to be almost perfectly round, with only a slight oblateness due to its rotation. It generates a respectable wind and seems to have radiated with little change for a long time. The only things it lacks are neutrinos and a companion star.

Because the Sun is the only star we can see in detail, it is the only star we can learn much about, the only one whose physics we can really probe. Although current fashion treats astronomy solely as the study of distant galaxies, they too are made up of stars, and to understand those stars we must understand the Sun. When we observe flares on stars, we turn to the Sun to understand what those phenomena might be, and when we observe radio bursts from flares we search the stars for similar events. They will not be the same; the great variety of stars means that phenomena a thousand time greater or weaker may be found. But to get some idea of what is observed, to make the first step in understanding, we should look to the Sun.

We can learn a lot about the Sun by examining a projected full-disk image from even a small telescope. As in Fig. 1.1 we see a limb-darkened disk, called the photosphere, covered with granulation and with occasional sunspots. The limb-darkening tells us that the temperature is falling with height at the photospheric level, since we must see higher in the atmosphere at the edge. Measurement with a light meter would show us that the limb darkening matches what one would get from the Eddington approximation (Chap. 6), so the photosphere must be in radiative equilibrium. The granulation structure suggests the presence of convection, because a purely radiative atmosphere would be homogeneous. So there must be a convective region underlying the photosphere. The presence of sunspots tells us there is some other factor affecting the structure, a factor we now know to be magnetic fields. If we were to make drawings on successive days we would see that the Sun rotates with a 27-day period, and if we did it for a longer time we would find that it has differential rotation. If there were spots near the limb we would notice that they are accompanied by bright plages which cannot be seen at disk center.

Everyone knows we are not supposed to look directly at the Sun; that requires specialized hardware. Normally we solve the problem by using dark filters or looking at a projected image. A more interesting way is to use an Hα filter which cuts out all but the light absorbed by H atoms in the second level. In this narrow wavelength range the atmosphere is no longer transparent and we see the chromosphere and prominences. Because the gas pressure drops faster than the magnetic field strength, the latter dominates, and the structure is much more complex (Fig. 1.2). Prominences stand out at the limb or form dark clouds at the surface. Bright plages and

1.1. A photograph of the Sun in white light at sunspot maximum, 8 April, 1980. Two bands of large sunspot groups cross the Sun in the northern and southern hemisphere parallel to the equator. The spots are still relatively far from the equator at this time. Bright faculae are visible near the limb where we see a higher and darker layer of the solar atmosphere. The faculae mark regions of magnetic field that are associated with the spots. The central dark area of each spot is the umbra; the surrounding grey region, the penumbra. The photospheric granulation (rice-grain pattern) is barely visible. (BBSO)

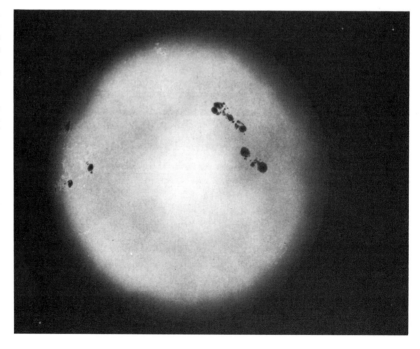

faculae can be seen everywhere on the Sun. The limb shows a thin forest of spicule jets shooting up into the corona.

Although we do not detect the neutrinos we expect (Chap. 6) we still believe the Sun's energy to arise from nuclear processes deep in its interior. To support the overlying mass the temperature at the center must be over 10 million deg; at the surface, direct measurement indicates a temperature of about 6000 K. The Sun's radius is 700 000 km, so that the gradient of temperature is only 14°/km. Since the mean free path – the average distance a particle travels between collisions – is of the order of a few cm (for either photons or particles), the material is comparatively well-behaved and can be described by the theoretical properties of a gas at equilibrium with a radiation field at the same temperature. So far as helioseismology has probed, the Sun has the same average rotation rate at all depths. It is made up of the same material, more or less, as the rest of the universe – hydrogen and helium, with about a 1% admixture of heavier elements, chiefly carbon, nitrogen, oxygen, neon, magnesium, silicon, and iron. There are about 10 times as many H atoms as He atoms.

The earliest observers of sunspots found that those near the equator rotated faster than those at higher latitudes. The differential rotation was confirmed by other observations; the rotation period Ω varies from about 25 days near the equator to 30 days near the pole. It is fitted by

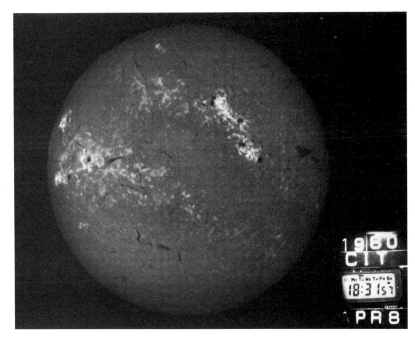

1.2. The Sun on the same day seen through a Lyot filter passing only Hα light. In this line the absorption is much higher than in integrated light, so we see the chromosphere. Dark objects are prominences in projection on the disk (filaments); the bright areas near sunspots are plages, regions of enhanced magnetic field weaker than that in the spots. The faculae far from the spots are remnants of old magnetic fields; as the cycle progresses they spread down to the equator. (BBSO)

$$\Omega = 14.42 - 2.30 \sin^2 \theta - 1.62 \sin^4 \theta \quad \text{deg/day}, \qquad (1.1)$$

where θ is the latitude.

We do not know if the differential rotation is a deep-lying phenomenon or a surface effect, nor if it varies with the solar cycle. Some models of solar activity suggest it generates the solar magnetic field by dynamo action.

The interior of the Sun is in radiative equilibrium, *i.e.* the energy is carried outward by photons. But in the outer envelope the opacity due to neutral helium and hydrogen impedes radiative transfer and convection occurs; the outer 20% or so is in convective equilibrium. The convective motions are thought to be responsible for many surprising surface phenomena.

One of these is the pattern of non-radial oscillations. The Sun oscillates in hundreds of normal modes of small amplitude. The dominant modes have a period of five minutes and velocities around 0.3 km/sec. The motion is a longitudinal sound wave thought to be excited by the convective envelope.

At the photosphere the photons can suddenly escape into space. What was a system in detailed equilibrium now loses half the photons emitted locally. The kinetic temperature continues to drop, but the radiation temperature drops even faster because of the photon escape. Since the magnetic field strength does not drop as sharply as the gas pressure, the material is strongly controlled by the fields; chaos is replaced by fibrils, jets and threads, ordered by the field, but moving and oscillating.

The photosphere reveals the convection below by two distinct patterns. On a scale of 1000 km it is covered by a uniform rice-grain pattern called the granulation. The rising granules are brighter, bringing energy up from below. There also is a cellular velocity pattern with a scale of 30 000 km, called the supergranulation; in each cell the flow is to the boundaries.

A few thousand km above the photosphere, the temperature rises sharply to a plateau at over a million deg, which we call the corona. The coronal density is very low, 10^{-12} times the Earth's atmosphere. Because the coronal temperature is high, the conductivity is high, and the temperature falls off only as the 2/7 power of the distance from the Sun. As a result the gas farther from the Sun cannot be contained, but flows out into space as the solar wind. We now know that many stars have winds, not all like the Sun's, which play an important role in the balance of the interstellar medium. On the Sun the outward flow is dominated by sources with open magnetic fields, called coronal holes, where no coronal emission is seen because the magnetic structure cannot hold it.

The corona is best seen during a total eclipse, when the moon blocks out the bright solar surface and allows us to see the faint coronal light, a million times fainter than the photosphere. Although the corona was known to the ancients, its remarkable physical nature was only recognized in 1940, when the coronal emission lines were identified as due to highly ionized atoms such as FeXIV. Modern space research has made possible observation of the X-ray and UV emissions from the hot outer layers without waiting for an eclipse, so they are much more familiar to us. Fig. 1.3 shows an X-ray picture of the corona with a coronal hole.

During an eclipse the transition region between photosphere and corona appears as a bright pink ring around the Sun, and is therefore called the chromosphere. The red color is due to the dominance of radiation in the red Hα line of hydrogen. The structure is dominated by a network of the same scale as the supergranulation and possibly produced by it. At the edges of the network are many small jets called spicules, which are particularly prominent at the limb. Inside the network cells is an amorphous region with strong oscillation.

Sunspots were known to Chinese astronomers of the first millennium and first observed through the telescope by Scheiner and by Galileo. The spots are about half as bright as the surrounding surface. Since the total rate of energy radiated by a star like the Sun is proportional to T^4 (the Stefan–Boltzmann law), the sunspot temperature must be 1000° lower than the

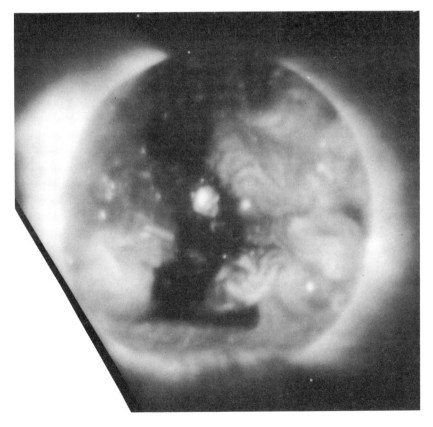

1.3. A picture of the Sun in soft X-rays taken from Skylab. In the center is a great coronal hole, with a few X-ray bright points. The rest of the Sun is covered with coronal X-ray emission, mostly loop-like structures connecting opposite magnetic poles, and the brighter regions are small dipoles. One small new dipole is in the center of the coronal hole. (AS&E)

surrounding 6000° photosphere. It is thought that the strong magnetic fields of sunspots suppress the energy coming from below.

In 1849 Schwabe discovered that the sunspot number rose and fell in an eleven-year cycle. A latitude variation was soon found; the first spots of the new cycle appear at high latitudes, and newer ones appear steadily closer to the equator till that eleven-year cycle subsides.

The solar constant is the amount of radiation that the Sun puts out, and, as is obvious from its name, is thought to be constant, at least over periods of years. Surprisingly, it was recently discovered that sunspots reduce the solar constant slightly; how or if the light lost in the spot gets out has not been determined.

When George Ellery Hale built the Mt. Wilson Observatory and began to study sunspot spectra, he found the spectrum lines to be split in accordance with the newly-discovered Zeeman effect. This meant that the spots had large magnetic fields, up to 4000 gauss in the big ones. Sunspots usually have bipolar structure (there are no magnetic monopoles), the same polarity preceding in all spot groups in one hemisphere, and following in the other. After studying several cycles Hale found that the polarities regularly reversed each eleven years, so the Sun has a 22-year magnetic cycle. Later research showed that the following polarities regularly drift to the poles and reverse the polar field. At a distance the effect of the local fields is small and the Sun appears like a dipole which reverses every eleven years.

The Babcock magnetograph permitted observation of weak magnetic fields outside the sunspots. As magnetic field elements break away from the sunspot dipoles, they cluster at the edges of the chromospheric network, perhaps because of the supergranular flow. Instead of cancelling with opposite magnetic polarity, the fields accumulate in extended unipolar regions trailing away from the active regions. It is these large-scale field agglomerations that drift toward the pole to produce the eleven-year field reversal. At sunspot minimum an irregular network of weak fields of mixed polarity remains all over the Sun. At the poles unipolar regions of opposite polarity remain, and the fields in the interplanetary medium reflect this simple structure.

There is a close correspondence between the magnetic field pattern and chromospheric structures. Sunspots, of course, are dark; but every other region of vertical magnetic field is marked by a bright plage or facula. Filaments occur at the inversion, or neutral line between regions of opposite magnetic polarity. If the lines of force looping between them are somewhat sheared, they will support a prominence. In other places where lines of force connect opposite fields directly, these contain smaller amounts of cool material and we see dark fibrils. Spicule jets come from the faculae of the network or the edges of plages. When the field is rapidly changing the $H\alpha$ brightness is enhanced.

In 1859 Carrington (1859) and Hodgson (Clerke 1903) independently observed a brightening of an area near a large sunspot group. Two days later there was a large geomagnetic storm accompanied by bright aurora borealis and widespread interruption in telegraphic communications. The phenomenon was what we now call a solar flare. Normally flares are transparent in the continuum and may be observed only in the line spectrum of hydrogen or some other abundant element. Carrington and Hodgson had observed an intense flare of the sort that occurs perhaps once a year and would rarely be seen in the cloudy conditions of England.

In Hα flares are brilliant transient brightenings accompanied by eruption of material in sprays and surges. The Hα brightness at least doubles; in bright flares it may increase tenfold. The Hα line can be broadened up to 20Å, suggesting internal velocities up to 1000 km/sec. Great quantities of high-energy nucleons and electrons are produced; it may be that all the flare energy input is in the form of these energetic particles. The electrons produce intense X-ray and radio bursts. Curiously, the highest-energy X-rays, above 10 MeV, are observed only from flares near the limb, which means they are preferentially emitted horizontally.

The nucleons from flares may be observed directly near the Earth; the nucleons of high first ionization potential are underabundant, and the same may be true in the corona. At the Sun the particles produce gamma-ray lines when they hit the ambient gas. The lines of electron–positron and neutron–proton capture are observed, as well as those produced by excitation of heavier nuclei. A few minutes after the flare starts these particles thermalize and form an extremely hot (40 000 000 deg) plasma which radiates X-ray lines of such ions as FeXXV and CaXIX. As that plasma cools, graceful loop prominences appear, draining down from the hot cloud. A limb flare is seen as an elevated brilliant structure a few thousand km above the surface.

Why do flares occur? We believe that magnetic energy builds up as a result of sunspot emergence or motion. The magnetic fields in this high conductivity plasma can only change slowly or cataclysmically; the stress builds up until an instability occurs and the energy is suddenly released. Almost all flares occur on magnetic neutral lines and involve filaments; the filaments mark the sheared boundaries of the magnetic fields. But details of the process have not been worked out.

There are flares without spots. Filaments, which one thinks would be likely to fall down, almost always erupt upwards. When this occurs in an old active region where the spots have died out, we can get a large flare. Filaments far from active regions also end their lives by erupting, and many of the filament eruptions are seen far out in the corona as coronal mass ejections (CME).

Solar activity has considerable effects on the Earth. The normal solar wind produces small changes in the Earth's magnetic field, but flares, CME and high velocity streams associated with coronal holes can produce a giant current system in the magnetosphere called a geomagnetic storm. These storms cause prolonged fluctuations of the order 1% in the Earth's surface field, and the field fluctuations produce particle acceleration in the magnetosphere, aurorae, and disruption of radio communication. The flux of high-energy particles during large solar flares may even reach dangerous levels in interplanetary space. The magnetic field fluctuations over large scales produce voltages in long-distance power transmission which can trigger circuit breakers. They generate interfering currents in long-distance communication lines and oil well logging wires, and other effects which appear as our world becomes more complex.

Finally, let us summarize some physical data on the Sun:

Table 1.1. The Sun's measurements.

Radius	R = 696 000 km = 6.96×10^{10}cm = 109 Earth radii
Volume	V = 1.412×10^{33} cm^3
Mass	M = 1.991×10^{33} gm
Mean density	ρ = 1.410 gm/cm^3
Gravity at surface	g = 2.74×10^4cm/sec^2
Total radiation	L = 3.86×10^{33} erg/sec
Distance from Earth	= 1 AU (astronomical unit)
	= 1.496×10^{13} cm = $214.94 R_\odot$

References

Allen, C. W. 1973. *Astrophysical Quantities*, 3rd ed. London: The Athlone Press.

Bruzek, A. and Durrant, C. J. 1977. *Illustrated Glossary for Solar and Solar-Terrestrial Physics*. Reidel: Dordrecht.

Noyes, R. W. 1985. *The Sun, Our Star*. Cambridge: Harvard.

2

Observing the Sun

2.1. Telescopes and auxiliaries

Solar observations present unique problems. Heating of the ground causes convective turbulence which degrades the image; orographic flow compounds the problem for mountain observatories. The Sun heats the telescope optics and any air inside it, causing distortion and internal air currents which further degrade the image. On the good side, however, the Sun is so bright that large images (long focal lengths) and short exposures are possible. Because there is a visible disk, the emphasis is on two-dimensional studies, particularly with monochromatic filters. The brightness of the Sun made it the first target of specialized instruments developed for ultraviolet, X-ray and radio imaging. In the radio band swept frequencies and very high time resolution may be used.

The most important ingredient of solar observations is a good site. The severe ground heating by the Sun, which produces convection, turbulence, and daytime cloud buildup, makes location in or near a body of water essential. This fact was originally discovered by Evershed and confirmed by the Aerospace Corporation site tests (Mayfield *et al.* 1964, 1969), which led to the location of the San Fernando Observatory in the upper Van Norman reservoir in the Los Angeles basin. The Caltech site survey by Leighton, Howard and Zirin (Zirin 1970) resulted in placing the Big Bear Solar Observatory in the middle of Big Bear Lake (Fig. 2.2). Both these studies determined that mountains are inferior sites because of orographic turbulence; even mountains near the sea are so affected.

Although the seashore itself may have good seeing, the frequent clouds and haze, as well as salt corrosion, make it unsuitable. Lakes are good for three reasons: the surface remains cool; the wind moves freely across the surface without generating turbulence; and if the water is cooler than the air, important refractive effects improve the seeing.

The actual "seeing" is due to temperature and pressure fluctuations along the line of sight and across the telescope aperture, which produce corresponding excursions in the refractive index. As a result the wavefront from the Sun is disturbed, just as if the glass in our optics had variable index. It is customary to evaluate such effects by using microthermal sensors to measure these fluctuations. While these are adequate for evaluating the local dependence, they cannot be used to compare sites because of the important role played by relative humidity. There is substantial literature on these effects in connection with laser propagation and other practical problems, largely unknown to solar and stellar astronomers. Wesely and Alcaraz (1973) define an atmospheric refractive index structure coefficient C_n:

$$C_n = C_T A_1 p < T >^{-2} (1 + 0.16\beta^{-1}), \qquad (2.1)$$

2.1. The McMath solar telescope of Kitt Peak National Observatory, the largest solar telescope in the world and the largest unobscured aperture (152 cm) of any telescope. The heliostat mirror reflects a rotating image down to a primary at the base of the tube. At right is a vacuum tower telescope for synoptic programs. The combination of a sunbaked mountain and long air path limits the seeing obtainable. (KPNO)

where C_T is a measure of temperature fluctuation, p is the pressure, and $< T >$ is the average of the temperature. Bowen's ratio β is the ratio of the thermal heat flux to the latent heat flux. Because the Earth's surface is either warming and evaporating or cooling and condensing, β is usually positive; over dry land it is typically 0.5. But when surface cooling and evaporation occur simultaneously, $\beta < 0$, C_n is decreased and the seeing is improved. Experimental verification is given by Friehe *et al.* (1975) as well as the empirical success of lake sites. The stability at any site may

be tested by looking at distant horizontal targets with binoculars or flying above it in a small airplane. But the best measurement is observation of the Sun. Photographic site surveys are hampered by problems of technique and the difficulty of testing many sites; an experienced observer can make accurate visual seeing appraisals at a number of sites fairly quickly.

Light coming from the Sun has a uniform wavefront before it enters the atmosphere; an objective in space would focus it to a diffraction-limited image. But the different paths leading to different parts of the objective have different optical paths due to pressure and temperature fluctuations, and these distort the wavefront. Obviously this effect decreases with height, so we look for the highest body of water we can find. It is important to locate in areas with strong inversions, usually the effect of cold coastal currents. Because the Sun is highest in summer, it is particularly important that the summer weather be the best possible and the many mainland areas with summer monsoons be avoided.

In recent years, extinct island volcanoes such as those in Hawaii and the Canary Islands have been used for solar work. In these cases one has the ultimate lake – the ocean. These mountains are rounded, which reduces wind-generated turbulence. But they are big and convective heating may be substantial. The Hawaiian sites appear not too good, the conditions in the Canary Islands are as yet not established. Hawaii has proven excellent for coronal studies; the Canaries suffer from Saharan dust in summer.

It is unfortunate that, before the problems of solar seeing were appreciated, most of the great solar telescopes were built on sunbaked mountain peaks capped by summer thunderheads. The peaks give dark skies, good for coronagraphs, but the natural daytime turbulence and cloud buildup makes them unsuited for solar work, even though they are good at night. A mountain lake such as Big Bear shares some of the orographic turbulence; due to general air flow, the local heating is reduced. A flatland lake such as Lake Elsinore in California has the most stable air (Zirin and Mosher 1987), but transparency is, of course, reduced.

Using a bigger telescope or a larger image does not help us see more, because the atmospheric seeing is usually the limiting factor. A large aperture must average over a larger air column and often gives a worse image. We find that telescopes 60 cm and larger will only achieve their limiting resolution a few days in the year. But when we build the telescope it should be carefully designed to have the best optics possible and avoid image distortion from internal heating.

Because solar telescopes have long focal lengths and need only point at part of the sky, they are often fixed, light being reflected into them by movable flat mirrors. In the McMath telescope (Fig. 2.1) a single mirror heliostat reflects the beam to a large primary mirror at the bottom of the polar axis, and the reflected image is then brought to a vertical spectrograph. This arrangement produces a rotating image; in the coelostat the face of the tracking mirror lies in and rotates about the polar axis, giving a non-rotating beam in a fixed direction. Since that beam is directed upwards a second flat is needed to send it to the telescope. As the Sun moves across the sky the second flat will shadow the coelostat, which must be moved to a different position. Since one wishes to avoid heat rising from the ground, the coelostat or heliostat is often placed atop a high tower, which also makes a long focal length and large image possible; however even the 150 foot Mount Wilson tower and the 130 foot Sacramento Peak tower (Fig. 2.3) are small compared to the mountain and offer only partial relief from ground heating. When continuous observations of the Sun are required, as for the study of solar oscillations, we must go to the South Pole (Fig. 2.4), where for a few summer months this is possible. Alternatively, a set of well-selected sites around the world can provide continuous observations throughout the year.

Both reflectors and refractors may be used, depending on the desired

2.2. The Big Bear Solar Observatory at an elevation of 2042 m in the San Bernardino Mountains. The lake is about 4 km long and 7 meters deep at the dome. Laboratories and living quarters are on shore. The main telescopes are a 65 cm vacuum reflector and 22 and 25 cm vacuum refractors. Powerful fans pull air through the dome slot and exhaust it behind the building. The tower is a single monolithic structure. The lake water stays cool so convection does not disturb seeing; also the smooth lake surface permits laminar airflow. (BBSO)

observational program. A lens reflects a few percent of the light incident on it, but absorbs very little of it; furthermore the tube is capped by the lens, which prevents chimney effects, and the light only passes down the tube once. While a singlet lens needs little focussing, it suffers from severe chromatic aberration and can only be used for monochromatic work. If we use an achromat, the overall power is the difference of that of the separate elements and will change rapidly with temperature, requiring endless focussing. A reflector is thus more suitable for spectrographic work, but the aluminized mirror absorbs 10–15% of the light and heats up; since the warm mirror is normally at the bottom of the tube and the light must go back and forth, almost all normal reflecting solar telescopes have terrible image quality. For this reason all modern reflecting solar telescopes are vacuum systems, so the entire optical path is free of disturbance. Then we require a front window, which must be carefully cooled and its edges shielded from the Sun. With the vacuum technique big reflectors can compete in resolution with refractors. A refractor usually need not be in vacuum; but the lens provides a window as well, so it is easy to evacuate it.

The resolution of a perfect optical surface is limited by the diffraction pattern of its aperture. If the root-mean-square (RMS) surface error is less than $\lambda/25$ (which corresponds to $\lambda/4$ peak-to-peak), there will be little deviation from diffraction-limited performance. The error of the total system will be the sum of the errors of all the elements as well as the detector, so each component must exceed these specifications. If we can manage it, the fewer optical elements, the better. The contribution to the error budget of elements near a focus or field stop is small, while elements near an image of the entrance pupil can be as important as the objective.

An error of 1/10 wave in a mirror surface produces an error of 1/5 wave in the resulting wave front, because the beam traverses this distance twice; but a similar error in a refracting surface only produces an error of 1/30 wave in the wave front, because it is the difference between glass and air paths $(n-1)/n$ (where n is the index of refraction) that counts. Thus for a given surface error a lens is superior to a mirror and preferred for monochromatic work.

If the telescope is equatorial a dome is mandatory to reduce wind shake and heating; but turbulence around the dome slot degrades the image. It helps to use large fans to pull the air into the dome and exhaust it elsewhere; at the Pic du Midi, the front end of one telescope sticks out of the dome as in popular cartoons. From this point of view the Sacramento Peak tower is ideal; the enclosed altazimuth mirror system at the top of the tower is relatively free of wind shake, and the entire optical path is in vacuum. At the Huairou Observatory in China the dome can be rolled away for domeless observing in calm weather.

Every non-optical surface exposed to the Sun must be painted titanium white to reduce heating. On older telescopes a separate inside tower protected by a decoupled shell was used, but it has proven impossible to remove seismic coupling in such structures and the two separate structures are invariably weaker than a single massive one.

For equatorial reflecting systems a short-focus primary is required, and the Gregorian (Fig. 2.5) system first introduced by ten Bruggencate at Locarno is particularly useful. Only the primary is exposed to a solar beam. The same system is used at Big Bear and has been proposed for solar telescopes to operate in space. The vertex of the Gregorian secondary must be carefully aligned to coincide with that of the primary. The field is limited and it is hard to change magnification, but the problem of solar heating of optics is solved. With modern low-expansion materials that may no longer be so important. In large systems like the Sacramento Peak or McMath

2.3. The solar tower at Sacramento Peak. This remarkable instrument, built by R. B. Dunn, has a vacuum tank 100 meters long. The Sun is acquired by a pair of altazimuth mirrors behind a window, avoiding wind shake. The clear aperture is about 75 cm. This is probably the best solar telescope in the world, but the site, a dry desert mountain, has mediocre seeing except for occasional spectacular moments. (SPO)

2.4. (right) The solar observatory of the Bartol Research Foundation of the Franklin Institute at the South Pole. In the Antarctic summer continuous solar observations for over 100 hours have been obtained. (Bartol)

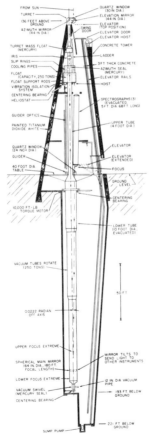

telescopes heating is less of a problem, because long-focus primaries are used.

The resolution limit of a diffraction limited system is 1.22 λ/D radians where λ is the wave length and D, the aperture; but at that point the modulation transfer function (MTF – Sec. 2.2) is essentially zero. The film and other parts of the system degrade the resolution further so that a point of this size cannot really be distinguished. For linear features we accumulate data over a number N of related points and the signal-to-noise ratio is improved by the square root of N. A larger image reduces the impact of detector MTF, but lengthened exposures introduce image jitter and other factors that limit resolution.

After good optical quality has been achieved we require good pointing stability or guiding. Since the solar surface varies from point to point we must be able to keep our telescope pointed at the same place for as long as our measurements require. Image jitter (seeing) and inaccuracy of the telescope drive make a servo-guider necessary. Photoelectric cells monitor the position of the telescope relative to the Sun and actuate servos in the drive to keep that relation constant. If there is no flexure in the telescope, the image will remain fixed. The guider need not have a long focal length, but simply must detect any deviation in pointing. A good guider may be made with a short-focus (say 50 mm $f/10$) lens projecting an image on a silicon quadrant cell, with an occulting disk that covers much of the image. Even if the image is out of focus any change in orientation will produce a detectable photon excess in the appropriate quadrants. The difference signal is then amplified and fed to the servos. Special precautions are required to compensate for passing clouds. The speed of response of the telescope drive is limited by the resonant frequency of the pointed element, which may be low if the whole telescope is moved. Therefore the main

2.5. The Gregorian system of the Big Bear Solar Observatory vacuum reflector. The primary mirror is mushroom-shaped for strength and lightness. A tilted cooled heat stop mirror at the first focus rejects all the light except what passes through a hole 3 arc min in diameter. This is enlarged by an ellipsoidal secondary and reflected out of the axis by a flat just behind the heat stop. The system is enclosed in a vacuum tank with a window.

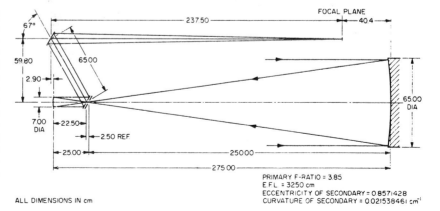

ALL DIMENSIONS IN cm

PRIMARY F-RATIO = 3.85
E.F.L. = 3250 cm
ECCENTRICITY OF SECONDARY = 0.8571428
CURVATURE OF SECONDARY = 0.021538461 cm⁻¹

telescope response is often filtered for slow guiding, while a lightweight mirror of high resonant frequency may be used to remove high-speed image jitter.

The opportunity to observe in the ultraviolet and X-ray regions of the spectrum has brought new challenges connected with the problems of reflecting that light. For wavelengths above 500 Å normal incidence mirrors with special coatings may be used; down to 150 Å images may be obtained with a single properly coated surface. Tousey (1973) at the U. S. Naval Research Laboratory (NRL) achieved remarkable images in this region with a concave grating at normal incidence acting as imager and disperser, producing (Fig. 2.6) a series of slitless spectrograph images in the different emission lines. That instrument, derisively called the "overlappograph" by

2.6. A slitless spectrogram of an erupting prominence made by the NRL instrument on Skylab, showing solar images in the emission lines (283–304 Å). The bright image is HeII 304 Å; the other images are coronal lines, in which the eruption does not appear. (NRL)

AUG 9, 1973   15^h 21^m U.T.

its detractors, produced some of the most remarkable solar images ever.

Below 150 Å the light must hit the surface at a grazing angle if it is not to be absorbed. In the X-ray region this has been achieved with the Wolter (1952) design, originally intended for X-ray microscopy. Light reflects at grazing incidence from a limited ring on a deep paraboloid. The extreme coma produced by this reflection is cancelled by reflection from a grazing incidence hyperboloid, concave in Type 1 and convex in Type 2. Wolter's third system, not yet used in solar physics, uses a hyperboloid and ellipsoid. Fig. 1.3 was made with a Wolter telescope on the Skylab mission in 1973.

At wavelengths below 10 Å the photons are absorbed even at grazing incidence; geometric devices must be used. The modulation collimator developed by Oda uses a fine grid rotating relative to one or more fixed grids and determines the position from the pattern of interruption. The aspect of the system relative to the Sun must be accurately known. This system was used successfully by the Japanese Hinotori spacecraft. A mosaic of geometric telescopes, aligned so that each detector sees only light from a distinct point, was flown on the Solar Maximum Mission (SMM) in 1980. The angular resolution of that system is the length divided by the grid spacing; intermediate grids remove secondary maxima. The rotating grids are superior in that they collect light over a large area, so they have been much more successful, but considerable computer processing is required. New X-ray telescopes designs (Crannell *et al.* 1986) use sets of grids designed to give various Fourier components of the source brightness distribution.

X-ray quanta are few, and collimators are limited in size and throughput. An interesting solution is the flying pinhole, in which the imaging objective is replaced by a "pinhole". The resolution is the ratio of the distance between pinhole and detector to the pinhole diameter. Thus a 10 cm diameter orbiting 10 km from the detector (a shield 200 meters diameter is needed to shade the detector from the Sun) gives resolution 10^5 (2 arc sec). All the flux incident on the 10 cm hole is detected.

The resolving power (RP) of a system is defined as the number of discrete elements into which it can separate what we are looking at. For a spectrograph that is $\lambda/\Delta\lambda$; for a telescope it is a radian (206 265 arc sec) divided by the angular resolution. For all interference systems the resolving power is the objective diameter divided by the wavelength:

$$\text{RP} = \text{maximum detector separation/wavelength} = D/\lambda. \qquad (2.2)$$

For imaging systems the maximum separation is the projected aperture. For a spectrograph it is the projected width between first and last rulings on the grating. A resolution of 100 000 will separate a unit (1 radian) angle into 2 arc sec elements. In X-ray collimators there is no interference and the resolution is the length of the system divided by the grid spacing; aperture is unimportant. A circular telescope mirror produces an image which includes all the Fourier transform components of the object down to the limiting resolution. An array of radio telescopes like the Very Large Array (VLA) may be thought of as a large circular dish with most of its surface covered except for apertures corresponding to the dishes, and the Fourier transform is limited to the spatial frequencies D_i/λ of the various antenna pairs. A mirror with two pinholes at either end of a diameter, a radio interferometer with two dishes, a Lyot filter with the thickest element only, or a grating with two lines will give only the Fourier transform corresponding to the highest spatial frequency. As we add dishes to the radio telescope, or lines to the grating, we add lower-frequency components which define the image and suppress the side lobes. As the Earth's rotation moves the object relative to the array, more Fourier components are added, and the image is improved. This is called aperture synthesis.

2.2. The modulation transfer function

It is easy to think of mirrors as perfect, images as sharp points and film as made up of very fine grains. But the image of a star is not a point but a spot, often blurred by atmosphere and poor optics, surrounded by a bright diffraction ring. The modulation transfer function (MTF) is used to describe the performance of optical systems more quantitatively (a complete description may be found in Smith 1966 or Dainty and Shaw 1974). The properties of the system can be described equally well by a point-spread function or line-spread function, but the MTF is often more convenient. It is defined as the ratio of the output modulation of a sinusoidal form with spatial frequency ω to the input modulation at the same frequency. To measure it, a regular pattern (such as bars) of a certain spatial frequency is imaged through an optical system and a new distribution is produced, either on fine grain film or with a photoelectric scanner. The ratio of amplitudes is the MTF for that spatial frequency.

If the system is perfect, the MTF is one. The MTF for the overall system is the product of the MTF's of each component: objective, film, atmospheric seeing, and any intermediate optics. However, one must be careful in treatment of systems where one element is designed to correct errors in another.

The system is producing a Fourier transform of the two-dimensional brightness distribution of the Sun; the MTF measures how each components is transmitted by the system. The MTF of an aberration-free circular dish is

$$\text{MTF}(v) = 2/\pi(\theta - \cos\theta\sin\theta)(\cos\theta)^k, \qquad (2.3)$$

where

$$\theta = \cos^{-1}(\lambda v f\#), \tag{2.4}$$

where v is the spatial frequency in cycles/mm, and $f\#$ is the ratio of focal length to aperture. To convert to ω (cycles/arc sec) we use

$$\omega(\text{cycles/arc sec}) = v \times \frac{F(mm)}{206265}. \tag{2.5}$$

The limiting resolution v_0 corresponds to the Airy limit:

$$\begin{aligned} v_0 &= \frac{1}{\lambda(\text{mm})f\#} \quad \text{cycles/mm} \\ &= 206265 \frac{\text{aperture}}{\lambda} \quad \text{cycles/arc sec.} \end{aligned} \tag{2.6}$$

A simple linear approximation to (2.3) is

$$\text{MTF} \approx 1 - \frac{v}{v_0}. \tag{2.7}$$

The effect of aberrations is to reduce the MTF, especially for v/v_0 from 0.2 to 0. Central obscuration lowers the MTF around $v = v_0/2$, but actually increases it at v_0. If the aperture is covered except for two small holes at the edges then we have a two-element interferometer, and the MTF is zero except for a spike around v_0. A central obscuration reduces the MTF in the middle ranges.

The MTF of film has a somewhat different form, fairly high for low spatial frequencies, then falling off steeply for higher frequencies near the grain size. But just as v_0 corresponds to MTF zero, the film has little modulation near its resolution limit. The commonly used film Kodak 2415, popularly said to resolve 100 lines/mm, has an MTF of only 0.25 at that spatial frequency. So we must enlarge the image if we seek the arc-second resolution everyone claims.

Consider a real telescope, the Big Bear 25-cm $f/15$ vacuum refractor. The focal length is 3750 mm, hence 1 arc sec is $3750/206265 = 0.018$ mm. The Airy limit is 0.66 arc sec and v_0 is 102 mm^{-1}, corresponding to 0.55 arc sec. Since the MTF is zero at this point we consider the MTF at 1 arc sec, $v = 55$ mm^{-1} and $v/v_0 = 0.5$; here the MTF is 0.40; the MTF of the film, 0.45, and the overall MTF is the product, 0.18. To improve this we use a secondary lens, which magnifies the image five-fold, so the film MTF is increased to 0.80, but now the secondary lens MTF, about 0.60, must be included, so there is minimal gain in the overall MTF. However, the contrast of the film is $\gamma = 3$, so the contrast transfer function, $\gamma \times$ MTF is about 0.50. The great brightness of the Sun makes possible alternatives not available in stellar work. Another technique now used for high-resolution imaging is to take many frames rapidly and select the best; this is done with video recording and digital image processing and gives remarkable improvement.

2.3. Spectrographs

The spectrograph is our basic instrument to analyze light from the Sun. It consists of a slit to limit the entering light; a collimator, which makes the light diverging from the slit parallel; a grating to disperse this light; and a camera to photograph the result. Various optical arrangements are used to achieve this result, depending on the goals of the particular spectrograph. All systems suffer from various off-axis aberrations, as the beams must go back and forth without hitting the other optics. The Ebert-Fastie design uses a single large mirror for collimator and camera, but the mirror required is large. In the Czerny-Turner design separate collimator and camera mirrors are used. Coma cancels between the two mirrors but there can be astigmatism. A Schmidt camera may be used as in the Palomar coudé to give wide spectral coverage and high speed. The grating is at the center of curvature of the camera mirror, and the focus is at half that distance. This system is fast and covers a wide wavelength range.

In the simple Littrow system, a single lens performs the function of both collimator and camera (Fig. 2.7). The slit is at the focus of the Littrow lens, which produces parallel light. This light is diffracted by the grating and refocused by the lens on the film or plate. An image of the slit at each wavelength is produced. This system has the advantage of great simplicity, symmetry and convenience in physical operation. Since the path length between the two ends of the grating is doubled, we get double retardation and higher dispersion. By combining BK7 and fluorite lenses a nearly achromatic system with good UV transmission is possible. Cross dispersers are often added to produce an echelle system, where the

2.7. The Littrow spectrograph, a simple arrangement utilizing a single lens for both collimator and camera. The slit and plate are both at the focus of the lens. Its symmetry makes it convenient for rotating spectrographs, but the use of a lens introduces chromatic aberration, which must be corrected by tilting the camera plane. The system is viewed (a): parallel to the slit and (b): perpendicular to the slit, which is always parallel to the grating lines. The f/ratio of the collimator must match that of the telescope.

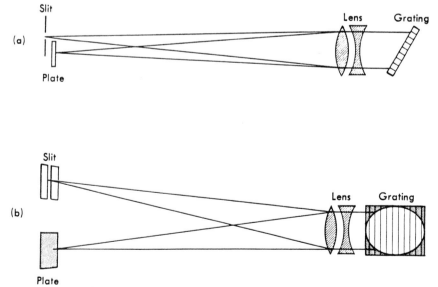

overlapping orders are separated and placed side by side. Good examples of the Littrow system are the thirteen-meter instrument at Sacramento Peak Observatory and the three-meter coudé at Big Bear.

Bartoe and Brueckner (1975) have developed a tandem Wadsworth spectrograph using two concave gratings placed along the Rowland circle to give high-resolution stigmatic spectra. Results with this remarkable instrument are given in Chap. 7.

Plane gratings diffract light according to the formula

$$n\lambda = d(\sin\theta + \sin\phi), \tag{2.8}$$

where n is the order, d, the separation of the lines of the grating, λ, the wavelength, ϕ, the angle of incidence, and θ, the angle of diffraction. It is because n and λ occur together in the formula that the orders cannot be separated; 6000 Å in the first order falls at the same point as 3000 Å in the second order. The inverse dispersion is the change in wavelength, $d\lambda$, per linear interval $fd\theta$ in the focal plane. Differentiating Eq. (2.8) we find (using the Littrow condition $\theta \approx \phi$, and $\phi = \text{constant}$)

$$d\lambda = d\frac{\cos\theta}{n}d\theta = \frac{\lambda}{2}\cot\theta d\theta, \tag{2.9}$$

so the inverse dispersion (in Å/mm) is (for a Littrow system)

$$\frac{d\lambda}{fd\theta} = \frac{\lambda}{2f}\cot\theta. \tag{2.10}$$

The dispersion, expressed in mm/Å, depends only on the focal length of the spectrograph and the tangent of the angle of diffraction. Since the focal length is usually a fixed physical structure, the only way to get maximum dispersion (if we want it) is to tilt the grating further and further over. Because the dispersion depends on the tangent of the angle of tilt, we get very large gains in dispersion at high angles. The factor two in Eq. (2.10) is due to the autocollimation aspect of the Littrow system. As

Table 2.1. Resolution of interfering systems.

System	Resolution	Biggest Value	Resolution
Optical Telescope	Aperture/λ	5 m telescope: 10^7	.02 arc sec
Diffraction Grating	Width/λ	25 cm grating: 5×10^5	.01 Å
Radio Interferometer	Baseline/λ	36 km VLA at 1 cm: 3.6×10^6	.06 arc sec
Lyot Filter	Retardation/λ	50 mm calcite/Hα: 1.3×10^5	.05 Å

in other interference systems, the resolution depends on the retardation in wavelengths between light diffracted from the two ends of the system. A two-groove grating has the same resolution as a complete one; the function of the other grooves is to separate orders.

For a typical high dispersion echelle-type grating, $\theta = 60°$ and $n = 13$, so the projected grating width is half the ruling width. For a given value of $\cot \theta$, various n, λ pairs are possible, so narrow band filters or predispersers must be used to isolate the particular λ we want. To fit the spectrograph to a telescope of focal length F, one must match the slit width to the desired optical and spectroscopic resolution. We define RP(opt) $= 1/\Delta\Theta$ as the optical resolution, and RP(spec) $= \lambda/\Delta\lambda$ as the spectroscopic resolution. The smallest linear element we can observe is F/RP(opt), and this

2.8. Optical diagram of the spectroheliograph. The spectrograph produces monochromatic images of the slit in all wavelengths. If the slit is moved across the Sun, images of the Sun in all wavelengths are built up. Slit 2 picks out a single wavelength, producing a monochromatic image of the Sun on the plate, which is held fixed. Pictures may be made in any wavelength, with arbitrary pass band. Exposures are long, because only a small fraction of the Sun is photographed at any time; but the exposure for any single element on the surface is short.

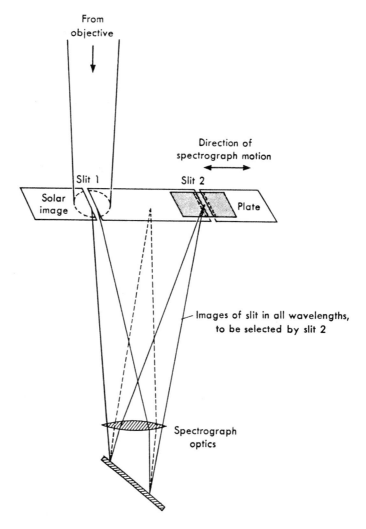

should match the slit size, which is $\Delta\lambda$ divided by the dispersion (Eq. 2.10). Therefore we can write

$$\frac{F}{\mathrm{RP(opt)}} = 2f\frac{\Delta\lambda}{\lambda}\tan\theta = \frac{4f}{\mathrm{RP(spec)}}, \qquad (2.11)$$

or,

$$\frac{F}{f} = \frac{4\mathrm{RP(opt)}}{\mathrm{RP(spec)}} \approx 4. \qquad (2.12)$$

Since vertical angles at the slit must be equal, the telescope aperture in this case can be up to 4 times the projected grating width or twice the actual width. The factor 4 comes from the doubling of retardation in the Littrow mode and the $\tan\theta$ term. Because of the limitation of Eq. (2.12), the speed of large telescopes for spectroscopy of point sources limited by the size of the gratings. Wavelength calibration may be obtained by placing a sealed tube with a few iodine crystals ahead of the slit.

In real life the exposure time with such a system may be too long to obtain optimum RP(opt). Spectroscopic exposures takes seconds, while seeing degrades exposures longer than 1/30 second or so. So one can choose to use a lower spectroscopic resolving power. Since we do not want to increase the slit size we simply reduce the grating angle θ; normally another grating with a different blaze must be used. If the telescope aperture is greater than 4 times the grating width a lower dispersion must be used, which is the usual procedure with stellar telescopes. The largest gratings available are currently $\approx$ 30 cm and their RP $\approx 10^6$. This restriction does not apply to other monochromators such as the Fourier Transform Spectrometer (FTS) or Fabry–Perot etalon. Both of these use the reflection between two plates to achieve high resolution. The FTS (Connes 1970; Brault 1979) uses a Michelson interferometer where the length of one arm is slowly varied and the output recorded. The system corresponds to a two-groove grating where the grooves are slowly moved to synthesize a many-grooved one. Because the mirrors may be far apart, enormous resolution may be achieved, but at present, FTS systems are only available to scan a single source. The entire spectrum is recorded at once and extremely high resolution obtained. Particularly in the infrared the FTS has made possible important advances. But the system is slow. Even higher resolution may be obtained in the IR by heterodyne mixing with stabilized lasers.

2.4. Monochromators

The "surface" or photosphere of the Sun is defined as the point at which it becomes opaque in continuous light. The overlying atmosphere is transparent in the continuum, but it is opaque in atomic lines and may be seen if we isolate the lines. The first device for doing this was the spectroheliograph invented by George Ellery Hale (Fig. 2.8) and used for his MIT B.A. thesis.

Whereas the spectrograph breaks up the Sun's light into the many lines of the spectrum, the spectroheliograph lets us make a picture in any line. The spectrograph produces an image in each spectrum line of the strip of Sun entering the slit. A second slit placed at the focus of the spectrograph limits us to the wavelength we desire. If we move the image of the Sun across the slit and simultaneously move the detector, we get a composite picture of the Sun in whatever line we have chosen.

The spectroheliograph has the advantage that it can be tuned to any wavelength with a narrow bandpass and high spectral purity. By slightly shifting the second slit, one can change the effective wavelength to compensate for and measure Doppler shifts due to local motions. But because at any time one looks only at the portion of the Sun subtended by the slit, the spectroheliograph is relatively slow. Changes in the sky transparency and guiding produce a ragged image as the slit moves across the Sun. With the electronic detectors in modern use one can produce complex combina-

2.9. A spectroheliogram in the center of CaII K made by Bruce Gillespie at KPNO. The bright plage of an active region is seen near the limb. In the lower part of the frame we see the chromospheric network with small intranetwork bright points. Because the K line broadens toward the limb, the upper part of this frame shows stronger absorption by spicules coming out of the network, which obscure everything at the limb. (KPNO)

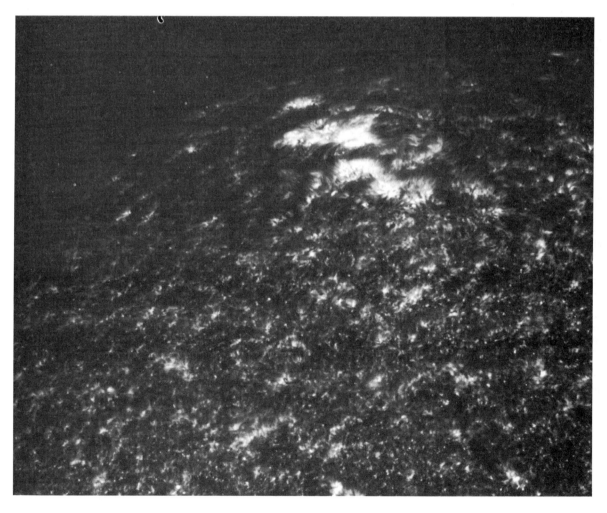

tions of parts of the line profile, such as the magnetic fields and Doppler velocities.

The Lyot filter, proposed by Lyot (1933) and independently by Öhman (1938), enables us to take instantaneous two-dimensional monochromatic images. If polarized light is passed through a birefringent crystal of quartz or calcite with the axis of the crystal parallel to the face, it is split into two rays, the ordinary and the extraordinary, of which the former travels faster. If ϵ and ω are the refractive indices for the two rays, the retardation when they have passed through the thickness d of the crystal is (since the velocity of a ray is inversely proportional to the refractive index)

$$n = d(\epsilon - \omega)/\lambda \qquad (2.13)$$

wavelengths. For calcite, $\epsilon - \omega = -0.17$; for quartz, $\epsilon - \omega = 0.009$. The plane of polarization rotates by an amount proportional to its thickness

2.10. Transmission curves for a three-element filter. Curve *a* is for the thickest, and *b* and *c* are for elements one half and one quarter as thick. Curve *d* shows the combined effect. The principal sidebands occur 2/3 of the distance to the next maximum of the thickest element. A suppressor element two thirds the thickness of the thickest element is often used to eliminate these. (After Evans 1949)

2.11. The Lyot filter. Thickest elements have the Lyot split-element design for wide field. The calcite axes are crossed and a half wave plate is put symmetrically in between. The back polaroids must be rotated twice as far as the front in line shifting. In the Evans design, the individual calcite elements of the split element are split once more, and rotated 90°; one of the thinner quartz elements is placed in between, with axis bisecting the other two, and the outside polaroid rotated 90°. This removes one polaroid per split.

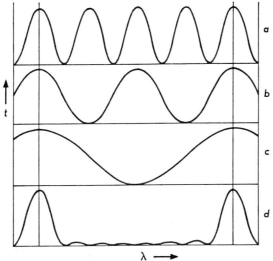

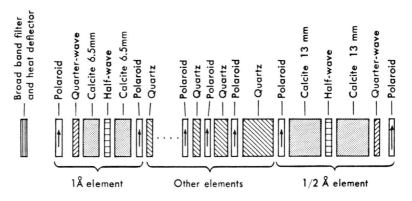

divided by the wavelength. Optical quality calcite is rare, so the wide band elements are made of quartz.

For some wavelengths n is integral, and the polarization is unchanged; for others we get circularly or elliptically polarized light; for 1/2 integral n the polarization is rotated 90°. A parallel polaroid is now placed in the emergent beam to pick out those wavelengths for which n is integral, and the transmission of the polaroid-quartz-polaroid sandwich is

2.12. (a) A large complex spot (7/25/81) in the center of Hα using a Lyot filter; (b) a video-magnetogram (VMG). Features marked include **S**, the main spot; **s**, the spot of opposite polarity; and **f**, the filament separating them. F is another filament and 1, 2, 3, t, **P** and **N** are plages. The rules for connecting magnetograms with Hα are given in Sec. 7.1 but are fairly obvious from these frames. Regions of strong polarity are bright in Hα, and dark fibrils either connect or separate those of opposite polarity. Please note that all the BBSO pictures in this book have W left, S top, while other observatories use E left, N top. This is a consequence of the optical setup. (BBSO).

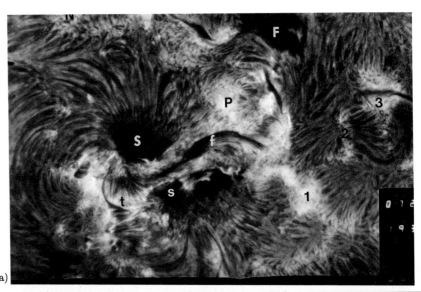

(a)

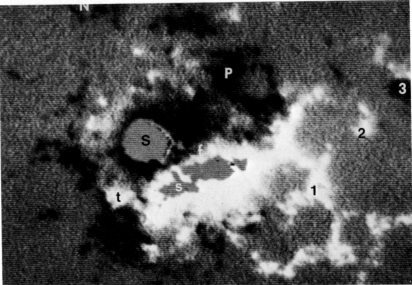

(b)

$$t_i = \cos^2 \pi n_i. \tag{2.14}$$

The transmission as a function of wavelength is given in Fig. 2.10. Now a series of quartz plates, alternating with polaroids, is added (Fig. 2.11). Each plate is twice as thick as the preceding one, so the total transmission of the stack is

$$T = \cos^2 \pi n_i \cos^2 2\pi n_i \ldots \cos^2 2^{k-1}\pi n_i, \tag{2.15}$$

where k is the number of plates. Typically, $n_i \approx 20$ and $k \approx 10$.

The transmitted light has $2^k - 2$ widely spaced maxima with weak side bands in between. The width of each maximum is determined by the thickness of the thickest element. If a broad-band filter is placed in front of the stack, only the desired line is passed. Usually the thickest elements are made of calcite, which has the highest birefringence but is hard to obtain in perfect form.

As we see in Fig. 2.10 there are a series of side bands, spaced at 1.6 times the half-width of the principal maximum. Their existence is easily seen by computing T for various n_i values. The side bands have a total intensity of about 15%, but since the main transmission band is often set in an absorption line (such as Hα) this uninformative light can amount to as much as 50% of that coming through the filter. To suppress the side bands and increase contrast one may add a "suppressor element" 2/3 the

2.12. (c) The same region at Hα − 1 Å in the blue wing.

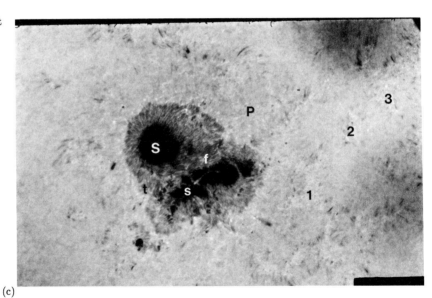

(c)

thickness of the thickest element. Or one may use a narrow band (≈ 10 Å) multi-layer filter.

Additional developments have made the Lyot filter the most powerful auxiliary for solar work. Lyot devised a wide-field version in which the element is split, the two halves rotated 90° and separated by a half-wave plate. This has the effect of making the optical axes symmetric and permits entrance of beams up to $f/15$ (4°wide). In addition, each element can be tuned in wavelength by placing a $\lambda/4$ plate before the second polaroid. Rotation of the polaroid shifts the wavelength between the normal peaks of the crystal sandwich. Evans (1949) showed that one could save a polaroid by placing one element, *sans* polaroid, between the split elements of a thinner one and crossing the end polaroids. The inner element is still wide field but the outer is not. However, if the outer is a thinner element, its field is wide anyway. It is possible to reduce the total polaroid number by four or five this way and nearly double the transmission. In this version the elements are no longer fully tunable.

Since each crystal can be tuned, a fully tunable filter in which every crystal group has $\lambda/4$ plate and rotatable polaroid is possible. The range is limited by the fact that a quartz crystal with $\lambda/4$ retardation at 6000 Å has a much different value at 5000 Å. This problem was solved by Beckers (1973) who employed "achromatized" wave plates made of MgF and quartz, which have opposite dependence of birefringence on wavelength, producing uniform retardation over a wide range in wavelength. Beckers' design led to the Universal Birefringent Filter (UBF), which is fully tunable from 4200Å to 7000 Å, the limitation being the transparency of the polaroids. The Evans split element design is, of course, not used in these. Newer designs have been developed by Title (Title and Rosenberg 1979). Outside of this range the performance of Lyot filters is limited by the performance of polarizers; for the CaII K line (3933 Å) crystal polarizers must be used.

Fabry–Perot type filters using a thin mica film and a multi-layer filter for blocking have become available in recent years. They are small, inexpensive and less temperature sensitive. However, they are limited to a narrow beam, about $f/20$, have low transmission (about 3 times lower than the Lyot), slightly poorer contrast, and cannot be tuned. The problem is that the pass band falls exponentially, while a Lyot filter element has real zeros in the transmission curve. But for many purposes they are most useful.

The measurement of solar oscillations has increased the interest in atomic resonance cells, which have a fixed wavelength reference in the atomic wavelength. A remarkable application of atomic resonance has been introduced by Cacciani (Agnelli *et al.* 1975), employing the Macaluso–Corbino effect. Light passing through sodium vapor in a strong long itudinal magnetic field undergoes resonant scattering in the σ transitions and is circularly polarized (Sec. 5.6). When the cell is placed between crossed polarizers, only the light absorbed and re-emitted in the σ transitions will have its plane of polarization rotated and pass through. Thus the filter isolates the wings of the Na D lines. A second cell placed in tandem is used to select the blue or red wing alternately.

Resonance cells may be made for any easily vaporized material; for example the potassium line at 7699 Å is used. To measure magnetic fields one only needs one cell, switching between right and left circular polarization. To obtain velocity observations one uses two in tandem, the second selecting one wing at a time. The cell is very stable because the atoms always absorb their proper wavelength; on the other hand, that means we can only look at that part of the Sun stationary relative to the observer; this changes with the Earth's diurnal and orbital motion. It is also difficult to keep the windows free of metallic deposits. But overall the Cacciani cell is the most stable and clean monochromator available.

2.5. Magnetographs

The study of magnetic fields on the Sun was revolutionized by the development of the magnetograph by Harold Babcock (1953). Until that time magnetic fields on the Sun could only be detected in sunspots; elsewhere the Zeeman splitting was too small. Babcock developed a photoelectric device to measure the weaker fields. As we will see in Sec. 5.6, the undisplaced π component is linearly polarized, but the two σ components are circularly polarized in opposite directions when viewed along the field. Since the σ components are shifted in wavelength, changing the polarization entering the system corresponds to shifting the line as we choose one or the other component.

Babcock used an electro-optic KDP crystal ahead of the spectrograph. When the proper voltage is applied to this crystal it produces a retardation of $\lambda/4$, which changes the circularly polarized σ components into two beams linearly polarized at right angles to one another. A polaroid placed behind the crystal lets only one of the beams through. As the voltage is alternated, the absorption line studied will shift back and forth slightly in the spectrograph as one and then the other σ component comes through. This shift is, of course, superimposed on a background of unpolarized light, which does not shift at all (this background is the result of light in other wavelengths and the π component, as well as scattered light and light emitted by atoms not in the magnetic field). At this point the whole spectrum is mixed up, but when the light is passed through a spectral analyzer, the intensity in the line wings is seen to be modulated.

Babcock placed a slit on each wing of the spectrum and measured the signal photoelectrically as the line shifted back and forth. The amplified signal was proportional to the longitudinal magnetic field strength. Further refinements include radial velocity compensators and null-seeking systems.

The Babcock magnetograph has operated at Mt Wilson for many years.

The Babcock system measured the field at a single point. Further developments concentrated on measurements over an area. Leighton (1959) introduced the techniques of making two-dimensional magnetograms and Dopplergrams using the spectroheliograph at Mt. Wilson. For Dopplergrams (Fig. 2.13) he took simultaneous photographic images in either wing of an absorption line, made a positive copy of one and superposed the two. Elements moving toward the observer are blue shifted and appear less intense in the blue; those moving away from us are brighter (darker on the negative). The red wing frame is opposite, so when we make a positive print the effect is reinforced. Superposition of the blue negative and red positive produces a uniform gray except where there is a Doppler shift. The superposition is necessary to eliminate the background brightness variation. In the case of a magnetic field, an analyzer is used to separate the Zeeman components, and subtraction as before gives an image with density depending on the magnetic field. Leighton's method was the first application of two-dimensional subtraction techniques; with it the supergranulation, the five-minute oscillation and the structure of the chromospheric network were discovered. With the coming of computers the tedious matching of photographs has been replaced by electronic subtraction.

2.13. Use of the subtractive technique for Doppler or magnetic measurements. The two bandpasses of the monochromator are marked, and the resulting intensity indicated below each. (from Mosher 1976)

Livingston *et al.* (1971) replaced the single or double element detector of the Babcock system with a double array of diodes along the slit of the KPNO spectrograph, so that the field at up to 512 points could be measured simultaneously. This powerful device produced the magnetograms of Fig. 2.15 and a remarkable daily record of the solar magnetic field; it still

DOPPLER SHIFT

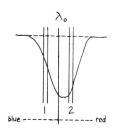

	(KDP+) Red Wing	(KDP−) Blue Wing
Rising Element	bright	dark
Falling Element	dark	bright

ZEEMAN SPLITTING

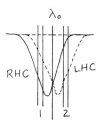

	Red Wing		Blue Wing	
	RHC	LHC	RHC	LHC
Positive Field	bright	dark	dark	bright
Negative Field	dark	bright	bright	dark
	(+)	(−)	(−)	(+)

produces the best full-disk magnetograms.

Leighton and Smithson (Smithson 1973) replaced the spectrograph with a Lyot filter and the diode array with a vidicon in the videomagnetograph (VMG) of the Big Bear Observatory (Fig. 2.14). Two-dimensional magnetograms could be obtained directly, without the slow process of adding up the linear results. A 1/4 Å filter is used to limit the light to the wing or wings of a single line (usually 6439 Å or 6103 Å, with the initial polarizer removed (Ramsey 1971) to permit a double bandpass. Ahead of the filter a KDP (potassium dihydrogen phosphate) crystal is modulated to give $\pm \lambda/4$. If it is positive, the left-circular Zeeman component σ_1 is linearly polarized in one plane and the σ_2 component is polarized orthogonally. When these two components enter the filter they are passed at $\pm 1/2$ the passband (or ± 0.125 Å) with the result that a given magnetic polarity produces the same effect in either passband. In the present system (Mosher 1976; Zirin 1986) the image is recorded by a CCD operating at video rates, digitized on-line and accumulated in a memory that is set to half intensity at the beginning of the sequence. The KDP voltage is now reversed and a new exposure is obtained, with reversed signal. The process continues at the 30 frame/sec video rate; for active regions good magnetic maps are produced in 32–128 frames, or 1–4 sec; for weak fields integrations of up to 4096 frames are used. Addition of a circular polarizer (polaroid + $\lambda/4$ plate) in the double pass band mode alternates the wavelength between red and blue wings of the line and we have a Doppler analyzer. This system

2.14. Scheme of the videomagnetograph, showing the transmission by the filter as the voltage on the KDP modulator is switched. (from Mosher 1976)

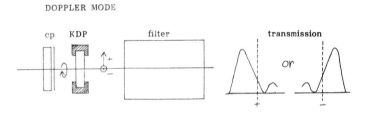

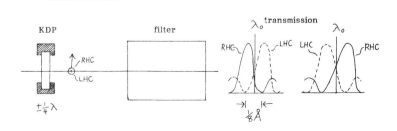

has produced all the videomagnetograms in this book, and a similar system produced Fig. 6.6.

The VMG is extremely sensitive because no scanning is involved and the images may be added up almost indefinitely. Leighton insisted on simple components operating at video rates; as a result pictures only 30 ms apart are subtracted and seeing effects reduced. Systems of this sort appear to yield better results than CCD designed for low light levels. The latter have better signal-to-noise ratio, but because they must be read out more slowly, the long interval between the subtracted images makes them much harder to match up. The many frames of the video system permit even higher signal-to-noise ratio. For deep magnetograms the brightness variation of the granulation becomes a problem. It can be removed by using the filter in a single bandpass on the blue edge of the spectral line where the blue shifted absorption due to the rising granule cancels its excess brightness and produces a more uniform brightness field.

This system is sensitive only to longitudinal magnetic fields, but if we insert a $\lambda/4$ plate the system we alternately see the linearly polarized π and σ components emitted at right angles to the magnetic field (Fig. 5.5). Modulation gives us the component amplitude of the transverse field in one direction. By rotating the $\lambda/4$ plate or modulating a KDP the polarization component along another axis is obtained and combining these two gives the direction and magnitude of the transverse field. Brueckner (Hagyard 1982) developed a vector magnetograph at the Marshall Space Flight Cen-

2.15. A magnetic map of the whole Sun made by J. W. Harvey at the Kitt Peak National Observatory. White field is North seeking, dark is opposite. These maps are made with a multi-element Reticon detector at the output of the spectrograph and measuring the shifts associated with the Zeeman effect (Sec. 5.6). The slit is swept across the Sun like a spectroheliograph. The active regions are all dipoles (no magnetic monopoles!) with opposite polarity in the North and South hemispheres. The magnetic fields remaining from old regions spread out into large unipolar regions, showing the small-scale structure of the chromospheric network. (KPNO)

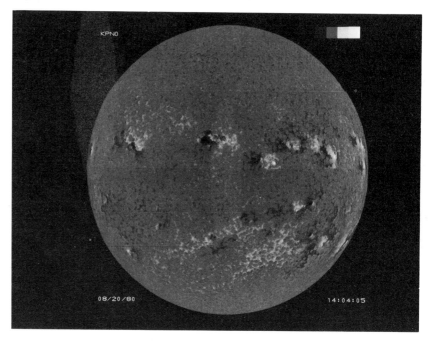

ter (MSFC) using such an arrangement. The sensitivity is not great, and the reduction, tedious, but the data are much prized in flare studies. With modern image processors these magnetographs should improve, although so far they tell us little that is not evident in a good Hα image.

CCD's or vidicons used in the video mode have a signal-to-noise ratio of about 100:1, so a ten second integration, which permits about 200 frames after overhead, yields more than 1000:1 signal-to-noise ratio.

2.6. The Coronagraph

As soon as astronomers became aware of the existence of the corona they sought means to observe it outside of eclipse. G. E. Hale climbed Mount Etna with a small telescope but volcanic dust and smoke stopped him. Anyway, since the corona is a million times fainter than the Sun, a system more free of scattered light than an ordinary telescope is required.

The problem was solved by Bernard Lyot, who (Lyot 1930; Evans 1953) understood that both the sky brightness and the scattered light in the telescope overwhelm the corona. He built a special low-scattering telescope which he placed on a mountaintop, where the sky was blue. For the coronal continuum the aureole must be fainter than 10^{-6} $I_\odot$ a few minutes of arc (the Sun is 30 minutes of arc in diameter) from the limb; for emission lines, scattered intensity below 10^{-4} $I_\odot$ is adequate. The best skies are about 10^{-5} $I_\odot$. A good way of testing the sky is to obscure the Sun with your thumb at arm's length. If the sky is deep blue up to the very edge of the

2.16. An active region photographed through a 5 Å wide multi-layer filter centered at the "G band", 4305 Å. In the blue part of the spectrum, contrast is enhanced, and low chromospheric structure can be recorded, especially if we pick one of the broad clusters of absorption lines and molecular bands. The limb of the Sun is at top. The dark spots are small sunspots associated with the group; the many bright faculae are the loci of magnetic fields. In between these we see the blue granulation, smaller than the white-light granulation. (BBSO)

Sun you are at a good coronagraph site. This is a good ploy to use if you should ever visit a solar observatory.

Lyot solved the scattered light problem through a number of clever precautions (Fig. 2.17). He selected an objective lens (O1) free of sleeks and bubbles and highly polished. By rubbing his finger at the side of his nostril and carefully spreading the nose oil over the lens, he obtained a surface free of static charge and dust. I personally always got a mess, which shows that the coronagraph business still has as much art as science in it; thus there is much dispute over the merit of the nose-oil treatment. The use of old diapers, washed many times, has its partisans.

Lyot removed the direct light of the photosphere with an artificial moon or occulting disk at the focus of the objective, which reflected away the bright solar image. Directly behind the occulting disk is a field lens (F1) which produces an image of the objective at a diaphragm (D). A screen placed at this point would show an image of the objective with a bright halo of light caused by diffraction at the outer edge, and a bright spot at the center produced by double reflection there. The halo is eliminated by making the diaphragm slightly smaller than the image, and the central bright spot is eliminated by placing a black dot at the center of the lens (O2) situated directly behind the diaphragm. Now all that remains is the image of the rest of the objective, which should contain no scattered sunlight except for that caused by dust or imperfections in the lens.

We do not want to look at the image of the lens but that of the corona. The lens (O2) forms an image of the eclipsed Sun on our analyzer. Coronal emission lines are picked out with a monochromator, either a spectrograph or narrow band filter. We still cannot see the coronal continuum, which is fainter than the sky.

In recent years space observations of permitted coronal lines have decreased the emphasis on coronagraphs, but they are still unsurpassed for prominence spectroscopy. Electronic cameras have permitted sky subtraction, which has greatly enhanced their sensitivity. Coronal lines may now be detected out to 0.5 $R_\odot$.

2.17. Scheme of the Lyot coronagraph. The lens F1 forms an image of the objective near O2, where the remaining diffracted light is removed.

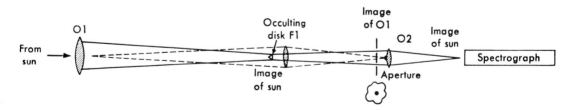

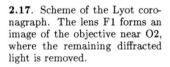

The detection of the continuous coronal emission required development of
the K-coronameter (Wlerick and Axtell 1957). This instrument is described
in detail by Billings (1966), and further development (Fisher *et al.* 1981)
has made possible observation of the K corona to some distance from the
Sun. Since the K corona, the continuous light from the corona, is less
than one millionth of the photospheric intensity (about 100 times fainter
than the emission lines and ten times fainter than a clean blue sky), these
instruments use the linear polarization of the K corona to separate it from
the sky, which is unpolarized near the Sun. This is usually done by an
electro-optic modulator, usually a KDP crystal. These crystals produce
phase retardation proportional to the voltage, so with a quarter wave plate
they give alternately 0 or 1/2 wave retardation, which has the effect of
choosing alternating orthogonal polarizations. These instruments must be
exceptionally clean optically.

The scattering of sunlight in the objective limits further improvement on
the ground, but the use of external occulting disks (Wagner *et al.* 1981) in
space coronagraphs has made possible observations out to five solar radii.
Several disks must be used in tandem to block out the diffraction from the
other disks.

2.18. Schematic of the externally occulted coronagraph built
by High Altitude Observatory
for the Solar Maximum Mission
(MacQueen *et al.* 1980). Three
occulting disks are located at
right, the second and third removing diffraction from the first.
A fourth disk is located at D4,
in front of the first field lens.
(HAO)

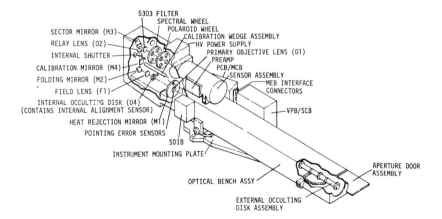

2.7. Solar radio telescopes

Solar radio telescopes require a short time constant and the ability to scan rapidly in frequency. The sensitivity can be lower but should be adequate to observe cosmic calibration. Interferometer arrays which represent partially filled dishes of very much larger size are most useful; they combine high resolution with rejection of the uniform solar background. The VLA, with three arms each 27 km long, synthesizes a colossal dish 39 km in diameter. The beam has many side lobes but it may be cleaned mathematically. Resolution of an arc second is possible in single frames, and 0.05" may be reached by observing all day.

Like all radio telescopes, their resolution is limited by the long wave length; for example the largest steerable radio dish is the Bonn 100-meter telescope, which has a resolution at 3 cm (the shortest wavelength at which the whole aperture is accurate) of

2.19. The VLA. Here the antennas were brought together in the closest (D) configuration; they can be located up to 22 Km from the center, but for solar work the smallest (C and D) configurations are most suitable.

$$\Delta\theta = 2.52 \times 10^5 \frac{\lambda}{D} = 2.52 \times 10^5 \times \frac{3}{10000} = 75 \text{ arc sec}, \qquad (2.16)$$

and the highest resolution dish in the world, Leighton's 10 meter Owens

Valley millimeter dish, has resolution 24" at 1 mm.

Because the Sun often shows intense transient sources against a complex background, interferometric observations are particularly valuable. They reject the larger scale resolved fluctuations and detect unresolved sources smaller than the resolving power of the system. From the phase of the signal it is possible to get fairly accurate lines of position for the source relative to the fringe nulls.

The solar interferometer at the Owens Valley Radio Observatory uses frequency-agile receivers which may switch to any frequency from 1 to 18 GHz. This makes it possible to obtain the spectra of bursts and also different lines of position; if the source position does not vary too much with frequency we can treat these data like a many-component map at a single frequency and get good positional data.

The meter-wave radio telescope at Culgoora built by Wild was a remark-

2.20. The solar disk in the wing of Hα, photographed through a half angstrom Lyot filter. From the rooftop of Robinson Laboratory in Pasadena where this small patrol telescope is located, many planes like this DC10 are seen cross the solar disk. (BBSO)

able array designed specially for solar work (Labrum 1972). It was a circle three km in diameter with 96 dishes which produced "pictures" of the Sun in left- and right-hand circular polarization every second at 40, 80, 160, and 320 MHz. The Culgoora array was unfortunately dismantled in 1984 but left a legacy of important data. A similar but more flexible system of this type has been built by Erickson at Clark Lake Radio Observatory.

A problem in calibrating solar radio observations is that the Sun dominates the antenna temperature and the gain must be suitably adjusted or kept constant as needed. For systems with a kind of automatic gain control, such as the VLA, it is necessary to measure the total solar power separately so that the true antenna gain can be determined.

References

Billings, D. E. 1966. *A Guide to the Solar Corona.* New York: Academic Press.

Dainty, J. C. and R. Shaw 1974. *Image Science.* New York: Academic Press.

Jenkins, F. A. and H. E. White 1957. *Fundamentals of Optics.* New York: McGraw Hill.

3

Plasmas in Magnetic Fields

If the Sun had no magnetic field, it might be a quiet, "classical" star–if such stars exist, with no corona, chromosphere, sunspots, or solar activity. Such phenomena are entirely due to the interaction of the ionized solar plasma with the magnetic field which this plasma has somehow produced. In this chapter we shall try to summarize the more important interactions that occur between plasma and field, but the actual behavior of the fields on the Sun is left for the later chapters; in fact that is what this book is largely about.

The term *plasma* refers to an electrically neutral material which is largely ionized. The existence of plasmas is made possible by the balance of ionization and recombination. If a gaseous plasma cools, the charged particles recombine, and it ceases to be a plasma; but in metals quantum-mechanical effects produce plasma-like behavior at even the lowest temperatures, because the electrons in the metal are detached from the atoms by pressure ionization and move about freely. Almost all of the Sun is a gaseous plasma, except for a thin unionized layer around the photosphere.

3.1. The properties of plasmas

Any plasma of large physical extent must be electrically neutral to great precision – otherwise strong electrostatic forces would arise and attract particles of the opposite charge, neutralizing the plasma.

The properties of a plasma are the sum of the properties of its constituents. The density is the sum of the densities of the various parts:

$$\rho = \rho_i + \rho_e + \rho_n = n_i m_i + n_e m_e + n_n m_n, \tag{3.1}$$

where the subscripts i, e and n refer to ions, electrons and neutral particles. If C_0 is the velocity of the whole parcel of gas, and C_{es} the velocity of an electron, then the peculiar velocity v_{es} of the electron s is

$$v_{es} = C_{es} - C_0. \tag{3.2}$$

Of course, the average peculiar velocity is zero.

The kinetic temperature of a constituent is defined in terms of the mean square peculiar motion:

$$\frac{3}{2}kT_e = \frac{1}{2}m_e\overline{v_e^2}, \tag{3.3}$$

where the mean square velocity is just the average squared velocity of all the particles. The mean velocities of the different constituents may be different because of the effect of electromagnetic fields and the like; but collisions between particles remove these differences rapidly and establish an equilibrium state in which the energy per particle will be the same for all varieties. The only exceptions are cataclysmic phenomena like flares, where high-energy accelerated particles are injected into cooler ambient plasma. Henceforth we will only use subscripts for particle varieties if they differ in properties. In the equilibrium state, the velocities of the particles are distributed according to the Maxwellian distribution $f(v)$. The number of particles/cm^3 with peculiar velocities between v and $v + dv$ is

$$N\,f(v)dv = 4\pi N \left(\frac{m}{2\pi kT}\right)^{\frac{3}{2}} v^2 e^{-mv^2/2kT}\,dv, \tag{3.4}$$

where N is the number of particles/cm^3. Similar distribution formulas hold for each type of particle. Any deviations from this distribution are rapidly wiped out by collisions. The peak of the velocity distribution is

$$v = \sqrt{\frac{2kT}{m}}. \tag{3.5}$$

In the case of a dilute plasma where collisions are infrequent, it is necessary to consider separately velocities in different directions. This might occur in interplanetary space, where particles are generally streaming from the Sun in one direction; or in a magnetic field, where the field restricts or alters the motion of particles across it. In these cases there may or may not be Maxwellian distributions in the motions in each coordinate, but when many collisions occur the velocity distributions in all coordinates must be the same. Liouville's theorem states that the density in phase space (that is, the number of particles at a point with a given velocity) is constant along dynamic trajectories. If it is possible for a particle to travel from one point to another, or to jump (via collisions) from one velocity in one direction to another velocity in a different direction, in equilibrium there will be an equal density and velocity distribution everywhere (except when potentials change the kinetic energy of the particles along the trajectory). An evacuated bulb will gradually fill through whatever minute leaks are present, and if we wait long enough, the final result will be the same no matter what the size of the leak is. In the velocity field, collisions between particles of different velocities play the same role as leaks between regions of different densities.

The plasma pressure is the average momentum carried across the surface by particles moving in each of the three directions; thus it is a tensor:

$$|P| = |nm\overline{v_i v_j}|. \qquad (3.6)$$

In each term we add the effects of all sorts of particles. In almost all real cases the pressure is a diagonal matrix with elements $nm\overline{v_i^2}$. If there are enough collisions, these are all equal, and

$$P_i = \frac{1}{3}nm\overline{v^2} \qquad (3.7)$$

in any direction i. When the magnetic pressure dominates the gas pressure, there may be a difference between the pressures across and along the magnetic fields. These are distinguished by writing $v_\perp$ and $v_\parallel$ for velocities across and along the field lines; the pressure tensor then has one diagonal term parallel to the field and two perpendicular.

The relaxation time is that in which equilibrium is established between the constituents of a plasma. Often accelerated particles are injected into a plasma made up of much lower energy particles, and we want to know how quickly they slow down to the ambient values. Because the Coulomb forces between charged particles are long range, this is a complicated business, described in detail by Spitzer (1962). But there is a simple formula (Takakura 1969) which can be used to compute t_e, the time for energy loss by fast electrons:

$$t_e \approx \frac{10^8}{nE(keV)^{\frac{3}{2}}} \quad \text{sec}. \qquad (3.8)$$

This energy-loss time is not the same as the deflection time t_D, the time for the particle to change its direction by 90°. The latter is shorter for high-velocity particles and longer if the injected particles are slower than the field particles.

For example, flares often inject streams of 40 keV electrons into the corona, where $N_e \approx 10^8$. In this case $t_e = 253$ sec. If the coronal temperature is 2×10^6 deg, the field particle energy is 0.17 keV and t_D is about 5 sec. These electrons travel with 1/3 the speed of light and easily escape the corona in this time unless restricted by magnetic fields.

The thermal conductivity of a plasma depends on the product of the mean free path $v_e t_e$ and the electron density n_e. Since n_e cancels in this product, the conductivity is independent of density; since each electron carries energy kT, the conductivity is proportional to $T^{5/2}$. In the solar atmosphere it is roughly

$$K \approx 5 \times 10^{-7} T^{\frac{5}{2}} \quad \text{ergs/cm/sec/deg.} \tag{3.9}$$

The electrical conductivity depends only on collision frequency and not the kT carried by each particle, hence

$$\sigma = 1.53 \times 10^{-13} \frac{T^{\frac{3}{2}}}{Z \ln \Lambda} \quad \text{emu,} \tag{3.10}$$

where $\ln \Lambda$ is the Coulomb logarithm, which allows for the cutoff of coulomb interaction due to screening. This logarithm, which is of the order 1–10 in the solar atmosphere, is tabulated by Priest and by Spitzer.

3.2. Motion of individual particles

We define the magnetic field **B** in terms of magnetic lines of force or the generating currents. The direction and density of lines of force at each point give the direction and magnitude of the field. If plasma motions or currents change the direction, we may consider them as moving or twisting the lines of force. But the force lines are not real; they only mark the direction and magnitude of the magnetic field.

Before we look at the gross behavior of charged particles in magnetic fields, it is valuable to consider the motion of individual particles. The force acting on a particle of charge e and mass m in a vector magnetic field **B** is

$$F = \frac{d\mathbf{p}}{dt} = \frac{e}{c} \mathbf{v} \times \mathbf{B}, \tag{3.11}$$

where **p** is the vector momentum and **v**, the velocity.

The cross product $\mathbf{v} \times \mathbf{B}$ is such that the resultant force is perpendicular to both **v** and **B**. As a result, a negative particle moves in a circular clockwise path as seen looking along the magnetic lines of force. The instantaneous center of the circle is called the center of gyration (Fig. 3.1).

3.1. A negative charge drifts in a right-hand spiral around the direction of parallel magnetic and electric fields. Its motion can be considered as a circle about a guiding center, which moves along the field line in the direction of $-\mathbf{E}$.

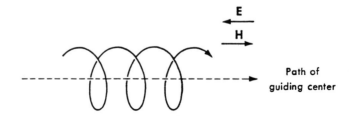

Since the effect of the magnetic field is a purely centripetal force perpendicular to the circular motion, the field does no work. The gyroradius, or radius of curvature ρ of the rotation is

$$\rho = \frac{p_\perp c}{eB} = 5.69 \times 10^{-8}$$
$$= 2.21 \times 10^{-2} \frac{T^{\frac{1}{2}}}{B} \quad \text{cm}, \tag{3.12}$$

where $p_\perp$ is the momentum in the plane perpendicular to the field lines and e the absolute value of the charge.

If we evaluate the gyroradius for a 40 keV electron in a one gauss magnetic field, typical of the corona, we find it to be about 7 meters; the electrons gyrate closely about the field lines.

The *gyrofrequency* ν_g, is

$$\nu_g = \frac{eB}{2\pi \gamma m c} = 2.8 \text{ MHz/gauss}, \tag{3.13}$$

where γ is a relativistic correction:

$$\gamma = \frac{p}{mv} = \frac{1}{\sqrt{1 - v^2/c^2}} = \sqrt{1 + \frac{p^2}{m^2 c^2}}. \tag{3.14}$$

The product

$$B\rho = \frac{c p_\perp}{e} \tag{3.15}$$

is the magnetic rigidity, which is of great importance in studying the trajectories of cosmic rays.

Alfvén introduced the concept of the "guiding center", the instantaneous center of gyration, which follows a trajectory determined by the nonmagnetic forces. For example, an electric field parallel to $\mathbf{B}$, exerting a force $-Ee$ on the electron, produces an acceleration $-Ee/m$ of the guiding center along the lines of force. The guiding center slides along them, while the particle traces a spiral along the outside of a cylinder of radius ρ. An electric field perpendicular to the magnetic field produces a drift of the guiding center in the direction perpendicular to $\mathbf{E}$ and $\mathbf{B}$, with velocity

$$v_{gc} = \frac{E_\perp}{B} = 10^8 E_\perp \quad \text{volts/cm}/B. \tag{3.16}$$

If the field is inhomogeneous, the radius of the circle about the guiding center varies with the field strength. The particle drifts at right angles to $\mathbf{B}$ and its gradient $\nabla \mathbf{B}$, because long half-loops at low $\mathbf{B}$ alternate with short half-loops at high $\mathbf{B}$. The drift velocity is

$$v_d = v_\perp \rho \frac{\nabla_\perp \mathbf{B}}{2\mathbf{B}}. \tag{3.17}$$

Drifts also arise in curved lines of force. In a gravity field perpendicular to **B**, the particles will drift perpendicular to that field.

When the magnetic field changes slowly with time, a series of *invariant* or almost invariant quantities is useful in determining the change of different properties. One such invariant is $W_\perp/B$, the ratio of the kinetic energy involved in the circling motion to the field strength. If we increase the field strength slowly, the energy of gyration must also increase.

Another invariant of importance involves the motion of the guiding center of the particles. If a charged particle is trapped between two regions of strong magnetic field ("magnetic mirror"), the integral of its momentum along the magnetic field through an entire circuit between the mirrors is an adiabatic invariant:

$$Q = \oint mv_\parallel ds = \text{const.} \tag{3.18}$$

This is true so long as the field does not change greatly through one circuit. If the two mirrors approach each other slowly, the particle moves back and forth between them, gaining energy all the time according to

$$T = \frac{1}{2m}\left\{ p_\perp^2 + \left(\frac{l_0}{l}\right)^2 p_\parallel^2 \right\}, \tag{3.19}$$

where l_0 is the original distance between the mirrors and l, the instantaneous separation. This process is the first-order Fermi acceleration mechanism; it is one of the likely ways in which high-energy particles are produced by magnetic fields. In this case, although the magnetic forces cannot do work on the particle, they provide a means by which the energy of many particles, acting in concert, may be conveyed to a single one. Since the mirrors are moved by large bodies of material, they are not slowed down appreciably by the energy losses to the particle.

3.3. Magnetohydro-dynamics

The actual behavior of the plasma in a magnetic field is governed by two series of equations: the Maxwell equations, which describe the relations between the electric field **E**, the magnetic field **B**, the charge density, and the current **j**; and the hydrodynamic equations, which interrelate pressure, density, temperature, and flow. In Gaussian *cgs* units* the Maxwell equations are:

$$\nabla \cdot \mathbf{E} = 4\pi c^2 \sigma = 4\pi ec(n_i Z - n_e) \tag{3.20a}$$

---

* A good discussion is in Priest (1982), Appendix I.

$$\nabla \cdot \mathbf{B} = 0 \tag{3.20b}$$

$$\nabla \times \mathbf{E} = -\frac{1}{c}\frac{\partial \mathbf{B}}{\partial t} \tag{3.20c}$$

$$\nabla \times \mathbf{B} = \frac{1}{c^2}\frac{\partial \mathbf{E}}{\partial t} + 4\pi \mathbf{j}. \tag{3.20d}$$

The first equation states that the divergence

$$\nabla \cdot \mathbf{E} = \frac{\partial}{\partial x}E_x + \frac{\partial}{\partial y}E_y + \frac{\partial}{\partial z}E_z \tag{3.21}$$

of the electric field is equal to the net charge. It means that electric field is conserved just like a fluid flow and a change in field in any direction is only produced by the existence of a net charge, which acts like a source or sink. The field near a point charge is divergence-free except in the region of the charge.

The second equation tells us there are no magnetic monopoles, or point sources of magnetic field. All lines of force must have two ends, and field can only be increased or decreased through the action of currents.

Eq. (3.20c) is the Faraday dynamo equation, and Eq. (3.20d) tells us that if there is any rotation in the field there must be a current. The common solar situation of nearby oppositely directed fields requires a current sheet in between to turn the field around.

If we combine the last two Maxwell equations with the generalized Ohm's law, which includes the Lorentz force in the resistance,

$$\mathbf{j} = \sigma(\mathbf{E} + \mathbf{v} \times \mathbf{B}], \tag{3.22}$$

we get the "induction equation"

$$\frac{\partial \mathbf{B}}{\partial t} = \nabla \times (\mathbf{v} \times \mathbf{B}) + \eta \nabla^2 \mathbf{B}, \tag{3.23}$$

where $\eta = 1/4\pi\mu\sigma$ is the magnetic diffusivity:

$$\eta = 5.2 \times 10^7 \ln \Lambda T^{-\frac{3}{2}} \qquad \text{m}^2 \text{ sec}^{-1}. \tag{3.24}$$

η is about 1000 in the chromosphere and 0.5 in the corona with $\log \Lambda \approx 10$. Eq. (3.23) describes the evolution of the magnetic field with time. The first term on the right can be transformed to show that the magnetic flux through a closed circuit moving with the field is a constant. The only change this term can produce is to move the field around; if the second term is zero, the field is "frozen-in" to the plasma and moves with it, expanding, contracting, *etc.* The second term on the right describes the diffusion of plasma through the magnetic field; if we replace the variables by their approximate scale we have

$$\frac{B}{T} \approx \frac{\eta B}{L^2}. \tag{3.25}$$

The time t_r for chromospheric **B** fields of scale L to change by their own magnitude through resistive diffusion is

$$t_r \approx 1000L(\text{km})^2 \quad \text{sec} \tag{3.26}$$

at the surface of the Sun (T = 6000°). The length scale of the magnetic field must be about 0.3 km for anything to happen in 100 sec. For fields the size of sunspots (10^4 km), $t \approx 10^{11}$ sec or 1000 years, and change in a field the scale of the Sun would take correspondingly longer, about 10^6 years. These long diffusion times have made it difficult to understand how the magnetic field could change fast enough to produce solar flares by reconnection.

If we look at the general solar field, inserting the size of the Sun for L gives a change time of about 10^{10} years. Imagine the surprise of solar physicists when Harold Babcock (1959) announced that the polar field reversed in 1957. Since then we have found that the general solar field reverses every 11 years, and takes less than a year to do it. L must be about 300 times smaller than the solar radius. The magnetograph has revealed that reversal involves small clumps of field around that size.

The magnetic field produces a pressure $B^2/8\pi$ transverse to the force lines and a tension $B^2/4\pi$ along the lines of force; but since the plasma moves freely along the field lines, it is unaffected by the tension. Across the field lines there must be a pressure balance

$$\frac{B_1^2}{8\pi} + n_1 k T_1 = \frac{B_2^2}{8\pi} + n_2 k T_2. \tag{3.27}$$

A sunspot with a strong magnetic field will reach equilibrium with its surroundings only if its temperature or density is lower. Hydrostatic equilibrium requires that the densities at a given height be roughly equal, or else the matter will sink until it does; sunspots must be cool to be stable.

If **E** and **j** are zero (no electric field, no current), Eq. (3.20d) gives:

$$\triangledown \times \mathbf{B} = 0, \qquad \mathbf{B} = \triangledown A. \tag{3.28}$$

So if there are no currents, **B** is given by the gradient of a scalar potential A. This is the normal "potential field" configuration we know from iron filings. If the first term in Eq. (3.27) exceeds the second – i.e. the magnetic pressure exceeds the gas pressure – any current must flow along the field lines; therefore,

$$\begin{aligned} \triangledown \times \mathbf{B} = \mathbf{j} = \alpha \mathbf{B} \\ \mathbf{j} \times \mathbf{B} = 0, \end{aligned} \tag{3.29}$$

where α is a scalar quantity. This is called a force-free field because the current produces no force. Along with the special case $\alpha = 0$ of the potential field, this equation describes most of the solar magnetic fields that we see. The energy is proportional to α^2. The current-free, or potential field has the lowest possible energy. The force-free field cannot exist in a vacuum; it must be anchored to a mass sufficient to balance the pressures it exerts, or it must be stabilized by currents.

Isolated plasmas in a gravitational field with their own self-enclosed magnetic field are diamagnetic; they move toward regions of lower magnetic field. If a diamagnetic bubble rises, the internal field expands to maintain equilibrium with the surroundings, and the gas may become still lighter. If the external field decreases more rapidly than the adiabatic gradient, the bubble will continue to rise until completely free of external pressure. Because of this diamagnetic effect, almost all prominences in the solar atmosphere erupt *upwards*. This is called the "melon-seed" effect, or magnetic buoyancy.

In the case of the force-free field the lower energy state is that of lower (or zero) α; in the diamagnetic case it is an uncompressed state of the bubble. The bubble field is not tied to anything else and may expand freely, but all other fields with which we deal are tied to the surface, and changes will require the crossing of field lines which is governed by the diffusion time Eq. (3.25). This time will only be short if the field breaks up into small scale structures or if the conductivity is lowered by phenomena like wave-particle interactions.

The stability of magnetic configurations is extremely important in controlled fusion, where hot plasmas are contained by magnetic fields. Much of this analysis can be taken over to solar physics. All magnetic configurations are unstable if they can jump to a lower energy state; for example, a field-free plasma contained in pressure by field lines curving concave inward is unstable because field and plasma can exchange positions – the field lines collapse inward and the plasma pops out. This kind of exchange is easier if field lines do not cross. However, some systems are stable against such transitions, and the stability can be determined by introducing perturbations of different scale into the equation and seeing if they grow or decay. The fact that the system is stable against one scale of perturbation does not at all ensure stability against others; but once any instability occurs, it will break up the system. Examples are discussed by Kulsrud (1967).

We often observe that solar magnetic fields rearrange into lower energy configurations, and elements which are connected to one pole reconnect to another. Indeed, if this did not occur, newly emerging sunspots would produce an endless increase in the field strength in the photosphere. For magnetic reconnection to occur, field lines must move across one another to connect to different poles. Crossing lines of force imply a double valued solution to Maxwell's equations which cannot exist. But as we saw in Eq. (3.25), rapid field diffusion can take place if the gradient (actually the Laplacian) of the field is large, or if the resistivity is great. Typically when an MHD instability occurs, the field breaks up into small elements which restrict current flow and increase the diffusion constant. This anomalous resistivity makes possible the field changes. A simple example of the kind of change that occurs is given in Fig. 3.2.

3.4. Wave phenomena in plasmas

Many different oscillatory phenomena can take place in plasmas, with the magnetic field or the particle interactions as the restoring force. Alfvén showed that if we disturb or displace a magnetic line of force, the disturbance will propagate along the lines of force as a wave with characteristic velocity

$$V_A = \frac{B}{\sqrt{4\pi\rho}} = 2.2 \times \frac{10^{11} B}{\sqrt{n_e}} \tag{3.30}$$

and time

$$\tau_A = L/v_A. \tag{3.31}$$

So if $n_e = 10^8$ (solar corona) and $B = 1$ gauss, v_A is 220 km/sec, and the time to cross the core of a flare 10 arc sec (7000 km) in diameter will be 30 sec. This disturbance propagates much like that in a plucked string; the restoring tensile force in the string is $B^2/8\pi$. The Alfvén velocity is characteristic for the propagation of all disturbances in the magnetic field; in particular, it governs the rate of growth of magnetic instabilities. The flare growth time will be t_A. The ratio

$$R_m = t_r/t_A \tag{3.32}$$

of the resistive diffusion time t_r from Eq. (3.25) to the Alfvén time is the "magnetic Reynolds number", used to measure the relative importance of these processes. Of courses, the Alfvén time is only important if instabilities occur. The Alfvén wave can also propagate perpendicular to the lines of force as a magnetosonic or "fast-mode" wave, with the restoring force

3.2. A new dipole with polarities p (preceding) and f (following) erupts near another dipole. The field lines between opposite polarity reconnect to form a lower energy configuration. Gold and Hoyle (1960) suggested this might be the energy source of solar flares. Resistive diffusive would produce reconnection in time t_r (Eq. (26)). But instabilities produce steep magnetic gradients that shorten the length L in Eq. (3.25) and reduce t_r.

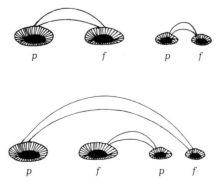

provided by gas pressure and magnetic tension. In this case the velocity is the Pythagorean sum of the squares of the Alfvén and sound velocities. If there is no magnetic field, this wave reduces to a normal sound wave. The exact propagation from a disturbance at a particular point will depend on the variation of field and density. Alfvén waves are characterized by frequency below the gyrofrequency and wavelength much longer than the interparticle distance.

If there is no magnetic field in a plasma, there are still different modes of wave propagation: electromagnetic and electrostatic. The electromagnetic waves are of the normal type (like light or radio waves), but their propagation is modified by the electrons of the plasma. The electrons can oscillate about the relatively stationary ions, with the electrostatic attraction as a restoring force. The characteristic frequency is the plasma frequency,

$$\nu_p = \frac{1}{2\pi} \left(\frac{4\pi n_e e^2}{m_e} \right)^{\frac{1}{2}}$$

$$= 8.97 \times 10^3 n_e^{\frac{1}{2}}.$$

(3.33)

Electromagnetic (em) waves of this frequency will resonate with the electrons of the plasma at ν_p. The wave energy will be transferred to the electrons and the wave will not propagate, being absorbed or reflected in a distance of one wavelength. The index of refraction is

$$n^2 = 1 - \left(\frac{\nu_p^2}{\nu^2} \right).$$

(3.34)

Since phase velocity is c/n and group velocity is nc, we see that if $\nu = \nu_p$, $n = 0$, phase velocity is infinite but group velocity is zero, so there is no energy propagation. If $\nu > \nu_p$, then $n > 1$ and $v_{ph} > c$, but $v_{gr} < c$. If $\nu < \nu_p$, n is imaginary and the wave is absorbed. If we divide the mean electron velocity $\overline{v_e}$ by n from Eq. (3.34) we can get the velocity of electrostatic waves.

Eq. (3.34) determines the depth in the solar atmosphere to which we see in radio wave lengths. The plasma frequency drops as the density decreases exponentially outward. Only high-frequency waves (above 3000 MHz) can escape from low in the atmosphere, but higher up, frequencies as low as 10 MHz escape. Since the opacity is greatest near the plasma frequency, emission comes mainly from that height.

If there is a magnetic field present, *em* waves propagate in oppositely polarized ordinary and extraordinary modes, depending on the relation of the polarization to the field. Moreover, two new characteristic frequencies are introduced: the gyrofrequencies ω_{ce} and ω_{ci} with which electrons and ions respectively circle around the lines of force

$$\omega_{ce} = \frac{eB}{m_e c} \qquad \omega_{ci} = \frac{eBZ_i}{m_i c}, \tag{3.35}$$

where $\omega_{ce} = 2\pi\nu_g$ from Eq. (3.13). These particles can resonate with the applied *em* field, giving zero propagation velocity; or at certain frequencies the velocity becomes infinite, so that total reflection, or "cut-off," takes place. These cutoffs and resonances are given by dispersion relations that give velocity V as a function of frequency ω. Eq. (3.36) is the dispersion relation for the circularly polarized ordinary wave that propagates parallel to **B** with electric vector along it, and therefore unaffected by **B**. The ordinary wave possesses a single resonance at the plasma frequency ω_p. The dispersion relation for the extraordinary mode is given by Spitzer as

$$\frac{c^2}{V^2} = 1 - \frac{\omega_p^2}{\omega^2 - \omega_{ce}\omega_{ci} + \frac{\omega^2(\omega_{ce}-\omega_{ci})^2}{\omega_p^2 - \omega^2 + \omega_{ce}\omega_{ci}}}. \tag{3.36}$$

When this is expanded in powers of ω_e/ω_i, two resonant frequencies and two reflections are found. The resonances correspond to oscillation by the electrons only and by electrons plus ions. The cutoffs, or reflections, are near the gyrofrequencies ω_{ci} and ω_{ce}, if the magnetic field is strong and those frequencies are large compared to the plasma frequency. If the plasma frequency is much the larger, they occur near it.

3.5. Shock waves

Although a shock wave is not a harmonic phenomenon, it is a wave in the sense that it has group and phase velocity, and propagates through a medium. A shock wave occurs when the velocity of a disturbance in a medium exceeds the normal propagation velocity. Then the normal propagation cannot carry away all the energy brought in, and there is a sharp change of physical parameters at the front of the wave. An analogy is the effect of a sharp highway constriction on traffic flow: in the reference frame of the traffic the effect moves with the general traffic speed, which is much higher than the relative velocity of the cars. A shock wave develops far upstream of the constriction, where there is a discontinuity as cars decelerate sharply when they feel the effect of the bottleneck. A piston moving in air faster than the molecules piles up gas in front of it. A shock wave propagates away from the piston.

Shock waves are best studied in a frame of coordinates in which the shock front stands still and the gas or magnetic field flows through it, with physical conditions changing sharply at the front. They can occur in almost any mode of wave propagation, as long as the macroscopic wave moves faster than the microscopic elements. A pressure variation moving faster than the sound velocity rapidly becomes a shock wave, characterized by an abrupt change in physical conditions. This change can be determined by solution of the Rankine–Hugoniot equations, which track the conservation of momentum, energy and matter across the boundary:

$$p_1 + \rho_1 v_1^2 = p_2 + \rho_2 v_2^2, \qquad (3.37a)$$

$$\frac{v_1^2}{2} + h_1 = \frac{v_2^2}{2} + h_2, \qquad (3.37b)$$

$$\rho_1 v_1 = \rho_2 v_2, \qquad (3.37c)$$

where the subscripts 1 and 2 refer to upstream and downstream, respectively, and h is the internal energy of the gas.

The solution of the Rankine–Hugoniot equations determines the properties behind the shock front, given the properties of the gas ahead of the front and the shock velocity. In compression shock waves, the density and pressure, as well as the velocity, increase behind the front. Rarefaction shock waves also occur. The properties of the shock are defined in terms of the Mach number M, the ratio of shock velocity to the sound speed and the ratio of specific heats γ. The pressure and density jumps are:

$$\frac{p_2}{p_1} = \frac{2\gamma}{\gamma+1} M^2 - \frac{\gamma-1}{\gamma+1}, \qquad (3.38)$$

$$\frac{\rho_1}{\rho_2} = \frac{v_2}{v_1} = \frac{\gamma-1}{\gamma+1} + \frac{2}{\gamma+1}\frac{1}{M^2}. \qquad (3.39)$$

In a perfect gas $\gamma = 5/3$ and the limiting density jump for large M is 4. However if the shock can radiate the extra energy behind it, it can become isothermal, in which case

$$\frac{\rho_2}{\rho_1} = M^2 \qquad (3.40)$$

and quite high density increases can occur.

If a magnetic field is present, the change through the shock front is constrained by the fact that magnetic flux is conserved, hence:

$$\frac{B_1}{\rho_1} = \frac{B_2}{\rho_2}. \qquad (3.41)$$

This puts a limit on the compression in the isothermal case, and reduces the compression in others.

An MHD shock wave will occur when a disturbance moves faster than the Alfvén velocity with which normal MHD waves propagate. These waves will be characterized by a jump in the magnetic field behind the wave, and the magnetic energy must be included in the Rankine–Hugoniot equations.

The thickness of a shock front in the absence of a magnetic field must be greater than the mean free path of the particles of the gas. This is obviously necessary for the transfer of "awareness" of the shock from perturbed to unperturbed particles. In an MHD shock wave, this is done by the magnetic field. Even if the density is low and the mean free path long, a sharp discontinuity in the field will occur. Such a collisionless shock front exists at the outer boundary of the terrestrial magnetosphere. At this point, solar plasma carrying a magnetic field impinges on the Earth's magnetic field with velocity around 400 km/sec, much higher than the local Alfvén velocity. Although the local mean free path is millions of miles, a collisionless MHD shock front occurs, with a thickness of only a few thousand km. This front, where the solar plasma becomes abruptly aware of the presence of the Earth, remains stationary with respect to the Earth, just like the shock front of a supersonic airplane or meteor; but of course it is moving relative to the solar plasma streaming by at 400 km/sec.

The dissipation rate in shock waves determines the temperature jump and of course the propagation. If a supersonic airplane produces a shock wave, it transfers energy to all the air by increasing density and temperature. This enormous amount of energy is not dissipation, but just what is required to get the shock going. But if the shock wave breaks someone's window, the energy imparted to the window is irretrievably dissipated. In plasmas, the rise in density and temperature behind a compression shock wave increases the rate of radiation, which escapes and constitutes an important dissipation mechanism. Although this dissipation can make the shock isothermal and increase the compression, eventually it limits the distance the wave will propagate.

3.6. Summary

Of the many properties of plasmas that we have discussed, there are a few that we shall return to over and over, and therefore summarize here:

1. The conductivity of a plasma is proportional to $T^{5/2}$; it is high in the chromosphere and corona.
2. The magnetic lines of force are frozen to the plasma at high conductivity. If the magnetic energy is large, the plasma must flow along these lines. If the plasma energy is larger than the magnetic energy, plasma motions will distort and twist the frozen-in fields.
3. Many kinds of waves propagate in the plasma. Disturbances in the magnetic field propagate with the Alfvén velocity $B/\sqrt{4\pi\rho}$. Propagation of electromagnetic waves is limited by various characteristic frequencies of the plasma.
4. A plasma plus magnetic field configuration is stable if no lower-energy states exist. If lower-energy states do exist, the system will pass over to them if *any* mode or scale of perturbation proves unstable.

References   Alfvén, H., and C. G. Falthämmer 1963. *Cosmical Electrodynamics. Fundamental Principles*, 2nd ed. Oxford: Clarendon.

Cowling, T. G. 1957. *Magnetohydrodynamics*. New York: Interscience.

Priest, E. R. 1982. *Solar Magnetohydrodynamics*. Dordrecht: Reidel.

Spitzer, L. 1962. *Physics of Fully Ionized Gases*. New York: Interscience.

Spitzer, L. 1978. *Physical Processes in the Interstellar Medium*. New York: Wiley.

4

The Interpretation
of Radiation

4.1. Some basic rules In 1859 Kirchhoff and Bunsen developed the science of spectrum analysis, showing that each element produced characteristic lines. Kirchhoff (1859) gave us a set of laws governing the production of the spectrum:

1. The ratio of emissivity to absorptivity is independent of the composition of the material and depends only on the temperature and wavelength.
2. An opaque body radiates a continuous spectrum.
3. A transparent gas radiates an emission spectrum that is distinct for each chemical element.
4. An opaque body surrounded by a gas of low emissivity shows a continuous spectrum crossed by absorption lines at the same wavelength.
5. If the gas has a high emissivity, the continuous spectrum will be crossed by bright lines.

The solar spectrum, like that of almost all stars, is a continuous spectrum crossed by dark lines (Fig. 6.19), so we know that the Sun is an opaque body surrounded by a gas of low emissivity. By measuring the wavelength of these lines we may determine the elements present in the envelope, and by studying the structure of the lines we may determine the abundances of those elements as well as the physical conditions there.

It is easy to understand the basis for Kirchhoff's laws. If we place two surfaces close together, they must each emit and absorb in the same ratio, or else one will rapidly get cooler than the other, and we will have built a magic refrigerator. This holds for any material, so the emissivity and absorptivity must depend only on temperature, so long as the surfaces are opaque. Because an opaque body is not transparent at any wavelength, it emits a *continuous spectrum* characteristic of that temperature. Kirchhoff was really talking about solid bodies in the laboratory, but the same is true of any opaque material. A transparent gas, on the other hand, will be opaque only at the frequencies corresponding to the energy difference between the atomic or molecular energy levels, and emit only in those frequencies. We get an *emission line* spectrum. Why is the Sun, which is too hot to be solid or liquid, opaque? Because as the density increases, various continuous absorption processes come into play and block the radiation at all frequencies. That fixes the position of the photosphere, the surface that we see in continuous light.

In an assembly of atoms there is a flow of energy back and forth between the kinetic, internal, and radiative energy fields; these interactions should lead to equipartition of these and all three should be specified by the same temperature. Collisions and absorption of photons increase the internal energy, while emission of photons and collisions of the second kind return that energy to the radiative and kinetic energy fields, respectively. The various processes are listed in Table 4.1, and we can tell where energy comes and goes from what goes into or comes out of each. Inside the star these processes are in equilibrium, because the temperature gradient is very small and the system is essentially closed. At the surface of a star, the photons pour outward without being re-absorbed, and half the photon emissions are not matched by re-absorption. We keep track of this more complicated state by the techniques of radiative transfer, which account for both the passing outward of radiative energy flowing from below, and the transfer of energy back and forth between the different forms.

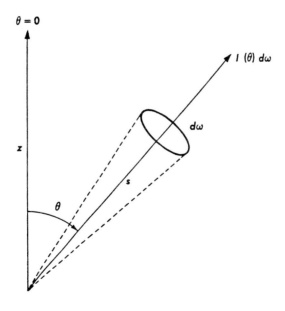

4.1. Diagram of the intensity $I(\theta)$ passing along s through the element of solid angle $d\Omega$. $I(\theta)$ is generally independent of the azimuthal angle ϕ, corresponding to rotation around the z axis.

4.2. Radiation transfer

The *intensity* $I(\theta, \phi)d\Omega$ (ergs/cm^2/sec/Hz/ster) is the radiant energy flux per unit frequency per unit solid angle $d\Omega$ passing through a unit surface perpendicular to the particular direction θ, ϕ. The total intensity in that direction (Fig. 4.1) is obtained by integrating over frequency:

$$I(\theta, \phi) = \int_0^\infty I_\nu(\theta, \phi)d\nu. \tag{4.1}$$

The intensity at angle θ to an element in the xy plane ($\theta = \pi/2$) is $I_\nu(\theta, \phi)\cos\theta$. The *flux* πF is the net intensity in all directions, which we obtain by integrating over all elements $d\Omega = \sin\theta d\theta d\phi$ of solid angle:

$$\pi F_\nu = \int_0^\pi d\theta \int_0^{2\pi} I_\nu(\theta, \phi)\cos\theta \sin\theta d\phi. \tag{4.2}$$

For isotropic radiation this integral is zero. Only when there is more energy going in one direction than the other, as at the solar surface, will there be a net flux.

The total energy emitted in all directions by an isotropic point source of light (such as an atom or a light bulb) is

$$S = \int_0^\pi \int_0^{2\pi} I(\theta, \phi)\sin\theta d\theta d\phi. \tag{4.3}$$

The total energy emitted by a star of radius R whose surface intensity is $I(\theta, \phi)$ is

$$L = 4\pi R^2 \int_0^{\pi/2} \int_0^{2\pi} I(\theta, \phi)\cos\theta \sin\theta d\theta d\phi \tag{4.4}$$

$$= 4\pi R^2 \overline{I}.$$

The apparent brightness of the star seen from a distance r is the intensity at point r divided by the solid angle subtended by the star:

$$b = \frac{4\pi R^2 \overline{I}}{4\pi r^2} \div \frac{R^2}{r^2} = \overline{I}. \tag{4.5}$$

Thus the apparent surface brightness is constant, so long as the source is not a point.

Finally, the radiation density u_ν at any frequency is

$$u_\nu = \frac{1}{c} \int_0^{2\pi} \int_0^\pi I(\theta, \phi)\sin\theta d\theta d\phi. \tag{4.6}$$

If I is isotropic,

$$u = \frac{4\pi}{c}I. \tag{4.7}$$

Along a path element $ds = \sec\theta dz$ along trajectory θ (we drop the ϕ-dependence; most cases with which we will deal are ϕ-independent), the change in intensity $dI_\nu(\theta)$ is the difference between absorption $\kappa_\nu I_\nu(\theta)ds$ and re-emission $j_\nu ds$:

$$dI_\nu(\theta) = -\kappa_\nu I_\nu(\theta)ds + \epsilon_\nu ds. \tag{4.8}$$

The quantities κ_ν and ϵ_ν are the absorption and emission per cm of path respectively. Because these are density dependent we define

$$\kappa_\nu(\text{cm}^{-1}) = k_\nu \rho \qquad \epsilon_\nu(\text{cm}^{-1}) = j_\nu \rho, \tag{4.9}$$

where ρ is the density in gm/cm^3. The units of k_ν are cm^2/gm/Hz, and of j_ν, ergs/gm/Hz/sec. For simplicity we assume j and k to be isotropic.

The optical depth $d\tau$ is the product of the absorption coefficient and the distance along the z axis:

$$d\tau = -\kappa_\nu dz = -\kappa_\nu \cos\theta ds \tag{4.10}$$

and Eq. (4.8) becomes

$$dI_\nu(\theta) = I_\nu(\theta)\frac{d\tau}{\cos\theta} - \frac{j_\nu}{k_\nu}\frac{d\tau}{\cos\theta}, \tag{4.11}$$

or

$$\cos\theta\,\frac{dI_\nu(\theta)}{d\tau} = I_\nu(\tau,\theta) - S_\nu(\tau), \tag{4.12}$$

where the ratio of emissivity to absorptivity

$$S_\nu = \frac{j_\nu}{k_\nu} \tag{4.13}$$

is the *source function*.

If we use a new variable,

$$x = \tau\sec\theta = \kappa_\nu s, \tag{4.14}$$

we may obtain a total differential form by multiplying Eq. (4.12) by e^{-x}:

$$e^{-x}\frac{dI}{dx} = Ie^{-x} - Se^{-x}, \tag{4.15}$$

$$\frac{d(Ie^{-x})}{dx} = -Se^{-x}, \tag{4.16}$$

where we have dropped the ν dependence because Eq. (4.13) holds for every frequency and the θ dependence is understood. Eq. (4.16) is easily integrated:

$$I(x,\theta)e^{-x} = -\int S(y)e^{-y}dy + c, \tag{4.17}$$

with the dummy variable y in the integrand. The constant of integration c is determined differently for the inward and outward flux. This evaluation gives

$$I_{\nu-}(x, \theta) = \int_0^x S_\nu(y) e^{-(y-x)} dy \qquad (4.18)$$

for the inward flux; and for the outward,

$$I_{\nu+}(x, \theta) = \int_x^\infty S_\nu(y) e^{-(y-x)} dy. \qquad (4.19)$$

Finally, the emergent intensity at the top of the atmosphere ($x = 0$) is

$$I_{\nu+}(0, \theta) = \int_0^\infty S_\nu(y) e^{-y} dy. \qquad (4.20)$$

We see that the source function is a measure of the energy contributed by each element along the trajectory, weighted by the fraction e^{-y} of the light emitted by that element which reaches the surface. If the source function is constant, we get the one-dimensional solution for the flux along $\theta = 0$ at optical depth τ:

$$I_\nu(\tau_\nu) = S_\nu(1 - e^{-\tau_\nu}). \qquad (4.21)$$

$$I_\nu \approx S_\nu \tau \approx j_\nu s \qquad \text{as } \tau \to 0. \qquad (4.22)$$

$$I_\nu \approx S_\nu \qquad \text{as } \tau \to \infty. \qquad (4.23)$$

So for small τ we add up the emissions along the path without worrying about subsequent re-absorption and for large τ, the emergent intensity approaches the source function.

4.3. Detailed balance: local thermodynamic equilibrium

We can determine the ratio of direct and reverse processes by the requirement of detailed balance in a closed system. Each and every transition (such as the emission of a photon of energy $h\nu$ by a jump between states B and A) is balanced by the inverse (absorption of a photon of the same energy in a transition from A to B). The medium is in detailed balance, and a state of local thermodynamic equilibrium (LTE) exists. The population of states depends only on temperature, and density changes affect only transition rates. If this were not the case, one part of the medium would get hotter and hotter than the rest. From Kirchhoff's first law the source function is a function of temperature only, which we define as

$$S_\nu = j_\nu/k_\nu = B_\nu(T), \qquad (4.24)$$

where B_ν is the Planck radiation function. In LTE the source function is the Planck function.

Consider the detailed balance between two states n (upper) and m separated by energy $h\nu$. Einstein showed that downward transition could take place either by stimulated or spontaneous emission. A radiation density $u_\nu = (4\pi/c)B_\nu$ produces stimulated emission at a rate $N_n u_\nu B_{nm}$ where N_n is the number of atoms per cm^3 in the n state. Even in the absence of a radiation field spontaneous transitions occur at a rate $N_n A_{nm}$. Stimulated upward transitions, or absorptions, occur at a rate $N_m u_\nu B_{nm}$. The coefficients A and B will be determined from the detailed balance

$$N_n(A_{nm} + u_\nu B_{nm}) = N_m B_{mn} u_\nu. \tag{4.25}$$

In most applications of interest to us the stimulated emission is negligible in comparison to the spontaneous emission. In thermodynamic equilibrium, the ratio of populations in upper and lower levels is given by the Boltzmann formula:

$$\frac{N_n}{N_m} = \frac{g_n}{g_m} e^{-(\chi_n - \chi_m)/kT} = \frac{g_n}{g_m} 10^{-\frac{5040}{T}(\chi_n - \chi_m)(eV)} = \frac{g_n}{g_m} e^{-h\nu/kT}, \tag{4.26}$$

where $\chi_n - \chi_m = h\nu$ is the energy difference between the states and k is the Boltzmann constant 1.38×10^{-16} erg/deg. g_n and g_m are the statistical weights of the levels, the number of indistinguishable states of the same energy in each level, or the number of electrons permitted in the level without violating the Pauli exclusion principle. In the second expression we put χ in eV for convenience.

Einstein showed that the transition probabilities are related by

$$A_{nm} = \frac{8\pi h\nu^3}{c^3} B_{nm} = \frac{8\pi h\nu^3}{c^3} \frac{g_m}{g_n} B_{mn} = 6.67 \times 10^{16} \frac{gf}{g_2 \lambda^2 (\text{Å})} \tag{4.27}$$

(gf is the same for upper and lower levels). If the populations are given by Eq. (4.26), the radiation density u_ν must be

$$u_\nu = \frac{4\pi}{c} B_\nu = \frac{8\pi h\nu^3}{c^3} \frac{1}{e^{h\nu/kT} - 1}, \tag{4.28}$$

and the Planck function Eq. (4.25) is

$$B_\nu d\nu = \frac{2h\nu^2}{c^2} \frac{1}{e^{h\nu/kT} - 1} d\nu \tag{4.29}$$

in the frequency scale, while in the wavelength scale

$$B_\lambda d\lambda = \frac{2\pi hc^2}{\lambda^5} \left(\frac{1}{e^{hc/k\lambda T} - 1} \right) d\lambda. \tag{4.30}$$

We must be careful of the differential factor $d\nu = -(c/\lambda^2)d\lambda$ which must be used as we transfer from the frequency scale Hz^{-1} to the wavelength scale cm^{-1}. The Planck function has two important asymptotic forms. At long wavelengths $(h\nu \ll kT)$, the denominator of Eq. (4.30) becomes $h\nu$ and we have:

$$B_\nu = \frac{2kT}{\lambda^2} \tag{4.31}$$

which is the Rayleigh–Jeans law. It tells us that when energy is not a factor, the radiation is proportional to the possible density of photons. For $(h\nu \gg kT)$, the exponential in the denominator dominates, and

$$B_\nu = \frac{2h\nu^3}{kT}e^{-h\nu/kT}, \tag{4.32}$$

which is the Boltzmann law from the fact that the distribution of higher-energy photons depends on the Boltzmann formula Eq. (4.26). Thus at short wavelengths the intensity falls off exponentially, more sharply than in the infrared, where all photon states are populated.

We have seen that in LTE, when the atomic states are populated according to the Boltzmann formula for temperature T, the radiation field will correspond to that same temperature. Logically the kinetic energy distribution of particles will be the Maxwellian distribution (3.4) for that temperature. Otherwise collisional excitation and de-excitation of the states would establish another temperature.

The Boltzmann formula Eq. (4.26) reflects the fact that in LTE the population of each state is proportional to its partition function

$$n_i = g_i e^{-E_i/kT}, \tag{4.33}$$

where g_i, the statistical weight or degeneracy, is simply the number of different atomic sub-states that are included in the state we are considering. In Chap. 5 we shall see that each atomic state of angular momenta L, S, and J can be split by a magnetic field into $2J + 1$ sub-levels designated by M_J. For all the J levels of a term LS there are

$$g(L, S) = (2L + 1)(2S + 1) = \sum_j 2J + 1 \tag{4.34}$$

different M_J sub-levels possible; for a hydrogenic shell n, there are $2n^2$ different M_J sub-levels possible. If a magnetic field is turned on, the Zeeman effect separates them.

The partition function for an atom is the sum of the Boltzmann factors (Eq. 4.26) of the individual levels:

$$u_i = \sum_{n=1}^{\infty} g_{i,n} e^{-\chi_n/kT}. \tag{4.35}$$

Because χ_n tends to a limit and $g \propto n^2$, this series is unbounded except that the atom is finite and the series must be cut off where the next atom starts. Unless excitation is high, it is sufficient to use the statistical weight of the ground state. Partition functions are tabulated by Allen (1981). As in Eq. (4.26), we can substitute for convenience:

$$e^{-\chi/kT} = 10^{-\frac{5040}{T}\chi(eV)} = 10^{-\Theta\chi(eV)}, \qquad (4.36)$$

where $\Theta = 5040/T$.

The partition function of a free electron is the density of states in phase space times 2 for the possible spin states:

$$u_e = 2\frac{(2\pi mkT)^{\frac{3}{2}}}{h^3}; \qquad (4.37)$$

and the ionization balance is given by the Saha equation, the ratio of partition functions of ion plus electron to that of the neutral atom, *viz.*:

$$
\begin{aligned}
\frac{N_{i+1}N_e}{N_i} &= \frac{u_{i+1}u_e}{u_i} \\
&= \frac{2u_{i+1}}{u_i}\frac{(2\pi mkT)^{\frac{3}{2}}}{h^3} \\
&= 4.83 \times 10^{15}T^{\frac{3}{2}}\frac{g_{i+1}}{g_i}e^{-\chi_i/kT}
\end{aligned}
\qquad (4.38)
$$

where g_i is the statistical weight of the ground states of the ith ion and χ_i is its ionization potential.

In detailed balance the population of all states is perfectly determined. But anything we see is leaking photons and not quite in detailed balance. So the Saha and Boltzmann laws are directly applicable only to the unseen gases below the surface of the Sun. In many cases the leaking photons make up only a fraction of the transitions. For example, the photospheric radiation comes from the level of $\tau = 1$; from that point only a small fraction of the radiation (less than $1/2e$) escapes, and LTE is not a bad approximation.

A special use of LTE is to determine the ratio of rates for inverse processes. We define a density-independent rate coefficient C_{LU} for collisional excitation from lower to upper state, such that $N_e C_{LU}$ is the rate of excitation per target atom. This coefficient is equal to the integral over the velocity space of the fraction $f(v)$ of electrons of a given velocity, their velocity v, and a cross-section $\sigma(v)$ which depends only on the process and the energy of the incident electron. The integration is taken over energies above the threshold energy of the transition:

$$C_{LU}(v) = \int_{mv^2/2=\chi}^{\infty} N_U N_e f(v)v\sigma_{LU}f(v)v^2 dv. \qquad (4.39)$$

If we introduce the Maxwellian velocity distribution (Eq. 3.4) for $f(v)$ and equate the direct and reverse rates we have:

$$N_L N_e C_{LU} = 4\pi N_e N_L \left(\frac{m}{2\pi kT}\right)^{\frac{3}{2}} \int_0^\infty e^{-mv^2/2kT} \sigma_{LU} v^3 dv = N_U N_e C_{UL}.$$

(4.40)

Requiring that the rates be in detailed balance as in Eq. (4.39) and using the Boltzmann formula Eq. (4.26) to relate the populations of the states, we find

$$\frac{C_{LU}}{C_{UL}} = \frac{N_U N_e}{N_L N_e} = \frac{g_U}{g_L} e^{-\chi/kT_e}.$$

(4.41)

The ratio of direct and reverse rate coefficients depends on temperature only. We still have to calculate the rate coefficients in one direction, but the opposite rates will easily be available from Eq. (4.41).

For ionization processes we use the Saha equation to relate the rates. Collisional ionization, for example, is proportional to

$$N_i N_e C_i(i, i+1) = N_{i+1} N_e^2 C_{i+1}(i+1, i)$$

(4.42)

Table 4.1. Atomic processes, direct and reverse.

Process	Incoming	Outgoing	Rate
Absorption	photon + atom	excited atom	$u_\nu B_{mn} N_m$
Stimulated emission	phot. + excited atom	2 phots. + atom	$u_\nu B_{nm} N_n$
Spontaneous emission	excited atom	photon + atom	$N_{nm} A_{nm}$
Photo-ionization	photon + atom	ion + electron	$u_\nu N_m B_{m\kappa}$
2-Body recombination	electron + ion	photon + atom	$N_e N_i A_{\kappa m}$
Dielectronic Recomb.	electron + ion	phot. + excited atom	$N_e N_i \alpha_{diel}$
Dielectronic Absorption (Auto-ionization)	phot. + excited atom	ion + electron	$N_n u_\nu \kappa_{diel}$
Thomson scattering	photon + electron	phot. + electron	$\sigma_T N_e$
Free-free emission (Bremsstrahlung)	electron + ion	elec. + ion + phot.	$N_e N_i \kappa \kappa'$
Free-free absorption	phot.+electron+ion	phot. + elec. + ion	$N_e N_i B_{\kappa'\kappa} u_\nu$
Collisional excitation	electron + atom	elec. + excited atoms	$N_m N_e C_{mn}$
Collisional de-excitation	elec.+ excited atom	electron + atom	$N_n N_e C_{nm}$
Collisional Ionization	electron + atom	2 electrons + atom	$N_m N_e C_{m\kappa}$
3-Body Recombination	2 elecs. + atom	electron + atom	$N_e^2 N_i C_{\kappa m}$

where the right side is the equal and opposite rate of two electrons coming together with the next higher ion to produce recombination, usually called three-body recombination. The ratio of the two opposite rate coefficients is

$$\frac{C_i(i,i+1)}{C_{i+1}(i+1,i)} = \frac{N_{i+1}N_e}{N_i} = 4.83 \times 10^{15} T^{3/2} \frac{g_{i+1}}{g_i} e^{-\chi_i/kT} \qquad (4.43)$$

using the Saha equation Eq. (4.37). This will hold whether or not the system is in LTE, because the reaction rate coefficients depend only on the velocity distribution and the cross-section. We must, of course, know the rates in one direction first; this is usually easiest for the upward transition. The actual rates in a given situation are then obtained by multiplying by the local densities of reactants. If the velocity distribution is non-Maxwellian, we need to find individual rate coefficients for each velocity using Eq. (4.37) with the actual *f(v)*.

For radiative processes the same procedure is followed. We assume the radiation field to be given by the Planck function which gives the LTE ratios, determine the rate coefficients in one direction, obtain the opposite rate coefficient and multiply either by the real radiation field for non-equilibrium rates. For fully ionized hydrogen $N_i = N_e$, so the *emission measure* $N_e^2 V$ is often used as a measure of emissivity.

4.4. Collisions and radiations

There are many atomic processes that enter into the equilibrium of atoms; the most important are summarized in Table 4.1. Only the hydrogenic and magnetic dipole transition probabilities can be calculated exactly, but approximations have been developed for the other interactions. At present our knowledge of conditions is sufficiently rudimentary that these approximations are satisfactory. This is particularly true in cases where excitation or ionization potentials are high compared to kT, so there is a strong temperature dependence and the error in rates is mainly due to our ignorance of the temperature.

Absorption of a photon and its inverse, stimulated emission, increase or decrease the photon density in response to the *em* field. The classical analog is an oscillator resonating with the electromagnetic field. The rates are the product of the number N_n of atoms in the initial state, the radiation density u_ν and the Einstein coefficient B_{nm}. The photon produced by stimulated emission is coherent with the incident radiation; this is the basic process of masers and lasers. The radiation field may also stimulate an electron and ion to recombine, but stimulated recombination is not significant in the cases with which we shall be concerned. We can distinguish the absorption probability B_{nm} from the Planck function B_ν by the fact that the former always has two discrete indices. Rates involving B_{nm} are always proportional to the radiation density and the number of reacting atoms, as in Eq. (4.25).

When $h\nu \ll kT$, as in most astrophysical contexts outside LTE, spontaneous emission is the most rapid downward process. It depends only on the number of atoms in the excited state and A_{nm}. It has no classical analog, but depends on the fact that the transition probability is proportional to the number of photons in the final state. So even if there are no photons around, the final state will have one, and there is a non-zero probability of spontaneous emission. The number of high-energy photospheric photons is small, so this is the dominant process at most wavelengths of interest. Formulae for calculating or estimating the A_{nm} are given in Chap. 5, and values are tabulated by Allen (1973).

Instead of the Einstein A's and B's, which have big values, the classical oscillator strength f (or f-value) is often used. It has the useful property that the sum of the f-values for absorption or emission of light from a given level is equal to the effective number of electrons – for example, the total of f-values for absorptions to all levels from the ground state (or any state) of hydrogen is 1. For the strongest spectrum line from a level one can often use the approximation $f \approx 1$. The continuum f-value $df/d\epsilon$ is simply integrated and included in the sum of f-values for discrete transitions. The product gf of f-value and statistical weight is symmetric in absorption and emission, *viz.*

$$g_1 f_{abs} = -g_2 f_{em} \qquad g_1 f_{12} = -g_2 f_{21}. \tag{4.44}$$

The Einstein coefficients are related to f by

$$gf = 1.5 \times 10^{-8} g_2 A_{21} \lambda^2 \quad (\lambda \text{ in microns, } A \text{ in sec}^{-1}). \tag{4.45}$$

Free electrons can have any energy and can be considered to have a quantum number κ with continuous imaginary values relative to the nearest nucleus. Bound-free transitions may involve any energy greater than the ionization energy of the bound state. If a photon with more than the ionization energy is absorbed, photo-ionization occurs; an electron and ion result. Conversely an ion and an electron may combine to form the next lower stage of ionization in two-body recombination.

For photo-ionization rates we use the absorption cross-section a_ν; it gives the absorption rate from an integrated intensity I_ν and is measured in cm^2 per target atom. For hydrogenic atoms

$$
\begin{aligned}
a_\nu &= 2.815 \times 10^{29} \frac{Z^4}{n^5} \frac{1}{\nu^3} g \\
&= 0.791 \times 10^{-17} \frac{Z^4}{n^5} \left(\frac{Ry}{h\nu}\right)^3 g \ \text{cm}^2/\text{bound electron in state } n,
\end{aligned} \tag{4.46}
$$

where g is the Gaunt factor, a quantum mechanical correction of order unity and Ry is the Rydberg constant, the ionization energy of hydrogen, 13.6 eV. Values of the Gaunt factor for hydrogen atoms are given by Karzas and Latter (1961) and energy-averaged values by Glasco and Zirin (1964). Averages over all angular momentum states yield $g \approx 1$ for most cases as do circular orbits (high angular momentum). But low angular momentum states may have quite high Gaunt factors. For complex atoms, one must allow for the number of electrons in the absorbing state. In terms of Einstein coefficient $B_{n\epsilon}$ or continuum f-value, we have

$$
a_\nu = \frac{4\pi}{c} B_{n\epsilon} h\nu,
$$
$$
= 8.067 \times 10^{-18} \frac{df}{d\epsilon} \quad \mathrm{cm}^2/\text{bound electron.}
$$
(4.47)

For isotropic blackbody radiation with $h\nu \gg kT$, the photo-ionization rate per atom per second is

$$
\int_0^\infty 4\pi \frac{I_\nu a_\nu}{h\nu} d\nu = \frac{8\pi}{c^2} \int_0^\infty \nu^2 e^{-h\nu/kT} a_\nu d\nu.
$$
(4.48)

Since recombination is the inverse of photo-ionization, it can be calculated by matching the two processes in LTE:

$$
N_n I_\nu \frac{a_\nu}{h\nu} = N_i N_e R_n(h\nu),
$$
(4.49)

where R_n is defined as the recombination rate to the nth level of the lower stage of ionization and we get the ratio N_n/N_i from the Saha formula. We must, of course, consider only those electrons that have the right velocity to give photon energy $h\nu$ upon recombination. When Eq. (4.49) is properly solved, we find the recombination rate to the nl hydrogenic level to be

$$
R_{nl}(h\nu)d(h\nu) = 3.26 \times 10^{-6} T^{-\frac{3}{2}} Z^4 \frac{2l+1}{n^5} e^{\chi_{nl}/kT} g e^{-h\nu/kT} \frac{d(h\nu)}{h\nu}, \quad (4.50)
$$

per unit emitted photon energy. The total recombination rate to each nl level is the integral of Eq. (4.50) over energy:

$$
R_{nl} = 3.26 \times 10^{-6} T^{-\frac{3}{2}} Z^4 \frac{(2l+1)}{n^5} e^{\chi_{nl}/kT} \bar{g} \left(-Ei\left(-\frac{\chi_{nl}}{kT} \right) \right), \quad (4.51)
$$

where $-Ei(-x)$ is the exponential integral

$$
-Ei(-\chi) = \int_\chi^\infty e^{-t} \frac{dt}{t}.
$$
(4.52)

For simple calculations for the total recombination on all levels, using $g \approx 1$, we may use the approximation given by Spitzer (1956). Let

$$\beta = \frac{h\nu_0}{kT} = 157\,000\frac{Z^2}{T}. \tag{4.53}$$

The quantity $T/157\,000°$ is the temperature in atomic units, since $kT = 1$ Ry at $157\,000°$. Then

$$R = 2.07 \times 10^{-11} Z^2 T^{-\frac{1}{2}} \phi(\beta) \tag{4.54}$$

where

$$\phi(\beta) \approx \beta^{-0.6} \qquad \text{for } \beta < 1,$$
$$\phi(\beta) \approx 1 + log\beta \qquad \text{for } \beta > 1. \tag{4.55}$$

Exact values are tabulated by Spitzer (1956). Recombination is particularly important for high levels and at low densities.

Auto-ionization and its inverse, dielectronic recombination (DR), involve atomic states which lie above the ionization limit and behave like dense parts of the continuum. For example, in all stable excited states of helium only one of the two electrons is excited. States with two excited electrons lie above the ionization limit and are unstable against auto-ionization. The electron may jump to a lower, stable state, leaving the He atom in a normal excited state, or make a radiationless auto-ionization transition to the continuum, leaving ionized helium in an excited state. Ionization thus takes place when a photon excites the auto-ionization state from an already excited atom. Because there are relatively few atoms or ions in excited states at any one time, the rate is not so great; but this process may be more frequent than ionization of the neutral atom and subsequent collisional excitation. DR is important when the temperature is relatively high. An electron with fairly high energy excites the electron of the ion and is itself captured into the auto-ionization state. If the subsequent decay is to the neutral atom, recombination has taken place. Because the auto-ionization transition is radiationless, it can be related to the continuum of the ionized state by the Saha equation.

Burgess (1964) solved a long-standing problem of the ionization equilibrium in the solar corona by showing that if $kT \approx \chi_{i+1}$, DR could be ten times higher than ordinary two-body recombination in hot gases. In these conditions there are enough electrons with sufficient energy to excite the auto-ionization state and the collisionless transitions are much more rapid than radiative recombination. The exact values are hard to calculate because so many states are involved. For example, in helium all states of $npnl$ with $n > 2$ can make transitions $npnl \rightarrow 1snl$ leaving He in an excited bound state. An approximate formula is given by Allen (1981); a more accurate form is (Landini and Monsignori Fossi 1971; Jain and Narain 1976):

$$\alpha_{\text{diel}}(X^{+Z}) = \frac{2 \times 10^{-4}(Z+1)^2}{T^{3/2}} \sum_j f_{\text{eff}} W_{Z+1}^{1/2} e^{-10.6 \times 10^3 W_{Z+1}/T} \text{cm}^3/\text{sec},$$

(4.56)

where the f-values for the different transitions are replaced by an effective f-value for excitation energy W_{Z+1} :

$$f_{\text{eff}} = a - (b/Z) + (c/Z^2).$$

(4.57)

The coefficients for f_{eff} are given in Table 4.2.

Calculations of ionization including DR are now available (Summers 1974, Jordan 1969) for coronal conditions. The principal effect, besides decreasing the ionization, is to produce an extended high temperature tail in the abundance of Li-like ions such as OV. Auto-ionization plays a particularly important role in exciting certain lines which would not otherwise occur, since it permits spin changes by exchange. As noted above, a large fraction of X-ray lines from highly ionized atoms are excited in this way (Doschek 1985).

The next three processes in Table 4.1 are Thomson scattering and free-free emission and absorption, radiation processes involving free electrons. If a photon strikes a free electron, the electron oscillates with the electromagnetic field of the photon and Thomson scattering occurs. The light is scattered in all directions without change of frequency or absorption. The cross-section of a single electron for scattering is

$$\sigma_T = 3.3 \times 10^{-25}(1 + \cos^2 \Theta) \qquad \text{cm}^2,$$

(4.58)

where Θ is the angle between incident and scattered photon. For simplicity, an angle averaged $\sigma_T = 6.6 \times 10^{-25} \text{cm}^2$ is often used. The electron oscillates with the electromagnetic field of the photon and scatters the light in all directions without change of frequency and without absorbing any of it. Thomson scattering is responsible for the continuous emission of the solar corona (see Chap. 6). This scattered light is about 10^{-6} as intense as the photospheric light, which means it has been scattered by about 10^{19} electrons; these are distributed along a path about equal to the diameter of the Sun, or 1.4×10^{11} cm, so the average coronal density close to the surface

Table 4.2. Coefficients for auto-ionization.

Isoelectronic Sequence	a	b	c
H	0.56	0.00	0.00
He	1.08	1.02	0.51
Ne	5.50	17.61	14.49

must be about 10^8 electrons/cm^3. (This value, so simply determined, is not far from the truth.)

Free electrons may jump between continuum states with emission or absorption of a photon. These are called free-free transitions and may only occur in the presence of a nucleus. For free-free emission by high energy electrons the term *bremsstrahlung*, which means "braking radiation," is often used, because it involves the deceleration of fast electrons passing through matter. The rate of free-free emission (ff) by an ionized plasma per cm^3 is

$$I_{\text{ff}} = 5.44 \times 10^{-39} Z^2 g e^{-h\nu/kT} T^{-1/2} N_e N_i \quad \text{ergs/cm}^2\text{/sec/ster/Hz;} \quad (4.59)$$

or, in all wavelengths,

$$I_{\text{ff}} = 1.435 \times 10^{-27} Z^2 T^{1/2} g N_e N_i \quad \text{ergs/cm}^3\text{/sec.} \quad (4.60)$$

The inverse process of absorption is obtained by using Eq. (4.13). The most important application is at long wavelengths, where we have:

$$k = 1.98 \times 10^{-23} Z^2 g \lambda^2 N_e N_i T^{-3/2} \quad \text{cm}^{-1}(\lambda \text{ in cm).} \quad (4.61)$$

In radio waves, where we will use Eq. (4.61) frequently, the Gaunt factor g deviates considerably from unity:

$$g = 10.6 + 1.90 \log T - 1.26 \log \nu - 1.26 \log Z. \quad (4.62)$$

The next four processes in Table 4.1 are connected with collisions. If an electron with sufficient energy collides with an atom or ion, it may excite it to a higher level or ionize it. If there is one atom in a cm^2 column, and one electron is fired at it, the probability that a passing electron produces the transition is the cross-section σ, which has the dimension of an area. The rate coefficient is the integral over the Maxwellian velocity distribution of the product of the cross-section and the electron velocity, which depends only on the process and temperature. The rate of the process is then the product of the densities of the interacting particles (usually atoms and electrons) and the rate coefficient.

Experimental measurements of collisional cross-sections are difficult and inaccurate. Theoretical calculations are similarly uncertain, particularly near threshold. Errors of a factor two are common. But the rate coefficients depend on the number of incident particles with energy above the threshold, which depends exponentially on temperature. Thus the rate will be much more sensitive to the assumed temperature than the cross-sections, and approximate values may be used freely. For permitted transitions the cross-sections are about $\pi a_0 = 0.87 \times 10^{-16}$ cm^2, the area of the innermost hydrogenic orbit. Tabulations may be found in Vainshtein *et al.* (1973) or Allen (1981).

For excitation Allen suggests the form

$$\sigma = 1.28 \times 10^{-15}(f/EW)\, b \quad \text{cm}^2, \tag{4.63}$$

where f is the oscillator strength and b, a correction running from 0 at threshold to 0.2 for $E = 3W$.

Allen also gives an approximation for the total collision rates:

$$C = 17 \times 10^{-4}\frac{f}{T^{\frac{1}{2}}\chi}10^{-5040\chi(eV)/T}P\left(\frac{\chi}{kT}\right) \quad \text{cm}^3/\text{sec}, \tag{4.64}$$

where P is tabulated. For those who require more detailed values, Van Regemorter (1962) gives empirical formulas for permitted and forbidden transitions. The accuracy of the various approximations is not known but may be within a factor of two.

For ionization we can use formulae given by Allen (1981) for $kT < \chi$:

$$C_i = 1.1 \times 10^{-8}nT^{1/2}\chi_{eV}^{-2}10^{-5040\chi(eV)/T} \quad \text{cm}^3/\text{sec} \tag{4.65}$$

for neutral atoms and

$$C_i = 2.1 \times 10^{-8}nT^{1/2}\chi_{eV}^{-2}10^{-5040\chi_{eV}/T} \tag{4.66}$$

for the corona. χ is the ionization potential in eV and n, the quantum number of the shell being ionized. For example, in the chromosphere the temperature is about $7000°$, and C_i for hydrogen (χ=13.6 eV) is 8×10^{-19}, so at a typical electron density of 10^{12} cm^{-3} a hydrogen atom would be ionized once every hundred days (but photo-ionization may occur more frequently). In the corona, $T = 1\,000\,000°$, and C_i for FeXIII ($\chi = 361$ eV), a typical ion, is 7.3×10^{-12}; at the coronal density of $N_e = 10^8$ it would be ionized three times an hour.

For forbidden transitions we can use a dimensionless collision strength $\Omega(mn)$ of order unity

$$\sigma(mn) = \Omega(mn)\frac{\pi a_0^2}{g_m\epsilon}, \tag{4.67}$$

where ϵ is the energy of the incident electron in Rydbergs. Integration over the velocity distribution gives

$$C_{mn} = \frac{8.63 \times 10^{-6}}{g_n T^{\frac{1}{2}}}\Omega(1,j)e^{-\Delta E/kT}\text{cm}^3/\text{sec} \tag{4.68}$$

for the collision rate. Values of Ω (which can sometimes be far from unity) are given by Seraph *et al.* (1968).

Bahcall and Wolf (1968) showed that collisional excitation of forbidden transitions by protons is a mechanism important at higher temperatures.

The proton must penetrate close to the nucleus to produce a reaction, and will only have enough energy to overcome the nuclear repulsion at high temperatures. This is a quadrupole process and obeys the same selection rules as electric quadrupole radiation (see Chap. 5); the cross-section is proportional, for hydrogenic ions, to

$$\sigma_p < r^2 > = \frac{n^2}{2Z^2}[5n^2 + l - 3l(l+1)], \tag{4.69}$$

where n is the principal quantum number; l, the orbital angular momentum quantum number; and Z, the net nuclear charge (1 for H, 2 for He II) on the orbital electron. The curve of excitation reaches a maximum for proton energies around the ionization energy, at which the proton rate is about twice the electron excitation rate.

The inverse of collisional ionization by electrons is three-body recombination; two electrons go in and only one comes out. Although detailed balance tells us that this process just balances the inverse in LTE, it drops sharply with N_e^2 and is rarely important in the low densities of stellar atmospheres.

Just as the processes of collisional excitation and ionization transfer energy from the kinetic energy field to a state from which it may be contributed to the radiation field, the inverse processes of three-body recombination and collisional de-excitation carry excitation energy back to the kinetic energy field. The difference in the energy transferred in either direction is the energy radiated.

4.5. Non-equilibrium conditions

In the absence of detailed balance and LTE the populations of different atomic levels must be calculated in detail. That is a big problem of great profit to computer manufacturers. But for certain common physical conditions considerable simplification is still possible.

The easiest case is the population of the high levels of an atom. Because the orbits are so large and the energy difference between levels so small, collisional transitions are much more frequent than radiative transitions; the population of upper levels relative to the next stage of ionization is given by the Saha equation.

We shall refer to cases where the excitation is given by an LTE formula multiplied by a simple dilution factor as pseudo-LTE. This occurs when only radiative or only collisional processes dominate, so detailed balance is not lost if the density changes or the electron and the radiation temperatures are unequal. This case applies two atomic levels (Fig. 4.2) between which radiative transitions are forbidden or very rare. The relative populations are determined by the ratio of the upward and downward collision rates, which are in the same ratio as in LTE, so the relative populations are still given by the Boltzmann formula Eq. (4.26).

Inside the star everything is in detailed balance. Each photon absorbed is re-emitted, and each photon emitted is absorbed nearby. The photon mean free path is so short compared to the temperature gradient that no strange distributions of photon energy or electron velocity are seen, and LTE prevails. But at the surface of the star half the emitted photons escape, and detailed balance is lost. The density drops rapidly, so just a short distance above the surface collisional processes are unimportant, and the excitation is dominated by the photospheric radiation field, which is still strong, but half its sub-surface intensity. Upward processes of photon absorption take place at half the LTE rate, whereas downward processes go on just as usual (the main component of downward transitions is the spontaneous term A; induced emission B is small for most cases we consider). Thus the ratio of upper to lower state populations above the surface must be half that in LTE for the photospheric radiation temperature T_r at the wavelengths connecting the levels. At more distant points we use the Boltzmann formula Eq. (4.26) multiplied by a dilution factor w^{-2}, the fraction of solid angle subtended by the illuminating star.

For ionization, the same holds, so long as the density is low enough so that radiative recombination, the inverse of photo-ionization, is the dominant process. The Saha equation is then multiplied by the dilution factor. If the upper levels cannot be excited by radiation and cyclical processes via the ion are unimportant, the excitation is determined by collisions and we put the electron temperature T_e in the Boltzmann formula. Since T_e usually exceeds T_r, metastable states can be overpopulated. The photospheric radiation field is so strong that most atoms in the solar atmosphere show a Boltzmann distribution corresponding to $T_r = 6000$ deg with dilution factor $1/2$.

When $T_r \neq T_e$ and both collisional and radiative processes are important we must compute the balance of atomic processes between various

4.2. Two-level equilibrium for a metastable level n, with A_{nm} small or zero. Transitions shown by dashed lines are infrequent. Since there is nearly detailed balance between upward and downward collisions, we have a Boltzmann distribution, and the ratio of populations is given by Eq. (4.41).

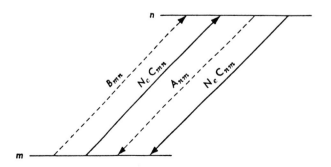

levels. The most important simplification is the assumption of a steady state where the level populations are constant in time. We can then match the transitions into a level equal to those out of it. A further simplification is the use of correction factors to LTE, which take out much of the temperature dependence. These were introduced by Menzel (1937); the coefficient b_n multiplied by the LTE value gives the true population of the nth state. In LTE the b_n are all unity.

The highly ionized atoms of the solar corona have atomic levels so high that the photospheric radiation field has almost no effect, and the corona is so rarified that there is little radiation. Excitation occurs only when those electrons with $E > \chi$ collide with an ion. Downward transitions are dominated by the Einstein spontaneous emission coefficient. The steady state equilibrium between two such levels balances collisions upward and radiation downward:

$$N_n A_{nm} = N_m N_e C_{mn}. \tag{4.70}$$

Since A_{nm} is very large, the ratio N_n/N_m is quite small, much less than that given by the Boltzmann formula. Although the rates in either direction balance, the individual collision and radiation processes are not in detailed balance. The emission will be correspondingly less than from a blackbody and in fact approximately the amount of energy transferred by collision processes to internal energy and thence to the radiation field. Thus, in addition to the Planck blackbody law, we have a second upper limit on the amount of radiation a body may emit: the energy it can transfer to excitation by collisions. This explains a famous contradiction in solar physics. The outer layers of the Sun are much fainter than the surface. Yet their spectra are replete with highly ionized and highly excited atoms. Although these regions are very hot, their density is so low that the rate of collisions, which is proportional to the square of the density, is even lower. Although the quality of the energy is high its quantity is low. The emission rate is

4.3. Two-level equilibrium for low density and weak radiation field showing transition probabilities up and down. Transitions shown by dashed lines are infrequent. A has typical values like 10^8/sec, whereas the collision rate may be only 1/sec (in interstellar space, 1/year).

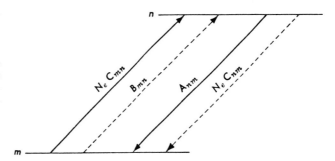

usually simply the total number of collisional excitations. When the path of escaping photons gets very long, they may be re-absorbed and lost to inverse processes. Then a detailed radiative transfer calculation is required. But in every steady state case the emission rate will be between the total number of excitations and the Planck function. In cases of a cooling plasma, such as that following a solar flare, we must adapt the previous arguments to the real situation.

References

Condon, E. U., and G. H. Shortley 1951. *The Theory of Atomic Spectra.* Cambridge: The University Press.

Thomas, R. N., and R. G. Athay 1961. *Physics of the Solar Chromosphere.* New York: Interscience.

Unsold, A. 1955. *Physik der Sternatmosphären*, 2nd ed. Berlin: Springer.

White, H. E. 1934. *Introduction to Atomic Spectra.* New York: McGraw-Hill.

5

Atomic spectra

We summarize those aspects of atomic spectra important to the interpretation of the solar spectrum. The atomic energy levels determine the structure of line spectra and the wavelengths and transition probabilities of lines, while the broadening or splitting of the levels by the physical circumstances of the emitting gas enable us to work backward and deduce the physical conditions from the observed effects.

5.1. Energy levels

Quantum mechanics tells us that as an electron circles the nucleus of an atom, the values of certain dynamic variables, in particular angular momentum and spin, may have only discrete, or quantized, values. Position and momentum are not uniquely determined but specified by a probability in accordance with the Heisenberg principle, while the energy of a state is almost unique, somewhat spread out by the Heisenberg principle according to its lifetime.

The atomic levels are bound, sp states of higher binding energy actually have lower total energy. The hydrogen ionization energy is 13.6 eV, or one Rydberg (Ry). The bound electron occupies discrete states of binding energy 1 Ry, $1/2^2$ Ry, $1/3^2$ Ry,...$1/n^2$ Ry, and so forth. No intermediate values are permitted. The levels are ordered in increasing orbital radius and energy from 1 Ry up by the principal quantum number n, which takes all integral values starting with one. For free electrons we use a continuous imaginary quantum number $n = i\kappa$, for which $1/n^2$ is negative.

The characteristics of the atomic state are specified by a wave function. The probable value of any dynamical variable may be determined from the integral of that variable between the wave function and its conjugate, a quantity we call a matrix element. For hydrogenic atoms the wave function is expressed as a product of radial and angular parts, *viz.*:

$$\Psi = R_{nl}(r)\Theta_{l,m_l}(\theta,\phi)\sigma_s m_s. \qquad (5.1)$$

In hydrogen the radial functions are known exactly; they are Laguerre polynomials, which are confluent hypergeometric functions that terminate because the arguments are integral. The Θ functions are spherical harmonics, solutions of the Legendre equation which describe patterns on a sphere. For these wave functions the angular momentum quantum number l has integral values up to $n-1$, and m_l, the projection of the angular momentum vector on the z axis, has integral values $|m| \leq l$.

The orbits corresponding to quantum number n may be considered in the Bohr–Sommerfeld theory as ellipses of semi-major axis $n^2 a_0$, where a_0 is the Bohr radius $= 0.527 \times 10^{-8}$ cm. The eccentricity of the ellipse depends on the angular momentum quantum number l, the orbit being circular for $l = n - 1$. The binding energy of state n is $Z^2 n^2$ Rydbergs (where Z is the charge on the nucleus) and the values of angular momentum and spin are quantized at $\mathbf{l}^2 = l(l+1)h^2/4\pi^2$ and $\mathbf{s}^2 = s(s+1)h^2/4\pi^2$ respectively. The quantity h is Planck's constant, 6.6×10^{-27} erg sec.

5.1. Energy level diagram for the lowest four shells of hydrogen showing the permitted transitions. The levels of the same n are very close, their energy increasing slightly with angular momentum L. Each term (except for the ^2S terms) is split into two levels $J = L \pm 1/2$. The levels converge to the ionization continuum at 13.59 eV. All the transitions between two shells merge into a single line; Hα consists of the three transitions between $n = 2$ and $n = 3$. The strongest transitions in each line are those furthest to the right; i. e., the strongest lines are those with the greatest L value. From a given level nL, the electron will most likely jump to the lowest possible n' with $L' = L - 1$.

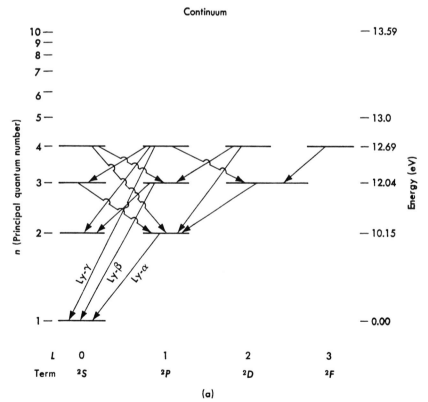

(a)

For $l = 0$ the orbit is a straight line passing through the nucleus. The probability that the electron is at distance r from the nucleus is proportional to $\Psi^2 r^2$; therefore the probability that it is at $r = 0$ is zero, and we need not worry about "collisions" with the nucleus. But because the $l = 0$ electrons penetrate close to the nucleus, they are less shielded from the nuclear charge in multi-electron atoms and hence more tightly bound than more circular orbits of higher angular momentum. In hydrogen the configurations of different l are split by relativistic effects. There are $2l + 1$ values of m for each l, $n + 1$ values of l for each n, and n^2 states of lm for each n. The spin function σ simply tells us if m_s is $\pm 1/2$. There are $2n^2$ possible substates $nlm_l m_s$ for each value of n, so the degeneracy or statistical weight of the n state is $2n^2$. Similarly the statistical weight of the nl state is $2(2l + 1)$ and of the nlm_l state, 2.

The energy of the state is modified by the spin–orbit interaction $\mathbf{l} \cdot \mathbf{s}$. This interaction depends on an atomic parameter ξ and the quantized values of the total angular momentum $\mathbf{j} = \mathbf{l} + \mathbf{s}$. The quantized magnitude of that angular momentum is given by

5.2. Principal lines and series in hydrogen. Wavelength in Å.

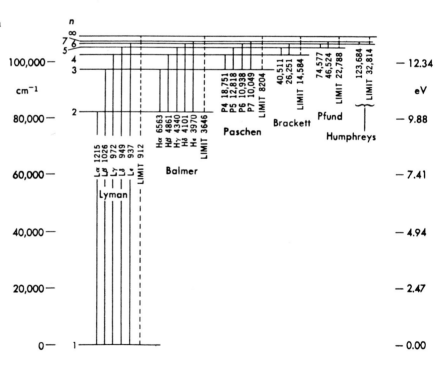

(b)

$$j^2 = j(j+1)\left(\frac{h}{2\pi}\right)^2, \tag{5.2}$$

where j ranges from $l+s$ to $l-s$. For an electron in a state $n=2$, $l=1$, we may have either $j=3/2$ or $j=1/2$ (since $s=1/2$). In fact, for all one electron levels $j=l\pm1/2$, except for $l=0$, where $j=1/2$ only. Such states with $s=1/2$ are called doublets.

In the many electron case we use the upper-case symbols L, S, and J for the momenta of the whole atom, since the lower-case symbols refer to individual electrons. The quantum numbers LS specify a basic state, the *term*, which is named by its multiplicity $2S+1$, and its angular momentum L (using the symbols $S, P, D, F, G,$* *etc.*, corresponding to $L = 0, 1, 2, 3, 4, \ldots$). The term $S = 1/2$, $L = 2$ is written ^2D and called "*doublet D*". The total angular momentum J specifies a *level* and we see that the multiplicity indicates how many J values are possible. $^2D_{5/2}$ is the level of "doublet D" with $J = 5/2$. These states are doublets because the electron spin can point up or down.

The energy levels for hydrogen are simple and are shown in Figs. 5.1 and 5.2. Energy levels are measured in cm^{-1}, the reciprocal of the wavelength in cm.† Because shielding is not important, the only significant energy differences are between states of different n.

The energy of a complex atom consists of the following main terms:

$$E = \sum_{i=1}^{N}\left(\frac{1}{2\mu}\mathbf{p}_i^2 - \frac{Ze^2}{r_i} + \xi_i(r_i)\mathbf{L}_i\cdot\mathbf{S}_i\right) + \sum_{i>j=1}^{N}\frac{e^2}{r_{ij}}. \tag{5.3}$$

The large bracket corresponds to the hydrogenic energy: the first term is the kinetic energy, the second is the nuclear attraction, and the third is the spin–orbit interaction. The second sum is taken over the electrostatic repulsion by the other electrons. If we have a single valence electron, the inner shells will cancel all but one of the nuclear charge, and we can fit the energies by a hydrogenic formula with $Z_{\text{eff}} = 1$. But often there is more than one valence electron and the outer electrons penetrate the inner shells, so the energy levels are determined by the degree of penetration of the orbits, i.e. the angular momentum. This is called the Russell-Saunders, or $L-S$ coupling scheme. Paradoxically this occurs when the $\mathbf{L}\cdot\mathbf{S}$ interaction is small, but means that the energy of a level is determined by the value of n, L and S (see, for example, Fig. 5.3) and the dependence on J is small. The matrix of $\mathbf{L}\cdot\mathbf{S}$ is a function of J and splits the $L-S$ terms accordingly; when this splitting is greater than the separation between the $L-S$ terms, the atomic levels are ordered according to J and we

---

* These letters come from the old spectroscopic designations *sharp, principal, diffuse, fundamental* for lines arising from those terms.

† 100 Å= 10^{-5}cm = $100\,000$ cm^{-1}, 1 eV = 8066 cm^{-1}.

have so-called $j - j$ coupling. If the spin–orbit interaction is comparable to the electrostatic repulsion, we have intermediate coupling, and neither LS nor J clearly define the levels. The multiplier, $\xi(r)$ of the spin–orbit interaction, is proportional to Z^4, so we find intermediate or $j - j$ coupling only in highly ionized ions. The noble gases Ne, A, *etc.* (but not He) are classified by a jl scheme introduced by Racah (1942a).

All of the configurations that we will consider obey $L - S$ coupling: the energies are ordered in terms by the LS value, but each is split by the spin–orbit interaction according to J. For terms in less than half-filled shells the levels are ordered according to J; for more than half-filled shells the levels of highest J lie lowest. If there is no configuration interaction, the spacing between adjacent levels is proportional to the greater J value. For example, the ground state of FeXII (Fig. 8.11) is ^3P, and because $S = 1$ and $L = 1$, the J values are 2, 1 and 0. The ratio of the spacing between the levels should be $2:1$, but the $J = 2$ level is perturbed and pushed down by the nearest term of the same J value ($^1D_2$) so the ratio is less. In this case the spin–orbit splitting is about a third the electrostatic interaction; if it gets much bigger, the levels will be ordered by J and we will have $j - j$ coupling. These matters are discussed in detail by Edlén and by Condon and Shortley.

The number of electrons in each nl shell is given by raising the l designator to that power: thus $2p^6$ means 6 electrons in the $2p$ shell. Because of the Pauli exclusion principle, no two electrons may have the same quantum

Table 5.1 Configurations of atoms.

Configuration	Atom	Configuration	Atom
$1s$	Hydrogen	$1s^2 2s$	Lithium
$1s^2$	Helium	$1s^2 2s^2$	Beryllium
$1s^2 2s^2 2p$	Boron	$1s^2 2s^2 2p^4$	Oxygen
$1s^2 2s^2 2p^2$	Carbon	$1s^2 2s^2 2p^5$	Fluorine
$1s^2 2s^2 2p^3$	Nitrogen	$1s^2 2s^2 2p^6$	Neon
$1s^2 2s^2 2p^6 3s^2 3p^6 4s^2$	Calcium		
$1s^2 2s^2 2p^6 3s^2 3p^6 4s^2 4p^6$	Iron		

numbers, so the values of l, s, m_l, m_s must be different for each electron. Since m_l has $2l + 1$ values ($m_l = +l, l - 1, \ldots -l$) and m_s has 2, $2(2l + 1)$ electrons may occupy each shell. By tradition, the designations s, p, d, f stand for $l = 0, 1, 2, 3$. The configurations of some important atoms are given in Table 5.1.

If we remove electrons from each of these atoms, we get the next lower configuration; a Roman numeral is appended for the order of ionization (*i.e.* FeI is neutral, FeII once ionized, *etc.*). This gives rise to isoelectronic sequences, series of ions of successively higher nuclear charge Z but with the same electron configuration. Li I, Be II, B III, C IV, N V, . . . Fe XXIV all belong to the Li I isoelectronic sequence. Note that because of penetration the 4s and 4p shells fill before 3d; that is why (Fig. 5.3) the $3d\ ^2D$ term in calcium is an excited state.

In LS coupling the energy levels cluster in terms determined by the values of LS permitted by the Pauli exclusion principle. The terms are designated $^{2S+1}L_J$ with $L = 0, 1, 2, 3$ given by S, P, D, F, etc. The quantity $2S + 1$ is called the *multiplicity* because it determines the number of J-levels into which the term may split. Because transitions between terms of different S are forbidden, we usually list all the terms of a given S together, ordering them by L. Table 5.2 lists the terms permitted in simple configurations. When two or more electrons have the same quantum numbers, they are called *equivalent electrons* because they are indistinguishable and must obey the exclusion principle. In two electron configurations $L + S$ must be even. But if the electrons occupy shells of different n, there is no such restriction. For example, for $2p^2$ $L + S$ must be even, but $2p3p$ has no such restriction and ^3S occurs. Filled shells make no contribution, n plays no role, and the configuration $nl^{2(2l+1)-k}$ is the same as nl^k except the J-levels are inverted.

Along an isoelectronic sequence the term separation increases as Z, but the J-level splitting increases as Z^4. Because the parity (evenness of the

Table 5.2. Spectroscopic terms from different configurations.

Configuration	Terms	Configuration	Terms	Configuration	Terms
ns	$^2S_{1/2}$			ns^2	$^1S_0$
np	$^2P_{1/2}$			np^5	$^2P_{1/2}$
np^2	$^1S_0$ $^1D_2$ $^3P_{2,1,0}$	np^3	$^2P_{3/2,1/2}$ $^2D_{5/2,3/2}$ $^4S_{3/2}$	np^4	$^1S_0$ $^1D_2$ $^3P_{0,1,2}$

sum of l) must change for permitted transitions, no strong lines occur between the terms of a configuration, and all the important transitions go to other configurations. In the CI isoelectronic sequence, for example, the configurations and the associated terms are $1s^2 2s^2 2p3d$ $^{1,3}PDFGH$, and $1s^2 2s2p^3$ $^{1,3}P, D$; $^{3,5}S$. The various sets of states and transitions between them are listed in Table 5.3.

5.2. Spectra of complex atoms

The easiest spectra to deal with are complex atoms with a single valence electron outside closed shells which behaves like hydrogen. They include the alkali metals – Li, K, Na, *etc.* and the ionized alkaline earths – Be II, Mg II, Ca II, Sr II.

Li nuclei are rare in the Sun; they are easily destroyed by nuclear reactions in the interior. The other alkali metals are not destroyed, but because Z is odd they have low nuclear binding energy and are not abundant. But the simple $s \rightarrow p$ transitions in their resonance lines require little excitation energy and are extremely strong.

Sodium (Na) has 11 electrons $(1s^2 2s^2 2p^6 3s)$; of these, 10 are in closed shells of zero net angular momentum which only serve to shield the nucleus. The spectral property of the ground state of neutral sodium is therefore determined by the $3s$ electron only and is $3s$ $^2S_{1/2}$. Energy differences between angular momentum states are greater than in hydrogen because of varying penetration of the inner shells. Instead of having the hydrogenic $n = 3$ binding energy, 1.5 eV, the $3s$ electron penetrates the inner shells and is bound by 5.14 eV, corresponding to $Z_{\text{eff}} = 1.84$. The $3d$ configuration is a circular orbit and has the hydrogenic binding energy. The transitions from $^2P_{3/2}$ and $^2P_{1/2}$ to $^2S_{1/2}$ form a strong doublet in the yellow region of the spectrum, at 5890 Å and 5896 Å, known as the D lines from Fraunhofer's classification of the photospheric spectrum. These lines are responsible for the yellow color of a sodium street lamp. They are a doublet because ^2P has two J-levels split by spin–orbit interaction, both of which may jump to the lower state. Mg II is isoelectronic with sodium and the corresponding doublet is at about 2700 Å. Since magnesium is an abundant element, these are the strongest absorption lines in the solar spectrum. The iron atom has 26 electrons; if we remove 15 of these to make Fe XVI, we again are in the sodium isoelectronic sequence. The analogous resonance doublet of Fe XVI is at 336 Å and 361 Å in the extreme UV spectrum and may be seen in Fig. 7.26. Each time we remove an electron, the effective unscreened nuclear charge is increased, the energy difference is increased, and the lines are at a shorter wavelength. Edlén (1964) terms such lines with no change in n "screening doublets". Their wavelength varies linearly with Z_{eff}.

There are several important one-electron isoelectronic sequences. That of Li I has Be II, B III, C IV, N V, O VI, and so on. If we look at the UV spectrum of the Sun (Fig. 7.8) we see the powerful resonance doublets of the more abundant of these ions, especially the nuclei with even Z.

The strongest absorption lines in the Fraunhofer spectrum are the resonance doublet of the potassium-like ion Ca II, called the H and K lines. As an even-Z element, Ca is abundant, and since it is easily ionized and excited, we get the strong lines of an alkali metal. The configuration is $1s^2 2s^2 2p^6 3s^2 3p^6 4s$; we shall in the future write only the valence configuration and assume the inner shells to be filled. As all freshman chemistry students are taught, the $4s$ shell fills before the $3d$ shell; the $4s$ electron penetrates deeper into the inner electron cloud and is therefore more tightly bound. Fig. 5.3 shows the lower energy levels of Ca II. Transitions may occur only when the parity changes, which here means $L - L' = \pm 1$ (The detailed selection rules are in Eq. 5.9). As a result there is no line between $4s$ and $3d$. The H and K lines are so strong that in their core we see the top of the atmosphere, resulting in an emission core in the center of the absorption line with absorption in *its* center; this is called double reversal. The MgII resonance doublet and most other strong chromospheric lines show this structure.

Since electrons in the $3d$ configuration cannot jump to the ground state it is called *metastable* because it behaves like a second ground state – the only way out is up. Actually electrons can leave the $3d$ state downward in de-exciting collisions, because collision processes do not strictly follow

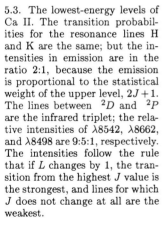

5.3. The lowest-energy levels of Ca II. The transition probabilities for the resonance lines H and K are the same; but the intensities in emission are in the ratio 2:1, because the emission is proportional to the statistical weight of the upper level, $2J + 1$. The lines between ^2D and ^2P are the infrared triplet; the relative intensities of $\lambda 8542$, $\lambda 8662$, and $\lambda 8498$ are 9:5:1, respectively. The intensities follow the rule that if L changes by 1, the transition from the highest J value is the strongest, and lines for which J does not change at all are the weakest.

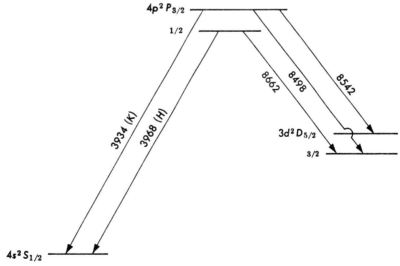

the radiation selection rules; but unless the density is high this process is much slower than radiation. Because J can change by 0 or 1 the $3d\ ^2D$ term is connected with the higher $4p\ ^2P$ term by three lines near 8500 Å, popularly known as the infrared calcium triplet.

Table 5.3 shows the hierarchy of atomic states and the transitions connecting them. Because of the prevalence of Russell–Saunders coupling most of the transitions we will deal with – such as the NaD doublet – are multiplets between terms, consisting of several lines between the different J-levels. These are classified in the Revised Multiplet Table (Moore 1959) and the term structure is illustrated by Merrill (1956).

A polyad consists of all the terms of the same multiplicity arising from the coupling of an electron of different nl to a term of a base configuration. In this case we added a $3p$ electron to an ion in the $2p^2\ ^1D$ state, and the terms results from combining the angular momenta $L = 2$ (from D) and $L = 1$ (the $3p$ electron). If we coupled to $2p^2 3p$ we would get different results; this is called *parentage*.

The helium atom has a number of interesting properties which, in addition to its great abundance, make it important to the astrophysicist. Helium is difficult to excite and is therefore not seen in the visible photospheric spectrum. But it shows many strong lines in the chromosphere where, in fact, it was first discovered in 1868 when Janssen obtained the first flash spectra at a total eclipse and was amazed to find emission lines of hydrogen and an unknown element which was named helium after its source. The two electrons of helium share the $1s^2$ ground state (Fig. 5.4). Because of the Pauli exclusion principle they must have antiparallel spins ($m_s = \pm 1/2$), so the total spin is $S = 0$ and the ground term is ^1S. In the excited states one electron remains in the ground state and the other is excited to $n > 1$, so there is no Pauli restriction and the spins may be antiparallel or parallel ($S = 1$). We have two complete sets of terms for singlet and triplet helium. Since radiative transitions with a change of spin are not allowed, these two systems of terms behave like two separate atoms.

The lowest term of the helium triplets, $2\ ^3S_1$, is a perfect example of a

Table 5.3 Classification of states and transitions.

State	Example	Transition
Sublevel	$^2P_{1/2}, M_J = -1/2$	Component
Level	$^2P_{1/2}$	Line
Term	^2P	Multiplet
Polyad	$2p^2(\ ^1D)3p\ ^2P\ ^2D\ ^2F$	Supermultiplet
Configuration	$1s^2 2s^2 2p^2$	Transition Array

metastable state. Because spin change is forbidden, there is no radiative transition to the singlet ground state, the only lower state. $2\ ^3S_1$ behaves like a second ground state. Exchange collisional transitions to and from the ground state are possible; the colliding electron sticks and one of different spin leaves. Because the dominant direct and reverse processes correspond, a pseudo-Boltzmann distribution at the local temperature occurs. This produces a higher population than other states, from which electrons drain downward by fast spontaneous emission. The 10830 Å line connecting to the next higher $2\ ^3P$ term is therefore strong, the only helium line visible on the disk (except for the ultraviolet resonance lines). In prominences it may be as strong as Hα. In the Fraunhofer spectrum, λ10830 is a weak absorption line, whose strength is increased in plages, prominences, or flares. The population of the singlet levels, by contrast, is determined by the ratio of the slow collisions up to the fast transitions down and therefore have an excitation temperature much lower than the triplets. The singlet lines are about ten times weaker.

Once helium is excited to the triplet states, upward transitions are easily produced by photospheric radiation, while downward jumps are still by

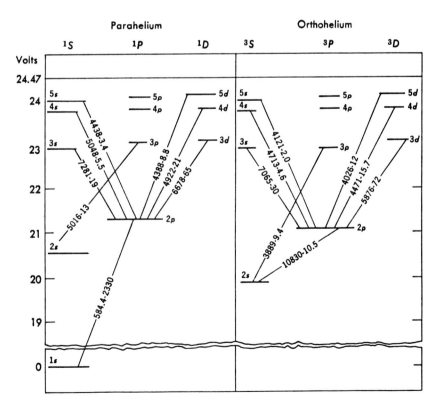

5.4. Energy-level diagram for helium. The first number given with each transition is the wavelength in angstroms; the second, the transition probability in 10^6 sec^{-1}.

spontaneous emission. So the excitation of the triplets relative to 2 ^3S is given by the photospheric temperature (6000°) with a dilution factor 1/2.

The helium-like ions are important in the chromosphere and corona, because it is easy to ionize the third electron in lithium-like configurations but hard to remove one of the two $1s$ electrons. So the lighter elements usually end up in the He-like ions, such as CV, OVII. This is not obvious from the spectra because the He-like ions have few lines, so the UV solar spectrum is dominated by strong screening doublets of Li-like lines such as OVI, NeVIII and MgX. In the X-ray range the He-like resonance lines are visible, and in solar flares these lines are strong. The He-like ion FeXXV is especially intense in the heart of flares.

5.3. Transition probabilities

The probability of spontaneous transitions between states of an atom is given by the integral over all space of the wave functions of initial and final states and the perturbation by the electromagnetic field; this kind of integral is called the matrix element of the em field between the two states. This is expanded in electric and magnetic parts (Condon and Shortley 1959); in order of descending importance these are: electric dipole (proportional to the matrix of $e\mathbf{r}$), magnetic dipole (proportional to $\mathbf{L} + 2\mathbf{S}$), and electric quadrupole (proportional to $e^2\mathbf{r^2}$).

The transition probabilities are written in terms of a symmetric *line strength* $S(2,1)$ between states 2 and 1; for example the Einstein A is given by:

$$A(2,1) = \frac{64\pi^4}{3\eta\lambda^3 g_2}S(2,1) = \frac{2.026 \times 10^6}{\lambda^3 g_2}S(2,1), \tag{5.4}$$

where g_2 is the statistical weight of the initial state 2. The line strength is proportional to the absolute value of the square of the matrix element of the dipole moment $e\mathbf{r}$ between the two states:

$$S(1,2) = S(2,1) = e^2|(2|\mathbf{r}|1)|^2. \tag{5.5}$$

Similar formulas for the magnetic dipole and electric quadrupole are given by Condon and Shortley. Formulae for evaluating the radial matrix element are given by Green *et al.* (1957).

The oscillator strength f (or f-value) is the effective number of electrons in the atom, and is asymmetric in absorption and emission, *viz.*:

$$g_1 f_{abs} = -g_2 f_{em} \qquad g_1 f_{12} = -g_2 f_{21} \tag{5.6}$$

(g is the statistical weight, discussed in Sec. 5.4; 1 is the lower, 2, the upper state). The absorption f-values are always positive and the emission f-values, negative. In terms of A and S

$$gf = \frac{0.03038\ S}{\lambda} = 1.5 \times 10^{-8} g_2 A \lambda^2 \ (\lambda \text{ in microns, } A \text{ in sec}^{-1}). \quad (5.7)$$

By a quantum mechanical identity the sum of the f-values for all transitions from a given level equals the number of valence electrons. For example in hydrogen the $\sum gf = 2$ and the individual values are $(g = 2)$:

$$gf(1 \to 2) = 0.83 \qquad gf(1 \to 3) = 0.16 \qquad gf(1 \to cont.) = 0.82. \quad (5.8)$$

In atoms such as sodium, $f = 1$ is an excellent approximation for the screening doublet, but where (as in hydrogen) the resonance line involves a change in n, $f = 0.5$ is good. Many f-values are tabulated by Allen (1981), who also gives sum rules.

The vector er can be broken up into terms $\mathbf{x} \pm i\mathbf{y}$ and $\mathbf{z}$, which correspond to Legendre polynomials $P_{1\,\pm 1}$ and $P_{1\,0}$, and also to Racah's (1942b) irreducible tensor operators (ITO) $T_{\pm 1}^{(1)}$ and $T_0^{(1)}$. The same is true of the electric quadrupole $e^2\mathbf{rr}$, which breaks down into elements of $T_{2,1,0}^2$. Racah showed that non-zero matrix elements only occur if L, L' and k (the order of the ITO, e.g. 2 for $T^{(2)}$) can form a triangle; thus $T^{(1)}$ can only connect L with $L' = L$ or $L \pm 1$ (but $L + L'$ must be ≥ 1) and $T^{(2)}$ can only connect L with $L' = L$, $L \pm 1$ or $L \pm 2$ (but $L + L'$ and $J + J'$ must be ≥ 2). The selection rules are:

$$
\begin{aligned}
&\text{electric dipole:} &&L' = L, L \pm 1 \\
&n, n' \text{ anything} &&J = J, J \pm 1 \\
& &&S = S'
\end{aligned}
$$

$$(5.9)$$

$$
\begin{aligned}
&\text{electric quadrupole:} &&L' = L, L \pm 1, L \pm 2; &&L + L' \geq 2 \\
&n, n' \text{ anything} &&J' = J, J \pm 2; &&J + J' \geq 2 \\
& &&S = S'
\end{aligned}
$$

The laxity in n is because r is not quantized.

All lines must obey the parity rule. The parity of a state is odd or even as the sum of the individual one-electron l's are odd or even; for an odd function like $\mathbf{r}$ the matrix element will be zero if the parity of the final and initial state is the same. For quadrupole transitions the parities of the two states must be the same. We cannot have electric dipole transitions between different terms of the same configuration (because Σl is unchanged), but electric quadrupole transitions that satisfy the triangular rule are permitted. Although $\Delta L = 0$ is permitted for dipole transitions, for one electron configurations (H, HeII) the parity rule requires its one electron to change angular momentum by one.

Transitions between M_J values are connected by the components of the ITO $T_q^{(k)}$ such that $M' = M \pm q$. Because of the geometry, T_1^1, which gives $\Delta M = \pm 1$ and corresponds to $\mathbf{x} + i\mathbf{y}$, gives circularly polarized radiation and T_0^1 ($\Delta M = 0$) gives linear polarization.

Because of the symmetry properties and the quantization of the angular momentum, the angular part of the wave function is known exactly, and that part of the line strength is easily calculated. The radial part, on the other hand depends on the solution of the radial wave equation, which, except for hydrogen, is known only by coarse numerical approximation. Line strengths are given as multipliers of the radial quantum integral and particularly useful for ratios of transitions with the same radial matrix elements. In other cases one can estimate relative values of radial integrals or use whatever values have been calculated by theoretical approximations.

For the strength of a line L between two J-levels:

$$
\begin{aligned}
S(L) &= S(\alpha S L J, \alpha' S L' J') \\
&= |(\alpha L \| P \| \alpha' L')|^2 (2J + 1)(2J' + 1) W^2 (L J L' J'; S'),
\end{aligned} \tag{5.10}
$$

where $P = \mathbf{er}$ and $(\alpha L \| P \| \alpha' L')$ is a "reduced" radial matrix element introduced by Racah (1942b) to remove the J dependence. It is valid for any of the lines in the $\alpha L \to \alpha' L'$ transition and is evaluated by calculating $\mathbf{S}$ for one pair (J, J) and using (5.10). The symbol α represents the other quantum numbers (e.g. n) and W is the Racah coefficient (Racah 1942a). For the sum over J' one finds

$$
\sum_{J'} \mathbf{S}(L) = \frac{(2J + 1)}{(2L + 1)} |(\alpha L \| P \| \alpha' L')|^2. \tag{5.11}
$$

The strength of the entire multiplet is

$$
\mathbf{S}(M) = \sum_{J, J'} (L) = (2S + 1) |(\alpha L \| P \| \alpha' L')|^2. \tag{5.12}
$$

In the case of a one-electron atom with angular momentum l:

$$
\mathbf{S}(\alpha l, \alpha' l \pm 1) = l > |(\alpha l | \mathbf{er} | \alpha' l')|^2, \tag{5.13}
$$

where the right-hand side is the actual radial matrix element squared multiplied by $l >$, the greater of the two angular momentum values.

By using Racah's methods (and additional rules given by Rohrlich 1956) for polyads we can easily estimate the relative strengths of lines and multiplets without facing the knotty problem of evaluating the radial integrals. The Racah coefficients are tabulated by Rotenberg *et al.* (1959) and by Edmonds (1957).

Without calculating transition probabilities, we can use some general rules to estimate relative intensities of lines. For transitions $nl \to n'l - 1$, such as the hydrogen Lyman series, the strongest line is that of the smallest

Δn. The D lines, for which $\Delta n = 0$, are the strongest of their series, and Lyman α, the strongest of its. In a transition array, such as $H\alpha$, the strongest component is that of the largest l value, $3d \rightarrow 2p$ for $H\alpha$. Lines such as $3s \rightarrow 2p$, for which l changes opposite to n, are much weaker than those like $3d \rightarrow 2p$ in which it changes in the same sense. For downward transitions with the same Δl the strongest is that of the greatest Δn. In a multiplet with $\Delta L = 1$, the strongest line is that from the highest J, and lines of $\Delta J = -1$ are strongest. But if $\Delta L = 0$, $\Delta J = 0$ gives the strongest line.

5.4. Statistical weight and degeneracy

We use the concept of statistical weight or degeneracy g to keep track of the number of indistinguishable quantum states in the current measurement. An atomic state is completely defined by a set of quantum numbers $nLSJM_J$, which has statistical weight one because it can only be occupied by a single electron. If the energy difference between states that produce a given spectral line is small we will see contributions from all those states. The transition probability to three indistinguishable states is three times that to a single one, and the same is true for the population. If the magnetic field is zero the $2J + 1$ levels of different M_J all have the same energy and are indistinguishable; the statistical weight of the LSJ level is $2J + 1$. In a single transition $J \rightarrow J'$, any of the M_J levels and the $M_{J'}$ levels may participate, so the statistical weights $(2J + 1)(2J' + 1)$ appear in (5.10). The degeneracy of an LS term is the sum of the degeneracies of all its J-levels, which is $(2S + 1)(2L + 1)$; for ^3P the statistical weight is 9. The statistical weight of a hydrogenic n shell is $2n^2$. This includes all possible sets of quantum numbers for a single electron in the shell n. The degeneracy appears in the Boltzmann formula because Liouville's theorem tells us that all states of equal energy are equally accessible.

This leads to an important rule for transition probabilities. Consider two states, m and n, of the same energy. In equilibrium there is a detailed balance between the transitions back and forth. F_{nm} is the transition rate per electron in the initial state and F_{mn}, its reverse. These are related by

$$N_n F_{nm} = N_m F_{mn}. \tag{5.14}$$

But according to the Boltzmann formula for $\Delta E = 0$,

$$\frac{N_n}{N_m} = \frac{g_n}{g_m} \tag{5.15}$$

and therefore

$$\frac{F_{nm}}{F_{mn}} = \frac{g_m}{g_n};\tag{5.16}$$

that is, *the transition probability is proportional to the statistical weight of the final state and inversely proportional to that of the initial state.* This makes sense, because when we multiply by the population of the initial state we will include the degeneracy of that state. This is incorporated in Eq. (5.8).

5.5. Forbidden transitions

There is no such thing as a completely forbidden transition. We use the term when the most probable transition, dipole radiation ($A \approx 10^8$ sec^{-1}) is impossible. When that happens, the much less probable "forbidden transitions," magnetic dipole ($A \approx 100$) or electric quadrupole ($A \approx 10^{-4}$ sec^{-1}) become important. Magnetic dipole transitions are calculated from the matrix element of $\mathbf{L} + 2\mathbf{S}$. Because this has only angular momentum terms it can only connect levels of the same configuration (no change in parity or in nl).

The selection rules for magnetic dipole radiation are simple:

$$S = S'; L = L'; J = J', J' \pm 1; nl = n'l'.\tag{5.17}$$

Thus magnetic dipole transitions can occur only between different levels of the same term – for example, $^3P_2 \to {}^3P_1$. This would seem to be an unimportant class, but in the highly ionized ions of the solar corona, the J-levels are split so widely (the splitting between the J-levels is proportional to Z^4) that jumps between them give rise to observable lines. All of the coronal emission lines in the visual region of the spectrum are magnetic dipole transitions.

The transition probability for magnetic dipole radiation is so simple that one does not have to be an expert to calculate it – although it took an expert (Shortley 1940) to figure out what the formula was in the first place. For transitions from J to J'

$$A(J \to J-1) = \frac{35,300}{2J+1}\frac{912}{\lambda^3} \times$$
$$\frac{(J-S+L)(J+S-L)(J+S+L+1)(S+L-J+1)}{4J}$$
$$A(J \to J) = \frac{35,300}{2J+1}\frac{912}{\lambda^3}J(J+1)(2J+1)g(SLJ)^2,\tag{5.18}$$

where g is the Landé g-factor:

$$g = 1 + \frac{J(J+1) - L(L+1) + S(S+1)}{2J(J+1)}, \qquad (5.19)$$

which is important in the Zeeman effect.

For the green line of Fe XIV the transition is $^2P_{3/2} \rightarrow {}^2P_{1/2}$; with $L = 1, S = 1/2, J = 3/2$, and $\lambda = 5303$ Å we find that

$$A = \frac{35,300}{4} \left(\frac{912}{5303}\right)^3 \times \frac{4}{3} = 60 \quad \text{sec}^{-1}. \qquad (5.20)$$

This compares with $A = 5 \times 10^7$ for Hα.

While the magnetic dipole rules forbid change in L or S, this can occur in intermediate coupling. This occurs when the spin–orbit interaction is great enough to perturb the individual LS states and mix with other states of different LS but the same J. A perturbation expansion of the wave function of a term will show each J-level to include a bit of other levels of the same configuration with the same J-value. The LSJ state A will include a fraction b of state B; and its wave function is a combination of the unperturbed states:

$$\Psi(A) = a\psi_0(A) + b\psi_0(B). \qquad (5.21)$$

The coefficient a is nearly unity, and the order of magnitude of b is given by the splitting of the J-levels (which is a measure of the spin–orbit interaction) divided by the distance between the levels. Because $\mathbf{L} \cdot \mathbf{S}$ only has matrix elements for $J = J'$, the mixing can only take place between levels with the same J values. The magnitude of the perturbation coefficient b is roughly the ratio of the spin–orbit to the electrostatic interaction.

The most famous example of this intermediate coupling is in the OIII nebular lines at 4959 Å and 5007 Å. The terms are $^1S_0$, $^1D_2$, and $^3P_{2,1,0}$. $^1D_2$ contains a bit of $^3P_2$, so magnetic dipole transitions can occur between it and $^3P_2$ or $^3P_1$. In this case, the forbidden line is a transition between different terms. In the solar corona the temperature is high, Z is much higher (as in Fe XIV above) and the J-levels are so widely split ($\propto Z^4$) that pure magnetic dipole transitions between them appear as lines in the visible spectrum. The transition probability is as given by (5.18) multiplied by b^2.

The $2\,^1P_1$ and $2\,^3P_1$ terms of helium in the $1s2p$ configuration are another example of intermediate coupling. The splitting of ^3P is small, so b is small; but strong dipole transitions from $^1P_1$ to the ground state $^1S_0$ may occur, so a weak transition is possible from $^3P_1$ to the ground state. A small fraction of $^3P_1$ has taken on the guise of $^1P_1$. The transition probability is b^2 times the normal transition probability from ^1P to the ground state. The electrons that arrive in ^3P almost all go downward to $2\,^3S$; but a few blunder into the "forbidden" transition to $1\,^1S$. In highly ionized helium-like ions b is much greater, and this transition becomes quite strong. In the configuration $1s^2 2s^2 2p^6 3s 3p$ there is no ^3S state to compete, so the line is even more important. The intermediate coupling makes possible a transition to the term $\psi_0(B)$ (which in this case is $2\,^1P_1$) by the magnetic dipole rules.

By any of these criteria the $2\,^3S_0$ term of helium remains rigorously metastable. Magnetic dipole and electric quadrupole radiation still require $S = S'$, and intermediate coupling requires the presence of another term with the same n, l, and J values. The singlet term $1s2s\,^1S_0$ has the same nl, but a different J value. Radiative transitions from $2\,^3S_1$ to the ground state have never been observed in the laboratory.

The selection rules for electric quadrupole emission given in Eq. (5.8) are quite low and these transitions are not normally observed in the Sun. Collisional excitation by protons, however, is a non-radiative process which only takes place through quadrupole transitions.

Forbidden transitions are principally of importance in tenuous gases, such as the corona or the interstellar medium. At high densities the rate of de-exciting collisions becomes so great that electrons return their energy to the kinetic field before they can radiate it by the slow forbidden mode, whereas permitted transitions are always efficient sources of radiation, when they can be excited.

Transitions that are highly forbidden to radiation may be produced by exchange collisions. Because electrons are indistinguishable the incident electron may be captured and one of the target electrons ejected in its place. No spin change takes place, but almost any final state is possible in these processes, which are most probable for incident energies near threshold.

5.6. Line broadening

One of the most fruitful sources of information about the physical conditions under which a spectral line is emitted is the broadening of the line, which is due to spreading of the atomic energy levels. If an atom is completely isolated from its neighbors and from all disturbing forces, it will radiate a sharp spectral line; the levels are broadened only by the Heisenberg uncertainty principle, which states that:

$$\Delta E \Delta t = h, \tag{5.22}$$

where h is Planck's constant. This is called natural broadening.

If an electron is in the ground state, it can go nowhere; Δt is large, ΔE, small, and the ground state is sharp. In an excited state, $\Delta t = 1/A$, where A refers to all rates of leaving the state. The uncertainty in the energy is hA ergs. For Lyman-α, $A = 4 \times 10^8 \text{ sec}^{-1}$, so ΔE of the upper level is 1.6×10^{-19} ergs. The energy of the $n = 2$ state is about 10 eV, or 1.6×10^{-11} ergs, and the wavelength of the Lyman-α transition is only changed by $1/10^8$. Although the effect is tiny, the effects of natural broadening may be great when a line becomes optically very deep.

The primary source of line broadening is the Doppler effect, which is due to the motion of the atoms. If a radiating source moves toward us with line of sight velocity w, the waves of a given frequency are crowded together and the wavelength shortened by

$$\frac{\Delta\lambda}{\lambda} = \frac{\Delta\nu}{\nu} = \frac{w}{c}, \tag{5.23}$$

where c is the velocity of light. If the source moves away, the shift is to the red. The displacement at Hα (6563 Å) is 1 Å for 50 km/sec. The broadening occurs because turbulent and thermal motion both produce random motions of the atoms along the line of sight. The distribution of the line of sight component w of the Maxwellian distribution is obtained from Eq. (3.4). The fraction $f(w)$ of particles of mass m between w and $w + dw$ is given by

$$f(w)dw = \frac{1}{w_0\pi}e^{-(w/w_0)^2}dw, \tag{5.24}$$

where

$$w_0 = \sqrt{\frac{2\pi kT}{m}}. \tag{5.25}$$

The line profile, or resulting distribution of emission in the line, is

$$I(\lambda) = I_0 e^{-(\Delta\lambda/\Delta\lambda_D)^2}, \tag{5.26}$$

where $\Delta\lambda$ is the wavelength separation from the center of the line, and

$$\Delta\lambda_D = \frac{w_0\lambda}{c} = \frac{\lambda}{c}\sqrt{\frac{2kT}{M}} = \frac{c}{\nu^2}\Delta\nu_D. \tag{5.27}$$

Obviously the profile is broader when the kinetic temperature is higher. We can write a numerical relation between the full width of the line at half-maximum, $FWHM$ (or, more simply, the half-width HW), and the temperature:

$$T = 1.95 \times 10^{12}M\left(\frac{HW}{\lambda(\text{Å})}\right)^2 = 5.2 \times 10^{12}M\left(\frac{\Delta\lambda_D}{\lambda}\right)^2 \text{ deg}, \tag{5.24}$$

where M is the atomic mass number. If the green coronal line of Fe XIV (5303 Å) has a measured half-width of 0.8 Å, we easily find (with $M = 56$) that the temperature of the emitting atoms is 2.48 million deg.

The absorption coefficient at the center of a Doppler-broadened line is, in terms of the f-value,

$$l_0 = \frac{\pi e^2}{mc} f \times \frac{1}{\sqrt{\pi}\Delta\nu_D} = \frac{.01497}{\Delta\nu_D} f \qquad (f = (37.7 h\nu/c)B_{mn}), \qquad (5.28)$$

where

$$\frac{\Delta\lambda_D}{\lambda} = \frac{\Delta\nu_D}{\nu} = \frac{1}{c}\sqrt{\frac{2kT}{Mm_{\rm H}}} = 4.4 \times 10^{-7}\sqrt{\frac{T}{M}}. \qquad (5.29)$$

Note that because $\Delta\lambda_D$ does not equal $\Delta\nu_D$, we have always to be careful to distinguish between normalization to the wavelength or the frequency scales. At $15\,000°$, $\Delta\lambda_D = 0.16$ Å for He atoms radiating at 5876 Å. The absorption coefficient anywhere else in the line is

$$l_\nu = l_0 e^{-(\Delta\lambda/\Delta\lambda_D)^2}. \qquad (5.30)$$

If we include the effect of natural broadening,

$$l_\nu = l_0 \frac{a}{\pi}\int_{-\infty}^{\infty}\frac{e^{-y^2}\,dy}{a^2 + (v-y)^2} \approx l_0\left(e^{-v^2} + \frac{a}{\sqrt{\pi}v^2}\right) \quad \text{for } a \ll 1, \quad (5.31)$$

where $v = \Delta\lambda/\Delta\lambda_D$ and the parameter a, which is usually small, is the ratio of natural to Doppler broadening. For small v the line shape is a Doppler core, but at greater v the exponential falloff of the first term leaves the second dominant, and in the wings of the line the natural broadening is preeminent. This behavior determines the shape of the curve of growth (Sec. 6.6). Because of the steep falloff of the absorption coefficient close to the line center, only the core is saturated at first and the line shape is dominated by the $1/v^2$ falloff in the wings. When a is larger the integral in Eq. (5.30) may be obtained from the tables of Harris (1948).

In terms of the Einstein A, the absorption coefficient is

$$l_0 = \frac{g_2}{g_1}\lambda^3\frac{A_{21}}{8\pi^{\frac{3}{2}}}\sqrt{\frac{M}{2kT}}. \qquad (5.32)$$

If there is macroscopic motion present it will broaden the lines of all atoms equally because they are all moving together. Since the thermal broadening depends on $\sqrt{M}$, we can separate Doppler and macroscopic motions by comparing the widths of spectral lines from atoms of different weight. The half-widths of lines from two different atoms are given by:

$$\begin{aligned}\frac{HW_1}{\lambda_1} &= \frac{1}{\sqrt{1.95\times 10^{12}}}\sqrt{\frac{T}{M_1}} + \frac{2\bar{V}}{C} \\ \frac{HW_2}{\lambda_2} &= \frac{1}{\sqrt{1.95\times 10^{12}}}\sqrt{\frac{T}{M_2}} + \frac{2\bar{V}}{C},\end{aligned} \qquad (5.33)$$

where $\bar{V}$ is the mean macroscopic velocity. If the two lines come from an identical volume, the value of T and $\bar{V}$ is the same in both equations. If we subtract one from the other, we eliminate $\bar{V}$ and solve for T:

$$T = 1.95 \times 10^{12} \left[\left(\frac{HW_1}{\lambda_1} - \frac{HW_2}{\lambda_2} \right) \left(\frac{1}{\sqrt{M_1}} - \frac{1}{\sqrt{M_2}} \right)^{-1} \right]^2. \qquad (5.34)$$

Although this looks like an elegant solution it is valid only if the two lines arise in the same portion of the emitting gas. Light atoms like H and He which are particularly sensitive to thermal motion usually arise in higher temperature regions than the heavy, slow atoms whose narrow lines fix the macroscopic velocities. So the H and He lines might be broad because they come from a different region with higher macroscopic non-thermal motion.

Perturbation by the surrounding particles will broaden the lines of an emitting atom. The effect of "collision broadening" is difficult to calculate and is usually treated in an average way so that we can make use of the theory of the Stark effect. The Stark effect is the broadening of spectral lines when the emitting atom is submitted to an electric field, and most of the collisional interactions may be approximated by electric fields. Because the hydrogen levels are degenerate, the perturbation matrix element is divided by a number near zero, and a big linear Stark effect occurs. In non-hydrogenic atoms the lines are only split by the quadratic Stark effect, which is much smaller. The broadening of the hydrogen levels increases with n^2, so the higher lines of the Balmer series grow broader and broader, till they finally merge with one another. This effect can be seen in solar and stellar flares.

The level for which the lines merge is related to the electron density by the Inglis-Teller (1939) formula, which, as modified by Griem (1964), is

$$n_s \approx \frac{1}{2} Z^{\frac{3}{5}} (a_0^3 N_e)^{-2/15}, \qquad (5.35)$$

where Z is the charge of the ion and a_0 the Bohr radius. A recent calculation by Kurochka (1967) gives

$$\log N_e = 22.0 - 7.0 \log n_s. \qquad (5.36)$$

In a magnetic field the energy levels of an atom split according to the quantum number M_j (written as M for short), which is the projection of the total angular momentum J on the direction of the magnetic field. This is called the Zeeman effect (Fig. 5.5). The line is split up into a number of components, depending on the M, M' values of the two levels. There are three groups of components corresponding to what is called the "classical Zeeman triplet," for $J = 1 \to 0$. The unshifted line for $\Delta M = 0$ is called the π component; it is seen when we look perpendicular to the magnetic field and is polarized parallel to the field. The two shifted components with $\Delta M = \pm 1$ are called σ_1 and σ_2 respectively; they are seen as left- and right-hand circularly polarized radiation along the direction of the magnetic field, and linearly polarized perpendicular to the field when we look perpendicular to the field. (The π and σ are from the German words *"parallel"* and *"senkrecht"*, referring to the orientation of the linear polarization relative to the magnetic field.) The net polarization is zero in any direction; perpendicular to the field the π component just equals the two σ components, and along the field the two σ components cancel.

The σ components are shifted in wavelength to the red and blue of the unperturbed line by

$$\Delta\lambda_H(\text{Å}) = 4.7 \times 10^{-13} g\lambda^2 H, \tag{5.37}$$

where H is the field strength in gauss, g is the Landé g-factor (Eq. 5.19) averaged between the two states, and λ in Å. The Zeeman splitting can be measured directly only in the center of sunspots, where the fields are thousands of gauss. This is how Hale detected the presence of magnetic fields in sunspots. But weak fields also broaden the lines.

Another important source of broadening is produced by merged fine-structure components. The $\lambda 10830$ line of neutral helium consists of the three lines produced by jumps from $^3P_2$, $^3P_1$, and $^3P_0$ to $^3S_1$. The first two of the ^3P levels are close and usually merged; the lowest gives a separate component of the line 0.5Å to the blue with 1/10 the intensity.

If the line broadening by other sources is of this order, the components merge in a broadened asymmetric line. The effect of the weaker component must be removed to assess the true line broadening. The strongest component is usually that of maximum $J \rightarrow J - 1$. The situation is even more difficult when the components are of nearly equal intensity, and the asymmetry is not immediately apparent. The Balmer lines of hydrogen are transition arrays made up of many multiplets merged into a single broad line. For Hα, for example, we have all the permitted transitions: $3s\ ^2S_{1/2} \rightarrow 2p\ ^2P_{3/2,1/2}$; $3p\ ^2P_{3/2,1/2} \rightarrow 2s\ ^2S_{1/2}$; $3d\ ^2D_{5/2,3/2} \rightarrow 2p\ ^2P_{3/2,1/2}$; seven lines in all. All the upper and lower levels are near-degenerate; the LS splittings are very small, and the seven lines of the supermultiplet are spread out over 0.1 Å. Because of the vagaries of excitation, the ratios of these components may also vary, so that each case has to be treated separately.

Thus by measuring line broadening it is sometimes possible, depending on the circumstances, to determine the electron density, temperature and magnetic field of the radiating gas. The presence of an electric field is unlikely, because the electrons move so freely that the conductivity is high, and electric fields are rapidly cancelled out.

5.7. Line ratios and optical depth

Many of the strongest lines in the spectrum are multiplets of known relative intensity with small separation. If the lines are weak, they will occur with the theoretical relative intensities. If they are strong, the stronger component will saturate, and we can estimate the optical depth from the observed ratio.

A good example is the NaD lines. The ratio of the transition probabilities is proportional to the statistical weights of the final states; the absorption coefficients will have a ratio of 1 : 2, just as in the H and K lines of ionized calcium. Applying Eq. (4.21) for lines of optical depth $\tau(D_2) = 2\tau(D_1)$, we have:

$$\frac{D_2}{D_1} = \frac{S_2(1 - e^{-2\tau})}{S_1(1 - e^{-\tau})}, \tag{5.38}$$

where D_2 and D_1 are the intensities and S_2 and S_1 the source functions of each line. In most cases $S_2 = S_1$. If τ is small $D_2/D_1 = 2$. In the limit of large τ the exponential disappears and the ratio of the lines is just unity. This is another way of showing that in the limit of a blackbody the radiation is a function of temperature alone, the Planck function. Between

5.5. The Zeeman effect. (a) Polarization of the Zeeman components viewed parallel and perpendicular to the magnetic field. Along the field we see only σ_1 and σ_2 components, with opposite circular polarization. Perpendicular to the field we see the π component polarized parallel to the field and the σ components linearly polarized in the perpendicular direction. The total radiation in any direction is unpolarized; the intensity of the transverse π component is twice that of either σ component.

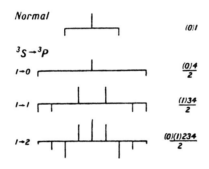

(b) Structure of the components of simple classical Zeeman triplet (top) compared with the "anomalous Zeeman effect" for both $J > 0$. In the latter cases the components cluster around the normal Zeeman triplet.

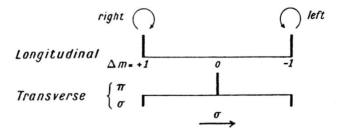

these limits the ratio of the lines will vary between 1 and 2.

In the photosphere the optical depth at the center of the *D* lines is great. Therefore the central intensities of the lines are virtually identical, differing in the wings. In prominences the temperature and ionization are higher, and there is little neutral sodium. The emission lines appear in the ratio 2 : 1 and also show the typical small τ Gaussian profiles shown in Fig. 5.6.

In the general case of two lines of theoretical ratio P and observed ratio R, we must solve the transcendental equation

$$R = \frac{1 - e^{-P\tau}}{1 - e^{-\tau}} \tag{5.39}$$

for the optical depth τ. Knowing the absorption coefficient per atom, we may then calculate the number of atoms along the line of sight.

As the optical depth of a line increases, the center of an emission or absorption line rapidly reaches an asymptotic value of S, whereas the wings keep on gaining with increasing optical depth (Fig. 5.6). The effect is to produce a squaring off of the peak of the line and an increase in the width at half-intensity. The line profile is then fitted by use of Voigt profiles which are tabulated by Allen (1981). The example shown is for no natural broadening, $a = 0$. With natural broadening the wings rise much faster. This calculation is for a constant source function, and the line profile in the

5.6. Profiles for an emission line with constant source function and pure Doppler broadening for various central optical depths τ_0, using Equation 5.23. For large τ_0 the line gets flat-topped as the emission reaches the source function. If damping is important for the line in question the wings will get very strong.

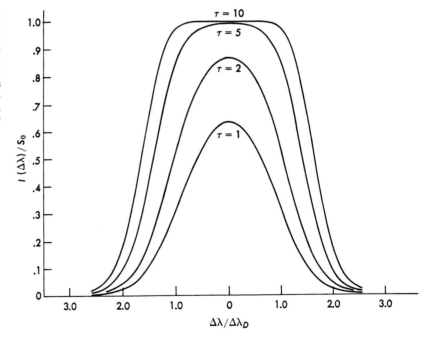

broad lines really maps out the source function as a function of increasing depth in the atmosphere.

One can use the line ratios to measure the Doppler width of pairs of optically-deep lines of theoretical ratio P. We simply measure the positions $\Delta\lambda_1$ and $\Delta\lambda_2$ at which the intensities of the two components are equal. Then, since $I_1 = I_2$,

$$1 - \exp\left[-P\tau_0\, e^{-(\Delta\lambda_1/\Delta\lambda_D')^2}\right] = 1 - \exp\left[-\tau_0\, e^{(\Delta\lambda_2/\Delta\lambda_D)^2}\right]. \tag{5.40}$$

We can solve for $\Delta\lambda_D$, finding

$$\Delta\lambda_D = \left(\frac{\Delta\lambda_2^2 - \Delta\lambda_1^2}{\ln P}\right)^{\frac{1}{2}}. \tag{5.41}$$

This method has been applied by Goldberg (1957) to find the Doppler broadening of the H and K lines at various parts of the profiles. Since we see much farther into the atmosphere in the wings of the line than we do near the line center, we can determine the height variation of the line broadening. The only trouble is that since the optical depths of the two lines differ by a factor of P, we are looking at different heights in each line, and the source functions S are not necessarily the same.

We have now completed our introduction into the mysteries of atomic spectra and the statistical mechanics of an astrophysical plasma. The time has come to look at the Sun.

References

Bethe, H., and E. E. Salpeter 1957. *Hdb. d. Phys.*, XXXV, 88.

Condon, E. U., and G. H. Shortley 1951. *The Theory of Atomic Spectra.* Cambridge: The University Press.

Edlén, B. 1964. *Hdb. d. Phys.* XXVII, 80.

Edmonds, A. R. 1957. *Angular Momentum in Quantum Mechanics.* Princeton: Princeton University Press.

Merrill, P. W. 1956. *Lines of the Chemical Elements in Astronomical Spectra.* C.I.W. Publ. 610.

Moore, Charlotte E. 1959. *A Multiplet Table of Astrophysical Interest.* N.B.S. Technical Note 36.

White, H. E. 1934. *Introduction to Atomic Spectra.* New York: McGraw-Hill.

6

Interior and Photosphere

The internal structure of stars is at once fairly well defined and virtually unknown. The Vogt–Russell theorem tells us that the structure of a star is uniquely determined by its mass and chemical composition. But the chemical composition changes steadily as hydrogen is converted into helium and is not always known. Further, we do not know if the helium generated in the center remains there or is carried to the outer layers by some mixing process. History may also play a role; it has been suggested that the Sun and other stars may still have rapidly spinning cores left over from the time of formation. Finally, the structure of the outer layers is modified by the existence of convection, which we have difficulty in modelling.

For these reasons there has been renewed interest in the solar interior, particularly because we can now probe it by solar seismology and neutrino observations. Since the Sun is the only star for which such observations are possible, these investigations are of the greatest importance for the understanding of all stars.

In Table 6.1 we give the relative abundance by number of the most common elements in the Sun as listed by Allen (1981). Although the exact numbers fluctuate, they are similar in the Sun and the stars. But we shall see in Sec. 7.4 that there is evidence for separation of the high first ionization potential (FIP) elements in the solar atmosphere and solar wind.

6.1. The solar interior

When stars form they collapse until the central pressure is high enough to support the outer layers. But the energy released by this contraction is not adequate to keep the star shining long in its present state. The energy of contraction would be used up in the Kelvin–Helmholtz (K–H) time, which is the potential energy E_G of the Sun divided by the luminosity:

$$T_{K-H} = \frac{GM_\odot^2}{R_\odot L_\odot} = 10^{15} \text{ sec} = 3 \times 10^7 \text{ yrs.} \qquad (6.1)$$

The virial theorem tells us that the thermal energy E_T is half the potential energy and of opposite sign. As the star contracts, half the potential energy is radiated away and half goes into internal kinetic energy, increasing the temperature until the nuclear processes ignite.

Because only a thin layer at the top is radiating into space, the inside of the Sun is close to thermodynamic and hydrostatic equilibrium. The mass and luminosity at a given radius are given by:

$$dM(r) = 4\pi r^2 \rho dr \qquad\qquad dL(r) = 4\pi^2 \rho \epsilon dr. \qquad (6.2)$$

These equations simply add up the mass and luminosity in successive shells, where ρ is the density and ϵ, the energy generation rate.

Two sets of nuclear processes convert hydrogen to helium: the proton-proton reaction and the carbon cycle. The first is a β-decay process – one of two protons in close approach undergoes inverse β-decay to become a neutron, and a deuteron is produced, with a positron and neutrino emitted. A free proton cannot β-decay to a neutron, but two protons can β-decay to a more tightly bound deuteron. The deuterons further interact with protons to form $_2\mathrm{He}^3$ and eventually $_2\mathrm{He}^4$, which is stable. The exact sequence is

$$e^- + \mathrm{H}^1 + \mathrm{H}^1 \rightarrow D^2 + \nu + 1.44 \text{ MeV},$$
$$D^2 + \mathrm{H}^1 \rightarrow {}_2\mathrm{He}^3 + \gamma + 5.49 \text{ MeV}, \qquad (6.3)$$
$$_2\mathrm{He}^3 + {}_2\mathrm{He}^3 \rightarrow {}_2\mathrm{He}^4 + \mathrm{H}^1 + \mathrm{H}^1 + 12.85 \text{ MeV},$$

where γ is a gamma ray and ν, a neutrino (the subscripts give the charge

Table 6.1. *The most abundant elements in the Sun.*

Element	Z	Relative Abundance	FIP(eV)
H	1	12.0	13.6
He	2	10.9	24.6
C	6	8.5	11.3
N	7	8.0	14.5
O	8	8.8	13.6
Ne	10	7.9	21.6
Mg	12	7.4	7.6
Si	14	7.5	8.2
S	16	7.2	10.4
A	18	6.8	15.8
Ca	20	6.3	6.1
Fe	26	7.6	7.9
Ni	28	6.3	7.6

and the superscripts, the atomic weight of each isotope). We see that, although only 1.44 MeV are liberated in the two-proton β decay itself, this step makes possible two further reactions giving about 25 MeV, c^2 times the mass difference between four protons and a He atom. The energy comes off in kinetic energy of end products, γ-rays and positrons which annihilate with electrons. An approximate formula for the energy production rate is:

$$\epsilon_{pp} = 2.5 \times 10^6 \rho X^2 \left(10^6/T\right)^{2/3} \exp\left[-33.8\left(10^6/T\right)^{1/3}\right] \text{ ergs/gm/sec},$$

(6.4)

where X is the fraction of hydrogen atoms by number.

Since each proton conversion gives about 6 MeV, the time to convert all the hydrogen into helium is simply the number of protons in the Sun divided by the luminosity:

$$T = \frac{2 \times 10^{33}}{m_H} \times 6 \text{ MeV} \div 4 \times 10^{33}$$
$$\approx 3 \times 10^{18} \text{sec} \approx 10^{11} \text{yrs}.$$

(6.5)

So the hydrogen burning can go on for a long time.

Although the proton-proton cycle is slow, it utilizes the most abundant material, protons, which also have the lowest electrostatic repulsion, and is probably dominant in the Sun. The carbon-nitrogen cycle involves collisions of protons with those elements to build up He nuclei; the higher nuclear charge of C and N produces a strong electrostatic repulsion and a much steeper temperature dependence. This restricts the CN cycle to stars hotter than the Sun.

Because the mean free path of a photon inside the Sun is a few centimeters, the photons produced in the core must follow an incredibly long random walk of about 10^{21} steps before they reach the photosphere. Even though they travelled at the speed of light, the photons emitted today were generated about 10 000 years ago. So surface phenomena occurring today should have no effect on the emission. But Fig. 6.24 shows that they do, for unknown reasons.

6.2. The neutrino problem

Since the conversion of four protons to one helium atom releases two neutrinos, the flux of neutrinos at the Earth should be

$$F(\nu) = \frac{2L_\odot}{25 \text{ MeV} \times 4\pi(AU)^2} \approx 3.5 \times 10^{12} \text{neutrinos/cm}^2/\text{sec}.$$

(6.6)

It is astonishing to contemplate such a huge flux passing through our bodies, no matter where we hide, and no doubt environmentalists will soon express concern over it; but it consists of low energy (0.42 MeV) neutrinos which cannot be detected with present techniques. There is an alternative chain to the third step of Eq. (6.4) which may produce detectable neutrinos:

$$_2\mathrm{He}^3 + {}_2\mathrm{He}^4 \rightarrow {}_4\mathrm{Be}^7 + \gamma$$
$$_4\mathrm{Be}^7 + p \rightarrow {}_5\mathrm{B}^8 + \gamma$$
$$_5\mathrm{B}^8 \rightarrow {}_4\mathrm{Be}^8 + e^+ + \nu \tag{6.7}$$
$$_4\mathrm{Be}^8 \rightarrow 2\ {}_2\mathrm{He}^4$$

Although this alternative only occurs in 0.015% of the chains, the B^8 decay neutrino has energy as high as 14.06 MeV and is the only one we can detect at present. Bahcall (1965) calculated the resulting flux of B^8 neutrinos at 2.5×10^7 /cm^2/sec or 40 SNU (solar neutrino units).

The neutrinos are detected by capture on Cl^{37} in a huge (4×10^5 liter) tank of cleaning fluid in a deep mine (Davis *et al.* 1968). Davis found a rate of 1.6 ± 0.4 SNU (1 SNU = 10^{-36} neutrino absorptions/sec/target atom). The theoreticians have sharpened their pencils and reduced the predicted rate to 4.7 SNU, which is three times the observed rate. Until recently (Bahcall 1985; Friedlander 1978) no solution had been found.

The B^8 neutrinos represent only 5×10^{-5} of the solar neutrinos. Since the calculated probability of their production may be subject to error, one would like to measure the low energy $p - p$ neutrinos directly, which

6.1. Standard solar model (Ulrich 1982). (a) density; (b) temperature.

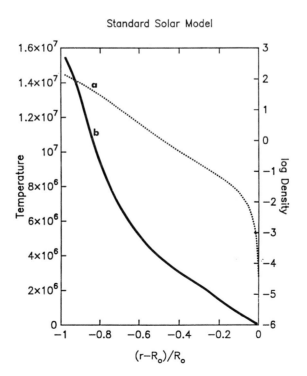

Standard Solar Model

can be done by measuring the capture of neutrinos on gallium to produce germanium. About 15 tons of gallium are required; the number of atoms converted is small but countable by special techniques. Other experiments are also in progress.

A number of implausible explanations for the low flux have been advanced:

1. The Sun's interior is fully mixed, so the central temperature is still that of the young Sun and the neutrino flux is low. This idea is supported by the long term constancy of the solar constant revealed by the geological record. The mixed model gives 1.5 SNU.

2. The Sun occasionally undergoes contraction and doesn't use its nuclear fuel and we are measuring at that rare instant.

3. The abundance of heavy elements in the solar interior is much lower than we think, hence the interior opacity and central temperature are lower, which produces less B^8.

The history and current state of the problem are reviewed by Bahcall and Davis (1982). The current neutrino measurements definitely rule out the carbon-nitrogen cycle (Bethe 1939) in the Sun, which produces 36 SNU.

In the past two years a new explanation of the neutrino deficit has won wide support. Wolfenstein (1978) proposed an oscillation between the electron neutrino, the kind produced in the Sun, and the muon neutrino, which is not detected by the chlorine experiment. Mikheyev and Smirnov (1986) suggested that in the high density gradient inside the sun, the phase shifts of electron neutrinos would be such as to amplify the Wolfenstein oscillation and convert them to muon neutrinos. The idea, now called the MSW mechanism, was refined and publicized by Bethe (1986). Calculations by Rosen and Gelb (1986) and Parke and Walker (1986) supported the conclusion. Bethe showed that the remaining flux of electron neutrinos are just about the flux found by Davis. It is hoped that the gallium detector will resolve the problem, which has great significance for theoretical physics.

6.3. Solar models

We have already given Eqs. (6.2) for the mass and luminosity of the star, the basic observable quantities that characterize it. In order to integrate these, we must know the variation of temperature and density throughout the star as well. This information is supplied by two additional equations.

First, we may assume the star to be in hydrostatic equilibrium. In that case the increase in pressure dP between point r and point $r + dr$ is simply the weight of the added material. That weight is the gravitational attraction $GM(r)/r^2$ of the mass of the star interior to r (the overlying material exerts no gravitational force on the material inside) on the density ρ :

$$\frac{dP(r)}{dr} = -\frac{GM(r)}{r^2}\rho. \tag{6.8}$$

This equation governs the variation of pressure throughout the star. We relate the density to the gas pressure by the gas law:

$$P_g = nkT = \frac{k\rho T}{\mu m_{\rm H}} = 0.825 \times 10^8 \frac{\rho}{\mu} T, \tag{6.9}$$

where k is the Boltzmann constant, $m_{\rm H}$ the mass of a hydrogen atom and μ, the mean molecular weight. This last quantity is the average weight of each individual particle in atomic mass units. Since ionized hydrogen gives a proton and an electron, $\mu = 0.5$. Helium has an α-particle of mass 4 and two electrons, giving $\mu = 4/3$. Heavy atoms with charge Z give $Z + 1$ particles when fully ionized, but have atomic mass of roughly $2Z$, so their mean molecular weight is 2. The mean molecular weight for a mixture with fractions X, Y, and Z (by weight) of hydrogen, helium and heavy atoms, respectively, is therefore

$$\mu = \frac{1}{2X + 3Y/4 + Z/2}. \tag{6.10}$$

For typical stellar mixtures μ is around 0.6.

From Eq. (6.9) we can derive the scale height, the e-folding distance in an atmosphere in hydrostatic equilibrium. If we let $GM/r^2 = g(r)$ we have

$$\frac{dP}{P} = -\frac{\mu m_{\rm H} g(r)}{kT} dr = -\frac{dr}{H}, \tag{6.11}$$

where

$$H = kT/\mu m_{\rm H} g \tag{6.12}$$

is the scale height. For short ranges where T and g change slowly we can integrate, finding

$$P = P_0 \exp -(r - r_0)/H. \tag{6.13}$$

The scale height is the measure of any barometric atmosphere. An identical expression applies for the variation of density and particle number, but we must remember that H is a function of T and r. If we integrate over height, we find that *the mass above any point is less than the mass in the next lower scale height*. So the integral from infinity down to some point x of a linear quantity like the absorption k is approximately $k(x)H$; for variables that vary with N^2, we replace H with $H/2$.

The run of temperature through the star will be determined by the efficiency with which the energy is transported from hot to cold, much as the temperature of an Eskimo in the arctic will be determined by the efficiency with which his parka transports the heat of his body outward. The Eskimo loses heat by conduction, radiation, and convection. In stars, the first process is important only in white dwarfs, while the latter two are important in all stars, depending on the circumstances.

The transport of energy will proceed by whichever process moves it most quickly and that process will establish the temperature gradient. If the Eskimo has a hole in his parka, air will tend to pass freely back and forth, and the transport will be by convection, since relatively little additional radiation will escape. Ladies' summer dresses, for example, are also designed for rapid convective transport with little radiative escape. If the Eskimo's parka is perfectly airtight, he still will lose heat by conduction through the parka, the inner shell of which is heated by radiation and convection from his body.

Inside a star, convective transport is possible only when a rising element of gas remains buoyant instead of cooling and falling back; this is convective instability. If an element of gas expands or contracts without exchanging heat with its surroundings, it is said to cool adiabatically, following the formulae:

$$P = k\rho^{\gamma} \qquad\qquad \frac{dT}{T} = \left(1 - \frac{1}{\gamma}\right)\frac{dP}{P}, \qquad (6.14)$$

where γ is the adiabatic constant, the ratio of the specific heats at constant pressure c_p and that at constant volume, c_v. For a perfect gas, $\gamma = 5/3$. If the ionization is changing rapidly, as happens near the surface of a star, heating will go into internal energy and c_v will increase. γ will be smaller, the rising element will cool less, and convection will be more likely. If we divide Eq. (6.14) by the distance element dr, we obtain a differential equation for the adiabatic temperature gradient:

$$\frac{dT}{dr} = \left(1 - \frac{1}{\gamma}\right)\frac{T}{P}\frac{dP}{dr}, \qquad (6.15)$$

where dP/dr can be obtained from Eq. (6.8). In some cases one has to allow for the radiative interchange with the surroundings and modify the adiabatic formula.

The temperature gradient for radiation alone is found by equating the energy flux into and out of a cell. The absorption coefficient per cm is $k\rho$, where k, the opacity, is measured in cm^2/gm. The energy absorbed per cm^2 in the shell 1 cm thick at radius r (area $4\pi r^2$) is $Lk\rho/4\pi r^2$. This amount must equal that contributed to the radial outflow, which is given by the radial gradient of the blackbody emission $d(acT^4)/3dr$. The factor 1/3 is due to the fact that we are interested in the radial flux only. Matching these two input and output rates, we obtain

$$\frac{Lk\rho}{4\pi r^2} = -\frac{d(acT^4)}{3dr}, \qquad (6.16)$$

or

$$\frac{dT}{dr} = -\frac{3}{4ac}\frac{k\rho}{T^3}\frac{L}{4\pi r^2}. \qquad (6.17)$$

Convective instability occurs if the absolute magnitude of the radiative gradient Eq. (6.17) is greater than the magnitude of the adiabatic gradient Eq. (6.15):

$$\left|\left(\frac{dT}{dr}\right)_{rad}\right| > \left|\left(\frac{dT}{dr}\right)_{ad}\right|. \qquad (6.18)$$

This is the *Schwarzschild criterion* (Schwarzschild 1906). If it is true, the rising gas element, cooling adiabatically, will still be warmer than its surroundings and remain buoyant.

Convection is so efficient that if the adiabatic gradient is only slightly less than the radiative, convection will carry all the extra energy. As a result the adiabatic gradient is the upper limit of the temperature gradient. Near the surface of a star, convection is less efficient because of the low density and temperature and the actual temperature gradient may be steeper than the adiabatic value. Even when convection sets the upper limit in the temperature gradient, most of the energy may still be transported by radiation, which is of course not turned off by the onset of convection.

The temperature gradient becomes steep enough for convection to occur in only two places in the star. In the center, the energy production may depend so steeply on temperature that high-temperature gradients arise. This usually occurs only if the C-N cycle is operative, because it has a high temperature dependence. In the outer layers of the star, atoms that were completely stripped in the interior recapture electrons and absorb radiation voraciously. The opacity k is increased, and the temperature gradient goes up. At the same time, the recombination of HeII and H provides internal energy for the rising gas elements to carry. This lowers the adiabatic temperature gradient significantly. The two processes lead to a deep convective zone. When the convection is connected with the ionization process, we talk about a hydrogen or helium convective zone.

The normal procedure in making a stellar model is to integrate Eqs. (6.2), (6.8) and either Eq. (6.17) or Eq. (6.15) simultaneously, starting from the surface with known values of mass and luminosity, and zero temperature and pressure. We have radiative equilibrium in a thin surface skin, then a convective zone, then radiative equilibrium down to the core. If the mass and luminosity come out to be zero at the center, we have a correct model for the Sun. We adjust the chemical composition and the depth of the outer convective zone until a model is achieved where M and L do vanish at the center. At this point we should have a good model for the Sun. One can also require models starting from inside and out to match in the middle. The procedure is described in detail in Schwarzschild's (1959) book, *Structure and Evolution of the Stars*. Its validity rests on the Vogt–Russell theorem: if opacity and energy generation depend on local conditions only, the structure of a star is uniquely determined by its mass and chemical composition.

An obvious source of uncertainty is the convective zone. The depth of the convective zone in the Sun is not known. Models of convection depend on the mixing length theory, the concept that each convecting element travels a certain distance (the mixing length) at a velocity determined by the unknown temperature overshoot. The energy transported depends on the square of the mixing length, which is usually taken as a multiple (one or two times) of the scale height. The depth of the convective zone can vary substantially with the choice of mixing length, so we will have to await better results from solar seismology to determine the true value. Recently Christiansen-Dalsgaard *et al.* (1985) have inverted solar oscillation spectra to set the base of the convective zone at $0.71\ R_{\odot}$. The details of turbulent convection are discussed by Spiegel (1966).

6.2. A "$k - \omega$" plot of spherical eigenvalue l (which corresponds to oscillation scale) *vs.* frequency in the solar brightness oscillations in a 6 Å band at the K line. These remarkable results were obtained by Harvey and Duvall in a 50-hr run at the South Pole. The modes have been averaged over the azimuthal eigenvalue m and the individual l mode frequencies are clearly separated. The abscissa (l) ranges from 6 (left) to 250 (right). The ordinate (frequency) ranges from 2238 to 4679 (top) μHz. The frequency resolution of 5 μHz is not good enough to resolve the modes of $l > 100$. (KPNO)

In Fig. 6.1 we exhibit a recent "standard model" computed by Ulrich and Rhodes (1977). The temperature at the center comes out around 15 million deg and the density, 150 g/cm^3. This model produces about 7 SNU neutrinos.

6.4. Solar evolution

As time goes by, the hydrogen in the center of the Sun is gradually transformed into helium; each four H nuclei are replaced by a single He nucleus, and the gas pressure at the center drops. The star shrinks, converting potential energy into internal energy by the virial theorem. But the nuclear generation is such a high power of the temperature that the core temperature cannot rise much, so the increased internal energy goes into an expansion of the envelope, resulting in a redder but brighter star. Model calculations show that the Sun (and almost all stars) does get brighter and redder as the helium content in the center grows. If for some reason the material at the center is mixed with the rest of the star, the change in composition is negligible and evolution does not take place. We know that other stars do evolve, because color-magnitude diagrams for star clusters show the color distribution one expects with no mixing.

Calculations by Torres-Peimbert *et al.* (1969) show that the Sun should be 1.7 times brighter after seven billion years than it started out, and should have brightened by about 10% in the last billion years. The Earth as now constituted would have been frozen under those conditions. Moreover the albedo of ice is so high that the solar output would have to be 30% greater before the ice thawed. Geochemical and fossil evidence indicate that terrestrial temperatures have changed little in the last billion years, and the temperatures were probably higher, not lower, before that. Possibly strong CO_2 blanketing kept the Earth warm in those days of a fainter Sun, and heating by decay of fossil radioactive elements may also have played a role. But these mechanisms are really not thermostatic, and the temperature stability has been remarkable. This stable temperature regime might be explained by a fully mixed sun, which would not evolve, because the He produced leaves the core. The fully mixed Sun produces about 1 SNU, in accord with observations. But alas, there are no known mechanisms for mixing, and other stars do appear to evolve by building up helium in their interiors without mixing.

In order to explain the advance of the perihelion of Mercury within a new picture of general relativity, Dicke (1963) suggested that the Sun might be oblate in shape because of a rapidly-rotating fossil core. The argument seems reasonable because conservation of angular momentum in protostellar clouds causes stars to form with sizable rotation speeds. As the stars age they slow down. Since the only way of slowing stars down to rotation speeds like that of the Sun is through mass loss from the surface, the core may still rotate rapidly. Dicke and Goldenberg (1967) built a special telescope with which they measured a solar oblateness of 0.048 arc sec (1/18 000 of the radius), 6 times greater than that produced by the 27-day rotation. The result was challenged by various authors. Chapman and Ingersoll (1972) argued that faculae produced a spurious signal; Goldreich and Schubert (1968) argued that a separately rotating core would be unstable; and Hill and Stebbins (1975) produced contradictory measurements. Dicke (1976) re-analyzed his data and found strong evidence for a rigid rotating core with a 12.2 day period. In 1982 the experiment was rebuilt by Kuhn, Libbrecht and Dicke, and observations in 1983 produced a null result (Libbrecht 1984). Further observations (Dicke *et al.* 1987) have continued to show little or no oblateness, so the Sun probably does not have a rapidly rotating core. Does the absence of a rapidly rotating core imply that there was mixing?

6.5. Solar oscillations

Oscillations can be seen in any good solar movie in any wavelength. They are particularly strong in the wing of Hα, the Ca K line and the blue continuum. The oscillatory motion was well-known in early movies of the chromosphere, and termed the "flagellant motion." It was given a place in the official literature by Leighton, Noyes and Simon (1962) who established that it was an oscillation with a five minute period. Evans and coworkers (Evans and Michard 1962; Evans 1963; Evans *et al.* 1962) made spectroscopic measurements and found the period to vary with height. This confirmed the result of Goldberg *et al.* (1961) that the rms velocity in a line increases with intensity, and therefore height. We see deepest where there is no absorption line at all, so presumably the granulation, which we see in the continuum, is even lower than the weakest lines. Therefore, if the curve is monotonic, the vertical velocities in the granulation must be smaller than the smallest observed in the absorption lines. The periods measured by Evans *et al.* are given in Table 6.2. Leighton *et al.* were able

Table 6.2. Oscillatory velocity amplitudes for various lines.

Line	Intensity	Element	Period (sec)	$\overline{V}$(km/sec)
3933.7	1000	CaII	150	2.00
6562.82	40	Hα	180	1.34
5172.7	20	MgI	—	0.61
5171.6	6	FeI	—	0.39
5168.7	1	NiI	300	0.31

to explain the variation of frequency with height in terms of the cutoff of acoustic waves propagating in the atmosphere, which allows only the higher frequency waves to travel upward.

The first hint of the organization of non-radial solar oscillations appeared in the work of Frazier (1968) who made the first two-dimensional plots of frequency versus wave number (popularly called $k - \omega$ diagrams) and found several peaks in amplitude. Ulrich (1970) suggested that Frazier's peaks matched the fundamental and first overtones of the solar envelope. Deubner (1975) obtained much better data, showing that the $k-\omega$ values of the modes formed a series of ridges, which matched the theoretical curves "to an embarrassing extent". He suggested that the long-wave (low k) extension of the ridges might be the radial modes predicted by Wolff (1972). Ando and Osaki (1977) calculated a whole spectrum of eigenfrequencies of p-modes and found them to correspond to Deubner's results. Deubner's work was later confirmed by Rhodes *et al.* (1977) and others. With modern techniques it is possible to separate the different modes, as in the $k - \omega$ diagram Fig. 6.2.

The amplitude (we follow Unno *et al.* 1979) of the oscillations can be characterized by a spherical function similar to the atomic wavefunction Eq. (5.1) which describes the normal modes of the Sun:

$$\Psi = R_n(r)P_{lm}(\theta)e^{im\phi}e^{i\omega t}. \tag{6.19}$$

For a given value of n we have a series of radial nodes, their position determined by the exact pattern of $R_n(r)$. The rest of the wave function gives the spherical distribution of the oscillation; together with n, the azimuthal eigenvalue m and the meridional eigenfunction l determine the frequency of a particular mode. For a given n all the frequencies of the nlm modes will lie along a ridge of the $k - \omega$ diagram.

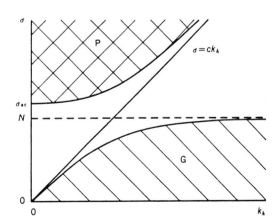

6.3. Diagnostics for a plane isothermal atmosphere. N is the Brunt–Väisälä frequency, σ_{ac} is the acoustic cutoff frequency, and k_h, the horizontal wave number. The vertical wave number k_z is real in the hatched (g-modes) and cross-hatched (p-modes) regions, but imaginary (evanescent waves) elsewhere. After Unno (1979).

The local propagation of waves is governed by two frequencies. The Lamb frequency

$$L_l^2 = \frac{l(l+1)c^2}{r^2} \tag{6.20}$$

(where c is the sound speed) reflects the angular momentum barrier as a wave propagates to smaller r. The Brunt–Väisälä (B–V) frequency is

$$N^2 = g\left(\frac{1}{\Gamma_1}\frac{d\ln p_0}{dr} - \frac{d\ln\rho_0}{dr}\right)$$
$$= \frac{g^2}{c^2}(\gamma - 1) \qquad \text{(isothermal case)}, \tag{6.21}$$

where

$$c^2 = \Gamma_1 p_0/\rho_0 \qquad \Gamma_1 = (d\ln p/d\ln\rho)_{ad}. \tag{6.22}$$

The B–V frequency is a measure of the existence of convective instability. Sound waves can only propagate with frequencies above both Lamb and B–V frequencies, and gravity waves must have lower frequencies (Fig. 6.3).

6.4. The B–V (heavy lines) and Lamb (light lines, l-value marked) frequencies as a function of radius in the solar interior, computed by P. Kumar. The photosphere is at log P = 5. *P*-modes propagate above both frequencies, so that they will exist between the Lamb cutoff at the left and the B–V peak at the photosphere at the right. There also is a region at upper right in the chromosphere where a *p*-mode resonance may exist. *g*-modes may propagate below both cutoffs, including the solar core, but there is no obvious way the waves should escape and there is no convincing evidence for their existence.

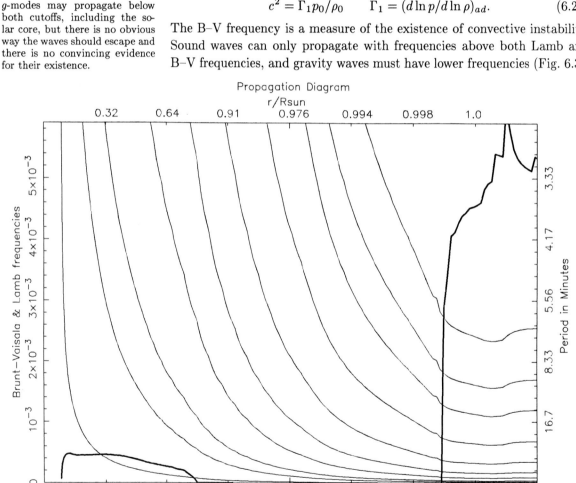

Propagation Diagram
r/Rsun

Gravity waves with higher frequencies would simply excite the convective instability and not oscillate.

Fig. 6.4 shows a plot of the B–V and Lamb frequencies through the Sun where the abscissa is the local pressure. There is a big peak in the Brunt–Väisälä frequency in the chromosphere and a smaller one in the core. The Lamb frequency, on the other hand, increases sharply inwards because of the r^{-2} angular momentum barrier. Each p-mode can propagate where its frequency exceeds both of these. The low l-modes are reflected deep in the Sun, while high l modes do not penetrate the convective zone. Gravity waves must have frequencies below both; hence they could only exist above the photosphere or in the core.

The dispersion relation for the wave frequency σ is

$$(\sigma^2 - \sigma_{ac}^2)\sigma^2 - (k_h^2 + k_z^2)c^2\sigma^2 + k_h^2 c^2 N^2 = 0, \qquad (6.23)$$

where σ_{ac} is the acoustic cutoff frequency:

$$\sigma_{ac} = \frac{c}{2H} = \frac{\gamma g}{2c} \propto T^{-1/2}, \qquad (6.24)$$

which is 1.02 times the B–V frequency for $\gamma = 5/3$. Since the acoustic

6.5. A $k - \omega$ diagram computed by K.G.Libbrecht based on 12 days of BBSO data, compared with frequencies predicted by Christiansen-Dalsgaard; there is excellent agreement, but the small discrepancies are thought to be significant. Some of the higher frequency data is based on the South Pole data (Fig. 6.2).

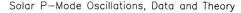

Solar P–Mode Oscillations, Data and Theory

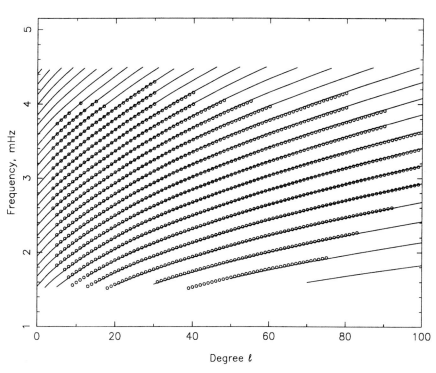

cutoff rises as the temperature falls in the photosphere, the low frequencies are filtered out, which explains the progression of higher frequencies in stronger lines found in Table 6.2.

We can rewrite Eq. (6.23) as

$$c^2 k_z^2 = (\sigma^2 - \sigma_{ac}^2) + k_h^2 c^2 (N^2 - \sigma^2)/\sigma^2. \qquad (6.25)$$

6.6. A Doppler picture of the Sun made with the 25mm helioseismology telescope, based on the VMG principle. Bright is approaching, dark receding. The network of bright-dark pairs must be cells in which the gas expands outward, first recognized by Leighton from his photographic subtractions. The shading across the frame is the solar rotation, 1.9 km/sec. (BBSO)

The zones of real vertical wave number k_z are plotted in Fig. 6.3 for frequencies σ greater than ck_h and N. The p-modes in the upper left lie above σ_{ac}; the g-modes at lower right, below N. If $k_z^2 < 0$, we have evanescent waves which decrease exponentially with distance. The line $\sigma = ck_h$ represents the Lamb waves or f-modes, which propagate horizontally as waves in the sea. Leibacher and Stein (1981) showed that if one treated the Sun as a resonant cavity, acoustic waves would be reflected from the point at which the sound speed equaled the horizontal phase velocity ω/k_h. This led to the relation

$$\frac{(n+1/2)\pi}{\omega} = T, \tag{6.26}$$

where T is the time of travel of the acoustic wave to the reflection point. Duvall (1982) reasoned that, since all the waves travel at the same speed, all the ridges on the $k - \omega$ diagram should collapse onto a single line when fitted to this formula. He found that this indeed occurred, not for the value in Eq. (6.27), but for that obtained by replacing $1/2$ by $3/2$. This supports the view that we are looking at a system of waves in a single cavity.

Duvall and Harvey (1984) measured the frequency spectrum of the 5-min oscillation and found radial modes of orders 2 to 26 for $l < 140$. They give the degree l as a function of the reflection radius r:

$$l = -1/2 + \left[1/4 + \frac{4\pi^2 \gamma^2 r^2}{c^2} \right], \tag{6.27}$$

and use this relation to match all the modes in a series of data, establishing them deep into the interior.

In order to determine the frequencies accurately one needs sharp peaks in the power spectrum. This is best obtained by long continuous observations of the Sun. A French group led by Fossat (Grec *et al.* 1980) was the first to utilize the South Pole Observatory (Fig. 2.4) for extended observations. Fig. 6.2 shows the results of South Pole brightness measurements by Harvey. The oscillatory velocities are 4–40 cm/sec, corresponding to amplitudes of less than 10 m in the solar radius. Eighty normal p-modes were identified. The frequencies correspond to spherical harmonics Y_{lm} of order $l = 1 \rightarrow 3$, and radial eigenfunctions of high n. No g-modes were detected by Grec *et al.*; although a number of peaks occurred at lower frequencies, their significance could not be established. While some workers have claimed detection of a 160-min period, the evidence is not strong.

Woodard (1983) obtained even sharper resolution by analyzing the long run of total solar brightness data obtained by Willson with a solar constant monitor on the Solar Maximum Mission. These data clearly show the rotational splitting produced by waves running with and against the solar rotation. Harvey and Duvall have measured this splitting in different modes from the South Pole and concluded that the solar rotation is constant as far in as 0.2 $R_\odot$.

The fact that the phase of the oscillations is fairly constant makes it possible to obtain good results from normal observatories, which offer much longer runs of clear weather than the Pole, albeit broken by the diurnal cycle. A problem with all observations is that we need to fit the oscillations with spherical harmonics, but we should have data over the entire sphere to determine the expansion coefficients. Allowing for limb-darkening, we see only one-third of the Sun; as a result our fits are filled with spurious frequencies (Fig. 6.5) which must be removed by cleaning techniques.

Since the waves are trapped inside the solar envelope, the $k - \omega$ ridges must depend on the nature of that envelope. Rhodes *et al.* (1977) parametrized the solutions and produced fairly stringent limits on the base of the solar convective envelope, a radius of $0.62\,R_\odot$ to $0.75\,R_\odot$ and a temperature between 1.7×10^6 and 3.2×10^6 deg. The upper limit is set by the observed existence of solar lithium, which would all be destroyed by nuclear reactions if the convection went too deep.

Ulrich and Rhodes (1977) studied the effect of a chromosphere on the normal modes. Using a homogeneous chromosphere 2650 km deep, they found modes with frequencies $\omega = 0.026$ and 0.035 sec^{-1}. Ulrich and Rhodes concluded that the chromosphere strongly affects the stability of the modes for longer periods. Since 1500 km is probably the greatest thickness for which a homogeneous model appears realistic, it would be interesting to see their results for that model. They found the eigenfrequencies of the normal modes depend on the entropy of the adiabatic portion of the solar convective envelope.

Most normal modes have lifetimes (i.e. no change in phase) of at least 5 days and many as long as 12 days (Libbrecht and Zirin 1986). Fig. 6.7 shows the distribution of power with frequency; there is as yet no convincing explanation why the peak power occurs around the 5-min period. A plot of power *vs* l gives much the same result; the energy rolls off above $l = 100$ (Libbrecht *et al.* 1986). The source of energy for the oscillation is thought to be the convection zone; Goldreich and Keeley (1977) have shown how turbulent convection may excite the oscillations. Libbrecht *et al.* find equipartition in the modes for lower l, which would agree with excitation by turbulence. There is no explanation for the high frequency roll-off, which might be due to viscosity or radiative loss or even interaction with magnetic fields. It is well-known that the oscillation amplitude in the chromosphere, at least, is sharply reduced by magnetic fields; it is possible that scattering off the magnetic network has the same effect globally.

There has been considerable interest in the measurement of solar oscillations of low l, because these modes may reach the core of the Sun. Hill (1980) reports diameter oscillations with amplitude > 10 m/sec and periods of 68 min and 45 min. Severny *et al.* (1976, 1980) and Brookes *et al.* (1976) report amplitude of 2 m/sec at 2^h40^m period. Others have disputed these results, and Goldreich and Keeley (1977) predict amplitudes less than 0.1 cm/sec for these modes. So at the moment only the short period p-modes in the 2–10 minute period range can be regarded as established.

6.6. Solar rotation

Since there are no markers, the details of the solar rotation rate are quite difficult to measure. But sunspot drawings early revealed that the low-latitude spots rotated considerably faster than those at higher latitudes. This remarkable unexplained fact is the basis of various models for the solar dynamo. Modern techniques have made possible considerable improvement in measurement of the differential rotation. The results are summarized by Schröter (1985).

Almost everyone who measures solar rotation for the first time finds their data does not agree with published values and then realizes that they have measured the synodic rate ω, the rate relative to the Earth, while we want the sidereal rotation rate Ω, which is the true rate. They differ by the Earth's orbital velocity of $0.9865°/\text{day}$. The synodic period is longer because the Sun must rotate a little further to face the Earth's new position. Rotation is normally fitted by the formula

$$\Omega = A - B\sin^2\theta - C\sin^4\theta \qquad \text{deg/day}. \qquad (6.28)$$

Various determinations are given in Table 6.3.

Newton and Nunn (1951) carried out the landmark modern measurement, using recurrent sunspots. Tang (1981) found a slightly different rate from high-latitude spots. Sunspot data is limited to the sunspot belts, and few returning spots are unchanged from the previous rotation. Magnetograph data cover a larger area and have many data points. Snodgrass (1983) used an autocorrelation technique on 15 years of Mt. Wilson data. Since there is a high persistence in magnetic data, this is probably the most accurate measurement of rotation. His result agrees closely with Newton–Nunn up to 30° latitude. Snodgrass suggests a small dip in the rotation rate at the equator. His values correspond to synodic periods of 26.9 days at the equator and 32 days at 60°. Beckers (1978) and Cram *et al.* (1983), searching for a possible polar vortex, found fair agreement with the published rates at 75° latitude, where the rotation period is about 35days.

One might think that spectroscopic measurements would be an accurate

Table 6.3. Solar rotation parameters (sidereal rate).

Method	A	B	C	Authors
Sunspots	14.38	2.96	0	Newton and Nunn (1951)
Sunspots	14.37	2.60	0	Tang (1981)
Magnetic fields	14.37	2.30	1.62	Snodgrass (1983)
Spectroscopic	14.19	1.70	2.36	Howard *et al.* (1983)
Coronal holes	14.24	0.4	0	Timothy *et al.* (1975)
Filaments	14.4	1.5	0	Adams and Tang (1977)

technique, but this is not the case. The problem is not so much in the spectrographs but rather the presence of substantial local velocities associated with the supergranulation and solar oscillation. There are also problems associated with scattered light and the limb shift of Fraunhofer line wavelengths. The spectroscopic results obtained by Howard *et al.* (1983) are 1.5% below those of Snodgrass.

Filaments are another good way to measure the rotation because they are long-lived. However, it is difficult to measure their position when they are not on the meridian because we do not know their height. Adams and Tang (1977) found an assumed height of 10 000 km resulted in correspondence with the spot data as well as the result given in Table 6.3.

Coronal holes should give an excellent measure of solar rotation; as we see in Chap. 8, they may last for five or six rotations and cover a great range of latitude. They show a near rigid-body rotation, quite different from the other measurements. The reason for this effect is not well understood. The results obtained by Skylab (Timothy *et al.* 1975) are heavily influenced by two large coronal holes. Shelke and Pande (1985) argue that there is differential rotation in all the holes measured in $\lambda 10830$ by Harvey at Kitt Peak; their data appears to me to show some holes with and others without differential rotation.

Schröter (1985) lists observations of filaments, prominences, *etc.*, which show differential rotation differing from the surface by as much as 0.5° per day at 45° latitude. This simply does not make sense. There is a perfect

6.7. The energy in each *nlm* mode as a function of frequency, calculated by Libbrecht.

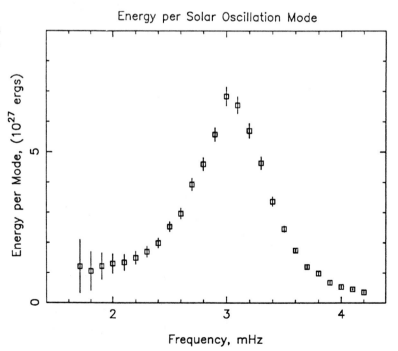

correlation between photospheric magnetic fields and overlying atmospheric features. We never see an active region with its accompanying coronal condensation trailing behind or ahead; the magnetic field holds it. Nor do we see filaments offset from the magnetic inversion lines with which they are associated. We must conclude that the data is faulty and there is little change in rotation with height, at least between photosphere and low corona.

Fig. 6.4 shows how the angular momentum barrier reflects waves of higher l before they reach the core of the sun. The frequencies of each mode should be split because we observe oscillations in both approaching and receding hemispheres; the splitting measures the rotational velocity and the

6.8. Granulation and an active sunspot group (Aug 17 1959) photographed by Stratoscope, a balloon-borne telescope built by Martin Schwarzschild of Princeton, which took the first high resolution granulation photos. At the edge of the large stable spot we see granulation elongated by the strong horizontal fields of the newly emerging satellite spots. The shear shows up in penumbral fibrils as well. (Princeton U Obs – Project Stratoscope)

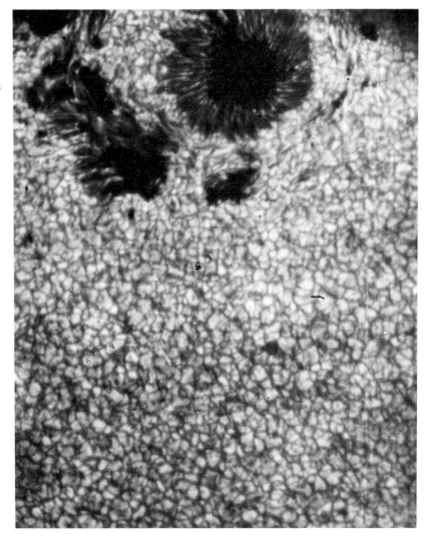

dependence of that value on l tells us how the rotation varies with depth within the Sun. All of the evidence so far points to rigid solar rotation to an accuracy of 10%. The data is as yet inadequate to determine the depth dependence of the differential rotation.

There is evidence for other large scale motions, all barely detectable and not completely certain. Howard and LaBonte (1980) analyzed spectroscopic velocity measurements obtained over 12 years and found alternating zones of fast and slow rotation, the difference from the mean being 3 m/sec (a fast walk or a slow run). They see these zones to move from the pole to the equator in roughly 22 years and propose that the Sun is a torsional oscillator with that period. They suggest that solar activity forms in a latitude zone marking the boundary of the fast and slow zones.

6.7. The granulation

The photosphere is easier to study than the unseen layers below and the transparent and variable regions above. It presents a fairly uniform surface roughened by granulation. Its density is high enough for collisions to set up equilibrium, so the radiation temperature is similar to the electron temperature, and the gas is not far from thermodynamic equilibrium. The radiation *intensity*, however, is diminished by the dilution factor 1/2.

The photosphere is defined as the outer surface of the Sun as seen in white light. Physically it includes the lower chromosphere and is intimately connected with the convective zone below. It is the point where the Niagara of energy flowing from the interior escapes into space with the enormous flux of 6.5×10^{10} ergs/cm^2/sec. The color temperature at 5500 Å is 6700°; this rapidly falls off at shorter wavelengths and overall the solar emission fits a 6000° black body.

Although the layers above $\tau = 1$ are in radiative equilibrium, the convective zone just below is manifested in the granulation, a universal pattern of small polygonal convecting structures. The granulation must be studied with photographs of the highest resolution, obtained from balloons, space, or sometimes from the ground. The best ground-based granulation observations are obtained by recording high resolution video images, where 30 frames catch occasional instances of good seeing, making up for the fact that the individual pictures are inferior to film.

Before the first modern high-resolution photographs were obtained by Schwarzschild (1960) with the Stratoscope balloon-borne telescope, the granulation was thought by some to be a perfectly symmetric convection – *i.e.*, the light and dark areas covered equal parts of the surface. Schwarzschild's pictures confirmed that the surface was mostly bright, with narrow dark lanes in between the granules. The mean granular size was measured to be about 640 km, a value later confirmed by Karpinsky (1977) using photos with the Soviet balloon-borne telescope. The granules are not round but irregular polygons. Since the width of the intergranular lanes is just about at the instrumental resolution limit, it is probable, as we can judge from Fig. 6.8, that higher-resolution pictures would bring out many more dark lanes, and we would find the granules to be considerably smaller. Because our best pictures are heavily filtered by the MTF of telescope and atmosphere, the true contrast of the smaller features could be considerably greater than appears on the photos.

Bahng and Schwarzschild (1962) found the rms temperature fluctuation in the granules (determined by the brightness fluctuation) to be about $\pm 92°$. Because the cool regions are so narrow, the contrast with the granules may be even greater. They defined the lifetime of the granulation as twice the time in which the brightness correlation function of successive exposures drops to $1/2$, and found this lifetime to be 8.6 min. The larger granules, naturally, last longer. Newer observations show that we must consider both growth and decay of granules, which leads to lifetimes of about 18 min. The velocity field of the granulation is not completely established. Lyot filter measurements (Bray *et al.* 1976) show a difference of 1.8 km/sec between granules and lanes. A good test is to examine fil-

6.9. A superb granulation photo by R. Müller with a 50 cm refractor at the Pic du Midi. Such photos can be made from the ground only by taking bursts of frames to catch the occasional good one. In addition, the telescope is built like a gun turret to avoid dome effects. Granulation is best visible with a green filter (5750 was used here); in the red the contrast is lower, and in the blue higher structure is seen. (PdM).

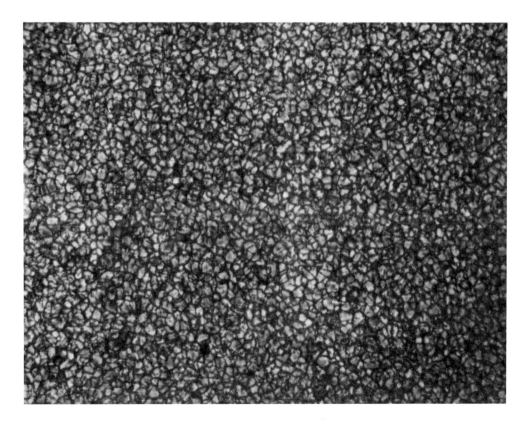

tergrams on either wing of an absorption line. On the red side Doppler motion shifts the absorption out of our band, increasing the local brightness; added to the granular brightening this gives enhanced contrast. On the blue wing the increased brightness of the granules is cancelled by the shifting of absorption into our band. Keil (1980) studied the granular velocity in a number of lines formed at different heights and, after eliminating the effect of oscillations, found a velocity amplitude of 2 km/sec at 100 km above $\tau = 1$ with a scale height of 80 km. But high-resolution velocity frames (Fig. 6.12) show a much larger scale pattern which is much different in appearance from the granulation. The amplitude of that pattern is well determined from oscillation measurements as 0.3 km/sec. So it is hard to understand the relationship between the velocity field and the granules.

Determined efforts at high resolution have brought new knowledge of the granulation. Rösch and coworkers (Carlier *et al.* 1968) found that granules often explode with a velocity of about 2 km/sec. LaBonte *et al.* (1973), who found a mean granular size of 1100 km and mean spacing of 1330 km (leaving 230 km for the lanes), concluded that destruction occurred by frag-

6.10. The area around a sunspot photographed at Big Bear by B. LaBonte. Kodak SO424, a very high contrast, almost grainless emulsion was used. The granulation is unchanged up to the boundary of the spot. Moore (1981) argues that the penumbral fibrils lie above the granulation; this is particularly evident at points A and B. This picture has been dodged to show both spot and photosphere. (BBSO).

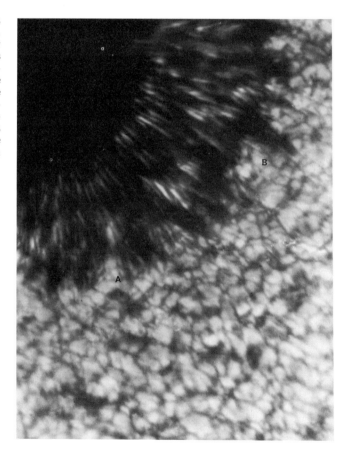

mentation (35%), decay (60%), and by merger (4%), but new granules arose only from fragmentation of old ones. They found a granule lifetime of 15 min when individual granules were tracked; the autocorrelation techniques used by others treats a positional shift as a change and gives a shorter lifetime. Recently sequences of excellent stability were obtained from Spacelab II by the Lockheed group; even though the telescope was only 25 cm aperture, the absence of atmosphere gave remarkable improvement. The new data show the significance of the exploding granule phenomenon. Although only 40% or so of the granules explode, the "explosion" covers an area of four or five granules, so most granules are affected by the phenomenon. The explosion is not just a phase effect; neighboring granules are pushed away by it. For a long time there has been evidence that the horizontal velocities in the photosphere were greater than the vertical; for example, the Fraunhofer lines broaden sharply as we approach the limb. Now it appears this is the effect of exploding granules. The important role of exploding granules is easily seen in movies made of selected video frames (Wang and Zirin 1987), which offer similar resolution but poorer stability. This technique can be used without waiting years for space opportunities, but cannot pick out the brightness subtleties visible from space. The presence of a larger scale pattern enhanced toward the limb was noted by Edmonds (1960) and by Bray *et al.* (1984).

In addition to the granulation, we see dark clouds in the continuum, features larger than the intergranular lane, which often enhance simultaneously over an area of one or two supergranules. There is indirect evidence that these are more marked near the limb, implying that they lie above the granules. They may be connected with the five-minute oscillation, although granules show that oscillation, too.

In summary, the granulation is convection with velocity ≈ 2 km/sec, scale of 700 km, and lifetime about 18 min, with a temperature fluctuation greater than $100°$. It is quite uniform over the solar surface, and is affected only by strong magnetic fields, particularly horizontal. These are usually associated with emergence of new flux. The brightness distribution of the granulation is typical of what in hydrodynamics are called Benard cells. Schwarzschild pointed out that the lifetime of a granule is just sufficient to move all the material one scale height. That the granules are indeed rising bright elements is confirmed by filtergrams in the wings of photospheric lines. But the great significance of the exploding granules as a real dynamical phenomenon will change our picture significantly.

The center-to-limb variation of granule contrast is important in understanding the height profile of the granulation. One should expect that close to the limb, when the height of granules becomes comparable to their separation, the contrast would decrease, and this is observed. On the other hand, the contrast associated with magnetic features increases near the limb. The heating effects connected with magnetic activity increase upwards while the effects of convection decrease.

The wavelength dependence of granular contrast reflects the combination of Planck function and height effects. Granular contrast is lower in the red because brightness varies linearly with temperature in the Rayleigh–Jeans region, while on the blue side it varies exponentially. Further in the blue we find the granulation contrast little changed down to about 4100 Å, where a smaller scale structure appears, as seen in Fig. 2.14. This structure is also seen in the *G* band, a cluster of lines at 4305 Å; therefore it must lie higher than the photosphere, and probably is identical to what we see in some absorption lines such as Na D. It is sinusoidal in form, with as much dark area as bright, smaller than the granulation and without polygonal structures. It probably is the sum of the contributions of the many lines in the spectrum in the blue, since it appears weak in spectroheliograms made in continuum windows (Fig. 6.10). This structure shows strong five-minute oscillation, and may be considered the upward extension of the granular structure. Velocity movies show a uniform cellular structure.

6.8. The supergranulation and the network

In Table 6.4 we show the "typical" values for the gas pressure NkT at various heights in the solar atmosphere, along with the magnetic field strength for which the magnetic pressure is equal. If the magnetic field strength exceeds the gas pressure, the material is constrained to move along field lines and should exhibit the regular structure characteristic of magnetic fields. We see from Table 6.4 that the magnetic pressure $B^2/8\pi$ exceeds the gas pressure in the photosphere only for fields above 1440 gauss, field strengths rarely found outside sunspots. By contrast any field above 8 gauss will freeze the motion in the chromosphere (in calculating NkT for the higher, ionized regions, one multiplies by two to allow for electron pressure). Over the long run magnetic and kinetic energy fields should reach equipartition; motions should stretch the lines of force, enhancing the field until equipartition is achieved. But in the steep density gradient of the photosphere the strong fields generated below dominate the more tenuous material above.

Because the fields are weak, we would expect to find no trace of magnetic structure in the granulation, and indeed it is uniform over the disk.

6.11. (opposite) An extremely high sensitivity videomagnetogram averaging 6144 frames (W top, N left; field 250 arc sec E–W). The apparent contours are a way of handling signal saturation; they represent field intensities 1,3,5, *etc.* times the saturation value. The sign of each pole is determined by whether it is dark (minus) or white (plus) outside the lowest contour. Even at this sensitivity only about half the area is covered with field, and some of that is due to smearing. The contoured areas are chromospheric network. This region is highly mixed, with many small dipoles, of which some are marked by arrows. (BBSO)

Danielson (1961a) found virtually no change in the granulation right up to the edges of the penumbras of sunspots. He did find the granules to be elongated near some spots associated with strong horizontal fields. I have seen this several times. Only horizontal fields change the shape of granular structure; vertical fields only affect the amplitude and velocity, as in umbrae. The field strength threshold for horizontal alignment is probably somewhat lower than equipartition, since the granular motion is across the field. In Fig. 6.8 we show a case of granule alignment associated with strong horizontal field in emerging flux.

There appears to be a sharp change in the photospheric structure at the edge of the penumbra. The entire pattern of granulation disappears abruptly and is replaced by a stable radial fibrous structure. Danielson (1961b) proposed that these filaments are due to roll convection, a mode that Chandrasekhar showed to be important in the presence of horizontal magnetic fields. The famous Evershed effect is an outward flow from the edge of the umbra of a sunspot. But our video selection observations show an *inflow* along the fibrils of 0.3 km/sec. Recently, Moore (1981) has proposed that the granulation in fact extends all the way to the umbra, and the penumbral fibrils are elevated structures hiding the granulation, which is in fact unchanged (Fig. 6.10). This might account for the opposite flows.

While the photosphere appears chaotic, the magnetic fields occur in a limited range of structures, as follows:

1. *Active regions*, characterized by sunspots, bright plages, and strong fibril structure.
2. *Unipolar magnetic regions*, made up of an "enhanced network" where the magnetic network is relatively strong, complete, and of one polarity. In centerline Hα and CaK the network is bright and in the wing of Hα a dark network of spicules is prominent. These are the accumulated remnants of active regions.
3. *Quiet magnetic regions*. The rest of the Sun, except for horizontal field regions, is covered by a pattern of weaker fields we shall call the quiet network. The magnetic elements are mixed in polarity and the network is incomplete. Sometimes (*e.g.* at the poles), these regions are unipolar.
4. *Horizontal field regions*, which separate unipolar regions of opposite polarity with filaments or horizontal fibrils and are devoid of network structure. The field lines may run parallel to or across the boundary.
5. *Ephemeral active regions* (ER), small dipoles with lifetime about one day and no sunspots which erupt everywhere on the Sun. They appear as small bright areas in chromospheric lines and X-rays, and are brighter than the network, but fainter than active regions.

Fields above 1000 gauss in active regions become sunspots, dark in the continuum, or plages, bright in the chromospheric and UV lines. Plages are also visible in the continuum with ultra-high resolution or near the limb.

6.12. A pair of VMG velocity and magnetic images showing the general relation of the supergranulation (a) to the magnetic network (b) at the South solar pole. Dark is receding. One can recognize the bright-dark pairs of the expanding velocity cells. They are similar to the network but do not match it. The oscillation pattern is superposed; the SG pattern is unchanged in the course of the day. (BBSO)

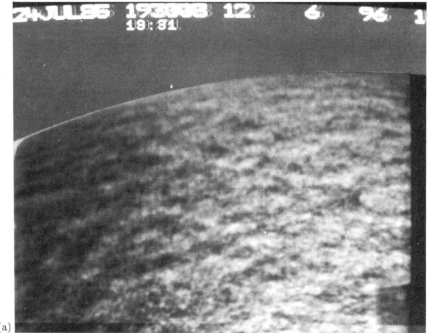

(a)

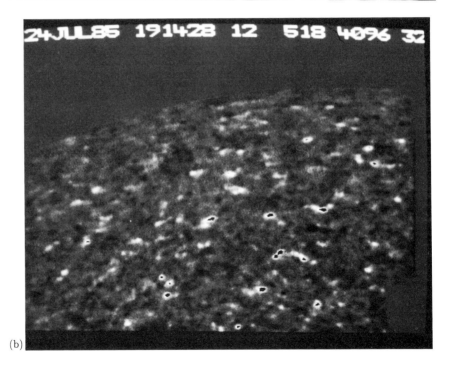

(b)

The bipolar components of active regions are separated by the scale of the active regions, about 60 000 km. The remains of complexes of active regions become an extensive unipolar enhanced network (Fig. 2.12 and 7.16) made up of 30 000 km cells, all of the same polarity. Inside the network cells there are mixed weaker fields which we call the intranetwork (IN) fields. There are weaker magnetic fields all over the Sun – everywhere we look. The IN fields are of mixed polarity and independent of the level of sunspot activity; but they too are arranged in what we call the quiet network. The IN fields in the mixed polarity quiet network are not different from those in the unipolar enhanced network. While the enhanced network clearly results from the sunspot cycle, the elements of the quiet network must be locally generated and have much lower field strength and shorter lifetimes than those of the enhanced network.

The bright regions of the network are variously referred to as faculae or plagettes. They are not much different from plages, except the latter are brighter and more extensive. The term bright flocculi was once frequently used for plages. We shall use the terms faculae or network elements.

While active regions are confined to the sunspot belts, between $\pm 40°$ latitude, but ephemeral regions erupt all over the Sun. The ER's are so numerous that they cannot be an extension of the sunspot cycle to lower size. Another remarkable characteristic of the magnetic fields is that even the weakest fields we observe are discrete, and we have not yet detected a diffuse magnetic background. In the chromosphere the magnetic network is marked by emission in the resonance lines and UV and jets called spicules.

Leighton, Noyes and Simon (1962) proposed that the chromospheric magnetic network is associated with a flow pattern apparent in velocity spectroheliograms (Fig. 6.10) which they termed the *supergranulation* (SG). The SG has the same scale as the chromospheric network; the bright-dark pairing we see in Fig. 6.10 means that each cell is a place of expanding horizontal flow. They surmised that the supergranulation was a large-scale convective pattern. In each cell the material flows outward from the center, sweeping magnetic fields to the cell boundaries, where downflow is sometimes observed. Simon and Leighton (1964) and Mosher (1977) presented further evidence that the edges of the SG indeed match the chromospheric

Table 6.4. Gas pressures and equivalent fields.

Region	Height	Density	Temp	NkT	B(gauss)
Photosphere	0	10^{17}	6 000°	8.3×10^4	1440
Chromosphere	1500	10^{12}	10 000°	2.8	8.4
Corona	3000	10^8	1 000 000°	0.028	0.84

6.13. Excerpts from a day-long series of sensitive magnetograms (Zirin 1986) showing the evolution of the weak fields. The reversals are due to wrap-around in memory, so each contour reversal is approximately a doubling of intensity.   The fields are concentrated in discrete elements to the lowest detectable levels; the wrapped around elements generally correspond to the network or ephemeral regions (ER). 1, 2: ER's which merge with poles of the same sign; 3: a reconnecting ER; 4: a rapidly separating ER; 7: a stronger pole forms from two intranetwork (IN) elements. 9: an unchanged network element. If we compare IN elements in left and right columns we see they are almost completely changed in 4 hrs, while the network elements are roughly unchanged. (BBSO)

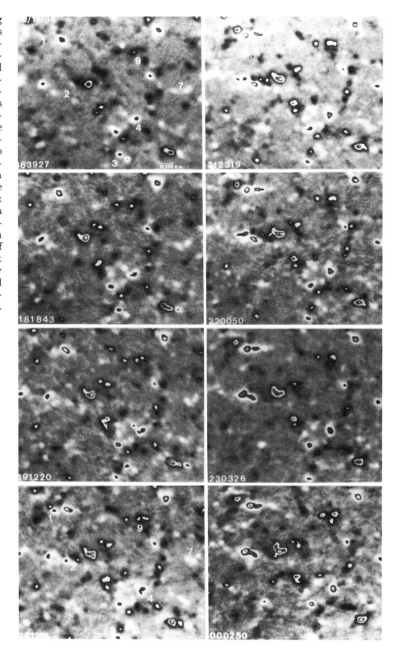

network. Because the spicule jets are localized in the edges of the network, it is the dominant pattern higher in the atmosphere.

Although the idea that the supergranulation creates the magnetic network is attractive, it is not proven. Fig. 6.12 shows images of the magnetic and velocity field recently obtained at Big Bear. SG and network are clearly visible, and of the same scale. Accelerated replays show that if we treat the field elements like wood chips, they do accumulate at the network boundaries. But this does not generate the network; since there are no monopoles, the SG flow carries both polarities to the cell boundaries. Clearly the flows can generate a unipolar network from the strong unipolar fields in the active region, and the net effect of interaction with the bipolar IN fields may simply maintain it in a way not yet clear. The lifetimes are not commensurate: the enhanced network appears to have a lifetime of several days while the supergranulation changes more rapidly. Another interesting fact is that the bright-dark boundaries in Fig. 6.12 should be parallel to the limb if they are simply expanding, but they are not. This could be caused by a slight rotational component in the supergranulation.

The lifetime of both the magnetic and velocity network elements has been reported by various authors as one day. This is uncomfortably close to the observational interval; but at least one measurement (Janssens 1970) used continuous Hα observations from the Arctic. The problem, as with all complex features, is to decide what constitutes the "death" of a magnetic feature. It is at least certain that the enhanced network near active regions lasts several days.

How are the IN fields related to the network? Can we see how these and other fields are formed? I studied the lifetimes and behavior of elements of the weak magnetic fields on several fine day-long videomagnetograph movies (Zirin 1986) obtained by Bill Marquette and Randy Fear at Big Bear (Fig. 6.13) and reached the following conclusions:

1. The quiet-sun network elements have a lifetime of about 50 hours. They must be regenerated by local processes. Their mean motion is 0.06 km/sec.

2. The intranetwork elements are discrete down to the lowest detectable levels. They move typically at 0.3 km/sec and live a few hours. In some places they flow into the network elements, seeming to follow the SG flow; in other cases they don't. IN elements may merge with others to form new network; they may "cancel" with elements of opposite sign; or they may be the debris of old network.

3. Ephemeral regions tend to come up near existing neutral lines. The elements separate at 0.3–0.6 km/sec, and produce a significant fraction of the changes in the quiet network.

We saw in Chap. 3 that the resistive diffusion time (Eq. 3.26) is about 100 days for magnetic elements 100 km across. So if nothing happens to

Fig. 6.14 A magnificent scan (Dunn and Zirker 1973) of the Hα line showing filigree in the wings, with elongated dark spicules coming out of the network elements. (SPO)

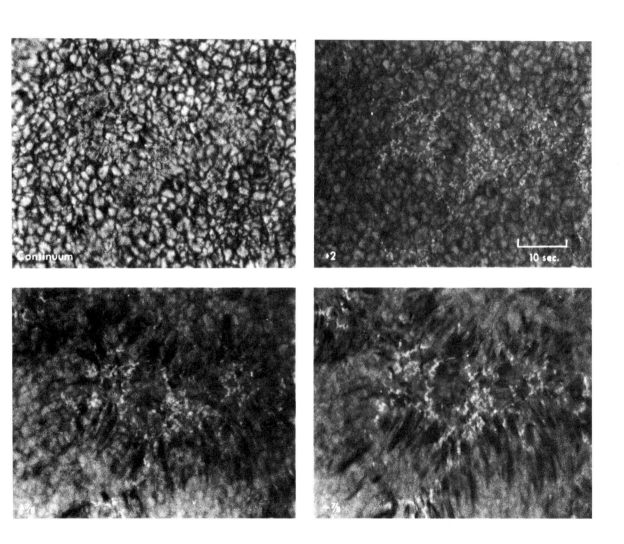

replenish them, the network fields will diffuse into a uniform unipolar blur, or, in the mixed polarity regions, disappear by reconnection. Whether the lifetime of the network elements is 24 or 50 hrs, after this period they must be replaced somehow. Since the network is always present, there must be an equilibrium between the creation and destruction of network elements. Creation may occur by fission of existing elements, emergence of ephemeral regions from below, or by merging of IN elements. Destruction results from merging, submergence or emergence from the surface, or simple diffusion. At this point we cannot say which is the dominant process.

Appearance of new dipoles by flux tubes coming down from the atmosphere has not been seriously considered. In the submergence or emergence, a flux loop must pass through the surface, and the dipole formed by its intersection with the surface must shrink or diverge.

Although there has been considerable use of the term "cancellation" with respect to this kind of flux disappearance, the term is confusing. Flux cancellation occurs only as the result of reconnection, which is a transition from force-free to potential field and represents a true loss of flux. Reconnection does take place when the elements shed their connections with their sibling poles and connect to others, but flux disappearance usually involves elements that are already well-connected. The field does not change, but vertical movement relative to the surface changes the points of intersection of the flux loop with the surface. When flux disappearance or emergence occurs there must be perfect symmetry; we lose a bit of plus field for every bit of minus. For this reason the net flux in any area changes slowly. Since there are no monopoles, every bit of plus field comes with a bit of minus. The only exception is the big unipolar regions, separated by the large-scale active complexes.

The difference in behavior of network and IN elements is striking. The former, like large beasts, are slow-moving and long-lived; the latter are fast-moving and short-lived. Yet one may be transformed into the other. The IN elements are only about 500 km across and move many times their own diameter during their lifetime without disruption. While they do appear to emerge as tiny dipoles, most of their existence they travel as if free of constraints. The only network elements that move with comparable speed are those spreading from an ER, or converging to submerge with an element of opposite sign.

Why do the IN field elements move so much faster than the network? What phenomenon generates these small point-like elements of flux? The supergranular flow speed is approximately that of the IN elements, about 0.3 km/sec. New magnetic field in any plasma of this sort is created by stretching and twisting existing lines of force. But the SG flow is rather systematic, and it is hard to see how it can generate little point fields. Rather, it is possible that the granule motions, which are of the same scale as the IN elements, are responsible for their creation. The increase of magnetic field by stretching existing field lines continues until equipartition of mass motion and magnetic energy occurs.

What is the source of the ephemeral regions? They have a scale somewhat smaller than the SG flow, but there is no other candidate in sight as their source. Since the ER's emerge fairly uniformly all over the Sun, we have to blame the supergranulation for them until we learn more.

6.9. Strong or weak fields?

Measurements of the weak magnetic fields in the photosphere give intensities of 5 or 10 gauss. This is somewhat surprising because Table 6.4 shows that the magnetic energy of fields less than a thousand gauss is less than the energy in photospheric motions and probably will be torn apart. But magnetographs are not high resolution instruments and integrate the brightness change due to the line shift all over their resolution element.

Simultaneous measurements in lines of different g-values could only be reconciled by the existence of strong unresolved fields. Frazier and Stenflo (1972) made magnetograms in two lines of different sensitivity and found the ratio of measured intensities to be constant. Because more sensitive lines should saturates easily, a wide variation in flux intensity would have produced a variable ratio, so they concluded that the magnetic field strength could be fitted only by discrete elements of high and constant field strength. Stenflo (1973) presented other arguments to support the conclusion that 90% of the magnetic flux intersecting the photosphere has intensity of one or two kilogauss. Tarbell and Title (1977) found high values in the network, but they only observed stronger elements, greater than 10^{18} mx. Since the magnetograph measures the average shifts in an area, these strong fields average with the weak or zero fields to give the low values measured. One would think we could confirm this by directly measuring the Zeeman splitting, but it is too small; for the widely used 5250 Å line ($g = 2.5$), the splitting (Eq. 5.18) in a 10 gauss field is only 3.2×10^{-4} Å. detection limit for small features. Only in sunspots could the Zeeman splitting be measured directly. Kurucz (unpublished) has contested the use of ratios, pointing out that high resolution spectra show that most of the lines used are blended and their effective g-factors are not really known.

High resolution photos of the Sun in the wing of Hα show the enhanced network and plages to be a network of fine bright points. Dunn and Zirker (1973) gave these a name, *filigree*, and a logic, pointing out that they must be associated with small-scale magnetic fields. On their best photos (Fig. 6.14) the filigree are not points, but u-shaped elements that they term *crinkles*. Dunn and Zirker's pictures show that the filigree is unrelated to (and presumably unaffected by) the granulation.

There is a big problem with the strong field models. Because of hydrostatic equilibrium, the density at any height in the photosphere must be the same everywhere. In the horizontal direction, therefore, the magnetic pressure and gas pressure inside the spot must balance the gas pressure in the photosphere outside:

$$nkT_{sp} + \frac{B^2}{8\pi} = nkT_{ph}. \tag{6.29}$$

The extra magnetic pressure inside can only be balanced if $T_{sp} < T_{ph}$, and when the temperature is lower we see a sunspot. The tiny 1000 gauss field regions would only be stable if they, too were dark. However, if the 1000 gauss field elements produce a magnetograph signal of 10 gauss, they must only fill 0.01 of the resolution element and the quiet Sun spots will be invisible, which in fact they are.

Like Lieutenant Kizh'e, the fictitious officer who required an invented life and death, the invisible pores have spawned a theoretical literature. Spruit

6.15. Tracings of the MgI line at 812 cm^{-1} (12μ) made with the KPNO FTS by Brault and Noyes (1983). The line shows the classic reversal of resonance lines, but at this great wavelength the Zeeman splitting is directly visible. (a) Spot umbra, absorption only; (b) penumbra, side toward disk center, σ components dominate; (c) penumbra, limb side, π component dominates; (d) and (e) plage near disk center, splitting about 500 gauss. (KPNO)

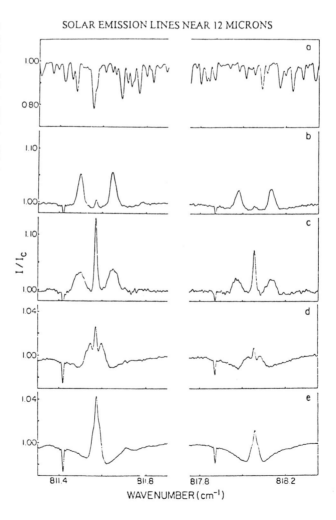

SOLAR EMISSION LINES NEAR 12 MICRONS

WAVENUMBER (cm^{-1})

(1976) and Deinzer *et al.* (1984) have calculated models of the center-limb variations of the faculae in which these invisible magnetic poles occur. The question is, why does a dark invisible spot give us a bright facula near the limb? It is observed that all magnetic elements of sub-sunspot strength are marked by small bright continuum faculae at meridian distances greater than 45°. The ingenious explanation by these authors is that while the pore is too small to be seen in disk center, near the limb the low opacity of the invisible spot allows us to see its wall, which is deeper and consequently brighter than the limb-darkened surroundings. This model predicts a peak brightening at about 60° from disk center, dropping off to zero at the limb. The observed brightness of faculae increases steadily toward the limb (Wang and Zirin 1986). This fact has not discouraged the believers.

There is also a problem with the integrated facular brightness. We know facular emission, or filigree, to be associated with field concentrations. These are observed near the limb with brightness contrasts as high as 20–40%. How can invisible dark spots produce such brightening? According to the invisible sunspot models, we see the bright walls of the invisible spots. Since the walls must be about the same scale as the spots, a filling factor of 0.01 would require walls twenty to forty times brighter than the surrounding photosphere, an unreasonable amount. By contrast, the highest photospheric brightness increase observed in huge solar flares is three or four times, so the magnetic elements would be brighter than the greatest flares. However, it could be argued that near the limb we look somewhat higher, and by then the flux tubes have spread apart. In any event the calculated model does not fit.

Another way to find out the strength of the quiet Sun fields is to measure the smallest fields you can. Wang *et al.* (1985) sought the smallest detectable solar fields by following the decay of small field elements and found field elements as weak as 10^{16} mx. Because the flux in maxwells is the product of field strength times area, this means that these elements could be 100 gauss and 100 km across, or 1000 gauss and 30 km. We believe that the smallest elements of structure in the photosphere cannot be less than the radiation mean free path, 30 km; otherwise the powerful radiation flux would smooth out any temperature inequalities. We must increase our sensitivity by a factor 10 before we really can rule out the 1000 gauss fields; but since it is unlikely that the Sun knows about the resolution of our telescope, I am confident that smaller elements will be found.

Brault and Noyes (1983) studied the Zeeman splitting of a series of newly identified (Chang and Noyes 1983) solar emission lines near 12 microns in the far infrared (Fig. 6.15), using the Fourier Transform Spectrometer (FTS). The lines are due to high ($n = 7$ to $n = 6$) transitions in neutral Mg and Al and formed in LTE. Because the free-free absorption increases as λ^3 we see fairly high in the photosphere at this wavelength. The lines show absorption wings as the temperature continues to drop upward; as we pass to the center of the line we see above the temperature minimum and there is an emission peak. Since the Zeeman splitting is proportional (Eq. 5.18) to the square of the wavelength, it is 600 times greater than in the visible for these lines. Exposures with the FTS are long, and the spatial resolution, low; but quite weak fields may be measured directly. Brault and Noyes (Fig. 6.14) found fields of 1600 gauss in penumbrae and 250–500 gauss in plages, but detected no splitting at all in the quiet Sun. Zirin and Popp (unpublished) obtained the same result. The evidence supports strong fields (but only 500 gauss), in plages, but quite weak fields elsewhere. It has been suggested that these relatively low values are the result of fields diverging with height, but plages near the limb do not give lower values, so there it is probably not a height effect. Evidently the plage fields are not sufficient to produce darkening, although seem to make an easy transition to plage and vice versa. The 500 gauss fields probably do not occur in the quiet Sun. It is odd that various observatories around the world have been measuring the fields in sunspots for many years by the Zeeman splitting and commonly list spot fields of 500–600 gauss. It may be that the measured splittings in these small spots are too low because of scattered light, but splitting can scarcely be affected by scattered light. There have been many, many such observations, and it is hard to discount them.

After we have criticized Stenflo's model, on *quantitative* grounds, there is no doubt that *qualitatively* it is correct, and it was this model that focussed our interest on discrete magnetic elements. The illustrations show that, to the limit of our sensitivity, solar fields are so clumped in small elements, and a uniform background field has yet to be detected. In the few cases where we have observed sunspot formation, we see individual pores which merge into a spot. An unknown stabilizing force keeps these elements discrete. We should not be unduly concerned by the field required by Table 6.4 to balance the kinetic forces in the photosphere; even stronger forces are required below the surface, yet the fields are stable.

Fig. 6.16 shows a high spatial resolution magnetogram in which the stronger fields may be seen. In taking this observation, I used the shortest possible (1 sec) exposure to reduce image drift. We see that these field elements are indeed quite small, about 1000 km or smaller. They are part of the enhanced network resulting from the breakup of old active regions. In the quiet Sun elements of the same apparent size are found, but their strength is about three times lower. Conversely, in active regions the plages reveal fields about two times more intense; the overall apparent ratio is of the order of five times. Since the fields in the plages are measured to be 100 gauss, and Brault and Noyes found 500 gauss from FTS data, the filling factor is about five, and in the weak network elements, 25–50. In the invisible sunspot hypothesis the filling factor would then be 100.

6.16. High resolution (a) VMG
and (b) Hα of an active region to
show the smallest scale elements.
The smallest elements are about
1 arc sec, or 700 km. At this resolution the close correspondence
between **B** and Hα breaks down;
the heating includes other factors. (BBSO)

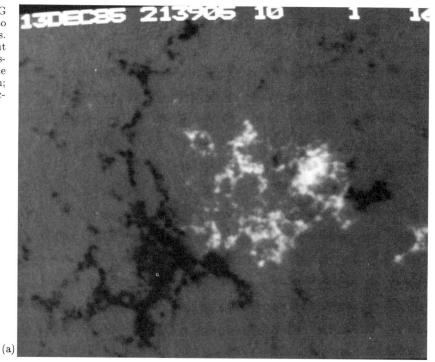

(a)

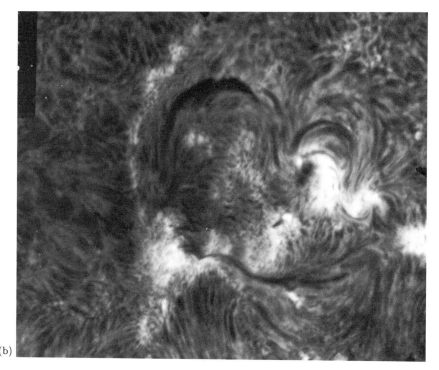

(b)

6.10. Photospheric models

The principal tool for building photospheric models has been the analysis of limb darkening as a function of wavelength. The two-dimensional (more often one-dimensional) brightness distribution across the Sun is mapped into the depth dependence of temperature.

The relation between position and intensity is best understood in terms of Barbier's (1943) relations derived from Eddington's work; they are

$$I_\lambda(\tau = 0, \mu) \approx S_\lambda(\mu) \tag{6.30}$$

$$F_\lambda(0) \approx S_\lambda(2/3) \tag{6.31}$$

for the intensity and the flux, respectively, where $\mu = \cos\theta$, the angle from the center of the Sun. Eq. (6.30) tells us that the emergent intensity at position θ is roughly equal to the source function at optical depth $\tau = \cos\theta$. The variation in brightness as a function of position maps out into the variation of source function as a function of depth. So at the limb we see radiation from the very surface ($\tau = 0$), and at the center of the disk the light that we see is produced mainly near $\tau = 1$. Eq. (6.31) tells us that the emergent flux F at $\tau = 0$ is equal to the source function at $\tau = 2/3$.

If we write Eq. (4.20) in angle-dependent form for the emergent intensity $I(\tau, \mu)$ at $\tau = 0$, we have

$$I_\lambda(0, \mu) = \int_0^\infty S_\lambda(\tau_\lambda)e^{-\tau_\lambda/\mu}\frac{d\tau_\lambda}{\mu}, \tag{6.32}$$

where

$$d\tau_\lambda = -k_\lambda\rho dx \quad \text{and} \quad \mu = \cos\theta. \tag{6.33}$$

We can see how the emergent intensity adds up the contribution from each height diminished by an absorption factor. The μ-dependence of the source function S_λ (which is taken to equal the Planck function B_λ for the temperature at depth τ) is fitted to the data.

The simplest treatment of limb darkening involves the application of the Eddington approximation, which assumes a fixed ratio of three between the mean intensity J and the radiative pressure K:

$$J = \frac{1}{2}\int_{-1}^1 I(\mu)d\mu = 3K = \frac{3}{2}\int_{-1}^1 I(\mu)\mu^2 d\mu. \tag{6.34}$$

From the relation between flux F and K we obtain (Mihalas 1978)

$$J(\tau) = \frac{3}{4}F\left(\tau + \frac{2}{3}\right),$$

$$B(\tau) = \frac{\sigma T^4}{\pi} = J(\tau) = \frac{3}{4}\frac{\sigma T_{eff}^4}{\pi}\left(\tau + \frac{2}{3}\right), \tag{6.35}$$

where B is the Planck function corresponding to the temperature at optical depth τ and is assumed to equal the source function S. If this is now inserted for S in Eq. (6.30), the result for the limb darkening is

$$\frac{I(0,\mu)}{I(0,1)} = \frac{3}{5}\left(\mu + \frac{2}{3}\right). \tag{6.36}$$

This gives a rather good approximation for the limb darkening, as we see in Table 6.5. From this fit Karl Schwarzschild first recognized that the photosphere was in radiative equilibrium. Pierce and Waddell (1961) used the expression

$$\frac{I_\lambda(0,\theta)}{I_\lambda(0,0)} = a_\lambda + b_\lambda\mu + C_\lambda[1 - \mu\ln(1 + \mu^{-1})]. \tag{6.37}$$

Fig. 6.17 shows the limb darkening observed by Pierce (circles) and by Peyturaux (dots). The latter is about 1% lower. The graphing by wavelength makes it easy to see the effect of different processes of continuous absorption, which vary the height (and hence the temperature gradient) that we see. The expression in Eq. (6.36) corresponds to fitting the depth dependence of the source function with

$$S_\lambda(T) = a_\lambda + b_\lambda\tau_\lambda + c_\lambda E_2(\tau_\lambda), \tag{6.38}$$

where E_2 is the second integral exponential function. Because this is continuous emission, we can assume that $S_\lambda = B_\lambda(T)$, the Planck function, which permits the determination of the temperature at each optical depth – which we want to convert to the actual geometrical height by inserting the yet to be determined k and ρ. But we can immediately obtain the wavelength dependence of the absorption by differentiating Eq. (6.33) to obtain

$$\frac{d\tau_\lambda}{dT} = -k_\lambda\rho\frac{dx}{dT}. \tag{6.39}$$

The derivative $d\tau_\lambda/dT$ is obtained directly from Eq. (6.35) and Eq. (6.37).

When the wavelength dependence of absorption was first obtained it corresponded to no known source of opacity. In a famous paper Rupert Wildt (1939) showed that the source was probably the negative hydrogen ion, which is formed by the weak attachment of a free electron to hydrogen. His idea was substantiated by calculations by Chandrasekhar (1945) and by Chandrasekhar and Breen (1946) as well as experimental measurements by Branscomb and Smith (1955). The most recent calculations are by Geltman (1962, 1965) and there is a convenient table in Allen (1981). There is so much hydrogen that H^- is quite abundant, and it absorbs strongly from its ground state. As we see in Fig. 6.18, there is an excellent match between the measured shape of the broadening and the theoretical data for H^-, so it must be the main source of continuous opacity in the photosphere.

Because the ionization potential of H^- is small, it is formed in LTE and its abundance is easily found from the Saha equation. The absorption coefficient is well-known, so is easy to calculate the opacity. Allen (p.102) gives a value of 2.8×10^{-26} cm^4 dyn^{-1} per hydrogen atom for the absorption at $6000°$ and 5000 Å. The opacity in cm^2 is obtained by multiplying by the electron pressure $N_e kT$. If we assume the gas is predominantly neutral except for the metals, which are easily ionized, we can estimate the electron density N_e as 10^{-4} of the hydrogen number density N_H. Then the integrated opacity down to $\tau = 1$ is

$$\tau = 2.8 \times 10^{-26} \times 10^{-4} N_H \times 1.38 \times 10^{-16} T N_H^2 H = 1$$
$$= 2.32 \times 10^{-42} N_H^2 H = 1, \tag{6.40}$$

where H is the scale height. Inserting $H \approx 2 \times 10^7$ cm from Eq. (6.12), we find that the density at $\tau = 1$ is $10^{17.5}$ atoms/cm^3.

Photospheric models are derived from limb darkening by a method given by Stromgren (1944), which combines Eq. (6.35) with the equation of hydrostatic equilibrium Eq. (6.11) to obtain

6.17. Limb darkening as a function of wavelength (Pierce and Waddell 1961). At 3500 Å and 13000 Å we see the effects of hydrogen continua. The limb darkening gets small in the IR as we reach the temperature minimum region.

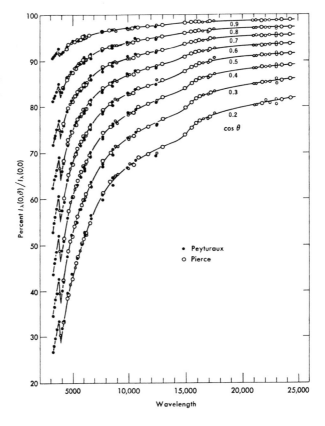

$$dP/d\tau = g/\overline{k}, \qquad (6.41)$$

where $\overline{k}$ is the mean absorption, obtained by weighting k_ν by the flux F_ν.

If the absorption coefficient is due to H^- only, we can use procedures similar to those above. The opacity k can be connected with the temperature by using the Saha equation to determine the amount of H^-:

$$\frac{N_H}{N_{H^-}} = P_e\phi(T) \qquad (6.42)$$

$$\log \phi(T) = -0.125 + 0.747\frac{5040}{T} - \frac{5}{2}\log T. \qquad (6.43)$$

The absorption coefficient per gram $\overline{k}$ is the absorption per H^- ion multiplied by the ratio Eq. (6.42) and divided by the mass of an individual hydrogen atom:

$$\overline{k} = \frac{\alpha(H^-)}{m_H} \times \frac{N(H^-)}{N(H)} = \frac{\alpha(H^-)}{m_H}P_e\phi(T). \qquad (6.44)$$

Combining Eqs. (6.42) and (6.44), we find

$$\frac{dP}{d\tau} = \frac{g}{\alpha(H^-)}\frac{m_H}{P_e}\phi(T), \qquad (6.45)$$

which can be solved for the dependence of P on τ if we can relate P_e to P. This Stromgren did by suggesting that since H and He are not ionized at $6000°$, all the free electrons present are due to ionization of the metals, which are easily ionized and in fact must all be singly ionized. If A is

Table 6.5. The Eddington approximation for limb darkening.

μ	I/I_0(Eddington)	I/I_0(observed, 5000 Å)
1.0	1.00	1.00
0.8	0.88	0.88
0.6	0.76	0.74
0.5	0.70	0.68
0.4	0.64	0.64
0.3	0.58	0.52
0.2	0.52	0.42
0.1	0.46	0.32
0.05	0.43	0.20
0.02	0.41	0.14

the fraction of metals (excluding C, N, O, Ne, and similar hard-to-ionize atoms), then the electron pressure P_e is easily related to the pressure:

$$P_e = AP, \qquad (6.46)$$

where the run of pressure is explicitly given by

$$P^2 = \frac{2gm_{\mathrm{H}}}{\alpha(\mathrm{H}^-)A} \int_0^\tau \frac{d\tau}{\phi(T)} . \qquad (6.47)$$

Once we know the pressure and temperature at every value of τ we can evaluate the opacity explicitly and connect τ with x with $A = 10^{-3} - 10^{-4}$.

Stromgren's method was developed for the precomputer days, and has now been replaced by great codes which keep track of all physical parameters and take into account such things as absorption by the hydrogen molecule and the ionization of hydrogen. The problem in using these codes is that one must understand the physics of the atmosphere, which we really don't. Access to the UV and IR has greatly extended the height range over which models are possible. In the IR, free-free absorption of protons and electrons becomes important, along with H^- free-free, and we see the temperature minimum. In the UV, photoionization continua are important and the brightness temperature drops to the same low temperatures. In simple approximations we assume LTE, apply the Boltzmann and Saha formulas, and use the Planck function for the source function. Since up to

6.18. (a) The observed logarithm of the inverse temperature gradient, $d\tau/dT$ (Eq. 6.36), as a function of wavelength for various inverse temperatures $\theta = 5040/T$, compared (Pierce and Waddell 1961) with (b) theoretical value based on H^- and other opacities.

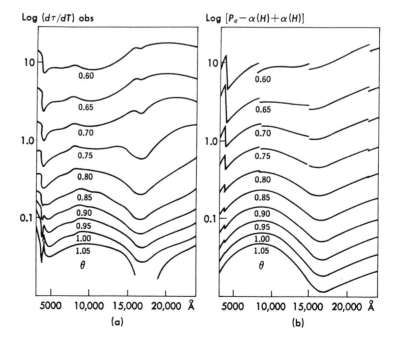

half of the photons escape into space in the outer layers, the photosphere is not in LTE, but allowances can be made.

In Table 6.6 we present an excerpt from the model of Vernazza, Avrett and Loeser (1976), commonly known as VAL. This model was computed with detailed attention to deviations from LTE in the ions which affect the UV, and gives a good fit to both limb darkening and the variation of brightness in the continuum. Note that the height scale is measured from $\tau_\kappa = 1$ at 5000 Å, where the temperature is 6423° and the density, $1.2 \times 10^{17} \text{cm}^{-3}$, about 10^{-4} of the atmospheric density at the Earth's surface. The weakness of models like VAL, which represent our present best current assessment of the state of affairs in the photosphere, is that they use observational data to extend the model from the low photosphere, where we probably understand what is happening (LTE, hydrostatic equilibrium), to the upper photosphere, where we do not understand what causes the upward increase

6.19. The Fraunhofer spectrum from 3800 Å to 7000 Å. At the far red are atmospheric bands; at the middle of the bottom line is Hα (6563). Also seen: 5890 Å, Na D lines; 5183 Å, Mgb; 4861 Å, Hβ; 3933 Å and 3969 Å, CaII H and K. (Mt. Wilson and Las Campanas Obs.)

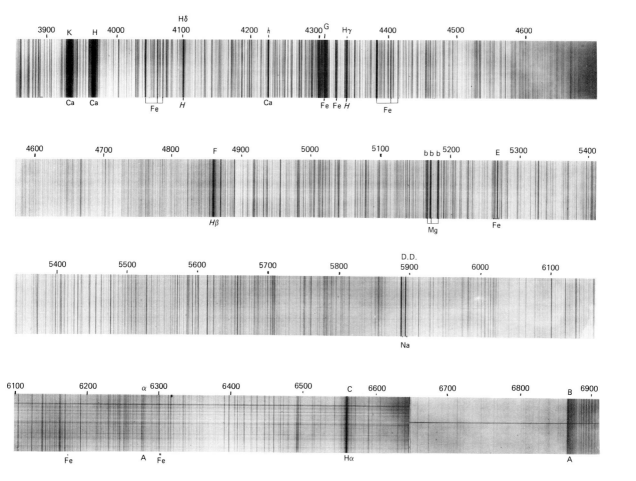

in temperature and where the atmosphere is quite inhomogeneous. So the photospheric models are pretty good around $\tau = 1$, but less dependable at greater heights, where the model is governed by uncertain brightness measurements in the UV and IR or the extreme limb, as well as unknown physics.

In the upper photosphere granular inhomogeneity may play a significant role. For many years the many limb-darkening measurements at the extreme limb coincided except for the sharp limb found by ten Bruggencate *et al.* (1950) as well as the eclipse measures of Lindblad and Kristenson (Kristenson 1951). Dunn (1960) used a large coronagraph to reduce scattered light and removed the instrument function with spicule observations. His data confirmed the minority view. Weart's 1972 eclipse data reproduced by Athay (1976) confirmed that result.

The VAL and other models are based on continuum data and are not valid above $\tau_{5000} = 10^{-2}$, where T = 4600°. The rest of Table 6.6 is really

Table 6.6. Model M from the VAL model (Vernazza *et al.* 1976).

h	T	$\tau_{0.5}$	n_e	n_p	n_H	ρ	P_{gas}	P_{total}
800	5360	2.20 −5	9.96 +10	7.78 +10	1.32 −14	3.09 −10	1.08 −2	1.11 +2
750	5150	2.77 −5	8.43 +10	5.39 +10	2.04 −14	4.77 −10	1.59 −2	1.63 +2
700	4890	3.53 −5	7.31 +10	2.89 +10	3.25 −14	7.59 −10	2.41 −2	2.47 +2
650	4600	4.73 −5	7.68 +10	1.03 +10	5.34 −14	1.25 −9	3.73 −2	3.81 +2
600	4350	7.35 −5	1.05 +11	2.67 +9	8.98 −14	2.10 −9	5.94 −2	6.06 +2
550	4170	1.45 −4	1.52 +11	9.52 +8	1.53 −15	3.58 −9	9.69 −2	9.87 +2
525	4150	2.19 −4	1.84 +11	8.30 +8	1.97 −15	4.61 −9	1.24 −3	1.27 +3
500	4150	3.36 −4	2.26 +11	8.21 +8	2.53 −15	5.93 −9	1.60 −3	1.63 +3
450	4200	8.08 −4	3.52 +11	1.10 +9	4.12 −15	9.63 −9	2.63 −3	2.68 +3
400	4330	1.92 −3	5.75 +11	2.32 +9	6.50 −15	1.52 −8	4.28 −3	4.35 +3
350	4460	4.45 −3	9.28 +11	4.90 +9	1.01 −16	2.37 −8	6.86 −3	6.98 +3
300	4600	1.01 −2	1.48 +12	1.10 +10	1.56 −16	3.64 −8	1.09 −4	1.11 +4
250	4750	2.23 −2	2.36 +12	2.86 +10	2.35 −16	5.50 −8	1.70 −4	1.72 +4
200	4920	4.79 −2	3.65 +12	8.19 +10	3.49 −16	8.17 −8	2.61 −4	2.65 +4
150	5120	1.00 −1	5.83 +12	3.05 +11	5.08 −16	1.19 −7	3.95 −4	4.01 +4
100	5445	2.04 −1	1.01 +13	1.72 +12	7.10 −16	1.66 −7	5.87 −4	5.96 +4
50	5840	4.23 −1	2.10 +13	8.84 +12	9.60 −16	2.24 −7	8.51 −4	8.62 +4
20	6153	6.93 −1	3.90 +13	2.39 +13	1.12 −17	2.63 −7	1.05 −5	1.06 +5
0	6423	1.00	6.60 +13	4.85 +13	1.23 −17	2.87 −7	1.20 −5	1.21 +5
−15	6690	1.38	1.08 +14	8.85 +13	1.30 −17	3.04 −7	1.32 −5	1.34 +5
−30	7040	2.00	1.98 +14	1.76 +14	1.35 −17	3.16 −7	1.45 −5	1.46 +5
−45	7460	3.07	3.85 +14	3.58 +14	1.39 −17	3.25 −7	1.58 −5	1.60 +5
−60	7880	4.94	7.04 +14	6.73 +14	1.43 −17	3.34 −7	1.71 −5	1.73 +5
−75	8320	8.16	1.25 +14	1.21 +14	1.46 −17	3.40 −7	1.85 −5	1.87 +5

extrapolation, and recent data (Fig. 6.22) shows that reality is probably quite different. That discussion is deferred to the next chapter.

There have been attempts to take the granulation into account by multi-stream models of the photosphere, with temperature differing according to the "observed" temperature difference between the granulation and background. Such models have many parameters and can fit the data, but as Pecker (1965) showed, different 2- and 3-stream models which all fit the same observations may vary wildly.

6.11. The Fraunhofer Lines

We have so far discussed what might be learned from the continuous spectrum, which is easy to understand and tells us the general structure of the photosphere. The Fraunhofer lines, by contrast, have a much more detailed story to tell, different for each element and for different excitation conditions. The problem of the production of the Fraunhofer lines is so complicated that we are never certain of our results. For example, one might think we can study photospheric velocities by Doppler shifts of the lines without understanding their formation. But velocities vary from line to line depending on the height of formation, so we must determine the height of formation to know what we are measuring. So we really have to understand the Fraunhofer lines to use them.

The abundance of the elements in stars the same age as the Sun is similar, and differences in spectrum are the result of temperature only. The hottest stars show lines of helium and hydrogen, and the A stars, at $10\,000°$, show little besides the Balmer series of hydrogen. The solar G2 spectrum (Fig. 6.19) shows that the atmosphere is cool enough for the metallic lines to appear. Most of the lines in the G2 spectrum are of neutral atoms, particularly those with low-lying levels that can be excited at $6000°$. Although hydrogen is the most abundant element, the Lyman series, which arises from the ground state (Fig. 5.1), is in the unseen ultraviolet. The only hydrogen lines in the visible are the Balmer series, which require excitation of $n = 2$, which is 0.75 Rydbergs, or 10 volts up. If we apply the Boltzmann formula Eq. (4.26) for $6000°$ with a dilution factor $1/2$ (because the photosphere fills only half the sky as seen from the atom) we have

$$\frac{N2}{N1} = \frac{1}{2} \times \frac{4}{2} \times 10^{-5040 \times 10.2/6000} = 2.7 \times 10^{-9}. \qquad (6.48)$$

So only a minute fraction of the H atoms are in the second level.

Fig. 6.19 shows the Fraunhofer spectrum in the visible, with various lines identified. The individual lines are classified by multiplets (Table 5.1) in the Revised Multiplet Table (Moore 1959). For a positive identification, all the stronger lines of a given multiplet should be present. The big lines were the features classified by Fraunhofer. Why are they strong? First, the elements involved are abundant, like hydrogen. Second, the lines may be resonance lines, i.e. absorption from the ground state, which has the highest population. This is the case for the NaI D lines at 5890 Å and the H and K lines of ionized calcium at 3900 Å. Although sodium is easily ionized, there are still many atoms in the ground state, and the f-value is one, so the lines are strong. The MgIb triplet at 5183 Å comes from a metastable state near the ground state of this highly abundant element. The situation of the lines and their occurrence can be understood by reference to Merrill (1956). Note that many abundant elements, such as the noble gases and C, N and O are absent because their lines in the visible require high excitation. Their resonance lines are prominent in the UV.

Fig. 6.20 shows the complex structure of the Fraunhofer lines when shifted by granule motions and solar oscillations. Much of the line width is determined by these motions. The brightening at the network edges produces what Sheeley calls "gaps", regions where all lines are weakened. The lines near the limb are quite different, nearly twice as broad as those in the disk center. This is probably due to the effect of exploding granules, which produce considerable surface motion. Usually people measure profiles of an averaged area with extraordinary care, ignoring the much more sizable point-to-point profile fluctuations. So it is quite difficult to evaluate work on the Fraunhofer spectrum, because we rarely know which elements of the atmosphere they were studying.

Why are the Fraunhofer lines dark? Kirchhoff tells us that this occurs when a cool gas is in front of a hot continuous source. This picture was embodied in the Schuster–Schwarzschild approximation, which considered that the photosphere radiates a continuous black-body spectrum, and the dark lines are produced by absorption in a cool, thin overlying region called the "reversing layer" in older texts. But the overlying layer could be hotter and still give absorption lines. If we substitute the phrase "lower excitation temperature" for "cool," we will have the right picture.

The Milne–Eddington model recognizes that absorption lines are formed because of the difference between the line and continuum absorption coefficients. That ratio $\eta_\nu = l_\nu / k_\nu$ is taken to be independent of optical depth. Then we can analyze the loss of photons from the line to the continuum as the radiation passes upward. More detailed treatment would have to allow for variation with depth of η_ν.

If we simply consider that at each wavelength we see the emissivity corresponding to the excitation temperature at the height to which we see. In the continuum we look deeper; in the line, higher. As in Eq. (6.30) at each wavelength we will see the local source function S_λ for that wavelength, and the absorption line maps out the height variation of S_λ. If S_λ decreases outward we will have an absorption line. Since S_λ cannot exceed the Planck function, we always have the *minimum* temperature at that point, if we can figure out where it is. With the line opacity we can get the exact height; in any event the gradient of $dS_\lambda/d\tau_\lambda$ is determined.

Three types of photon transfer in the lines are generally recognized: coherent and noncoherent scattering and pure absorption. When radiation is absorbed and re-emitted, the wavelength of the new photon may or may not remain the same. If, before the absorbing atom undergoes any collisions, it jumps back to the same level and re-emits the photon in the same direction, then the frequency will be unchanged. This process is called *coherent*, or pure, scattering; it also occurs when radiation is scattered by free electrons. If there is no emission from the gas itself, coherent scattering should eventually lead to a perfectly black line, because each photon is absorbed from one direction and re-emitted into 4π solid angle, where it may be captured by whatever continuous absorption is present (say, H^-) and lost to the continuum. There must always be some continuous absorption present to produce an absorption line in coherent scattering; otherwise all the photons eventually escape. Coherent scattering typically occurs in

6.20. A beautiful spectrogram from 3880 Å to 3889 Å. Each bright horizontal streak, which marks a granule crossing the slit, is marked by a blue (up) shift in the absorption lines. (SPO)

low-density gases in resonance lines where the probability of returning to the same level is great.

In a real atmosphere we must consider *noncoherent scattering*, in which the photon is re-emitted in the same line but in a new direction. Because the velocities of the atoms along different axes are independent, a photon absorbed in one frequency will be re-emitted in a different frequency in another direction along which the atom has a different velocity. We can see that this redistribution prevents the occurrence of completely black absorption lines in coherent scattering, because redistributed photons can always turn up in the center of the line.

The most important process of radiation transfer in a line is the so-called *pure absorption*, in which the absorbed photon has no control whatever over the activities of its progeny. For example, if we excite an iron atom, there are so many lower levels to which the atom may return that the probability of re-emission of a photon at the same frequency is small. If we have

6.21. Tracing of the solar spectrum near the Mgb lines showing varying shapes of weak and strong lines.

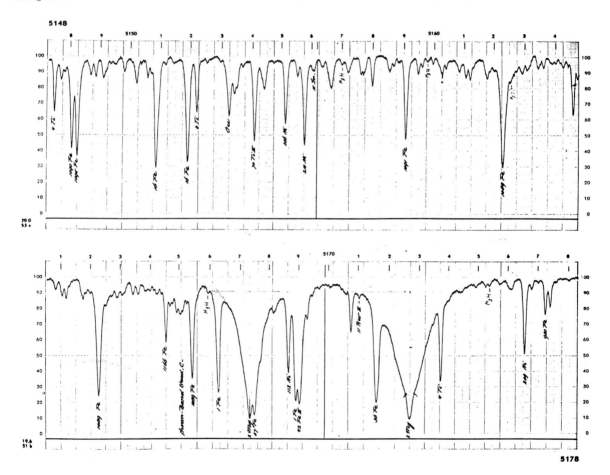

many collisions, this probability is further reduced. The coherent effect of stimulated emission is reduced because $h\nu > kT$ in the most important cases. It is also possible for the photon energy to be returned to the kinetic energy field via collisions of the second kind. In any event, the re-emitted energy is no longer determined by the flux of radiation in the line but by the microscopic circumstances of the gas.

In pure absorption the re-emission is governed by Kirchhoff's law; the source function (Eq. 4.13) is the measure of creation and destruction of photons. The net emission in the line can never exceed the rate of photon creation by collisions; if this does not balance the net loss due to scattering, absorption results.

In the ultraviolet the combination of line and continuum opacity prevents us from seeing much below the temperature minimum, where $T \approx 4200°$. The strong resonance lines are all formed at much higher temperatures in higher levels, so that only emission lines are seen in the spectrum below 1400 Å. In principle there could be weak absorption lines in the spectrum below this point, but none have been detected.

A photon at the wavelength of an Fe line near 5000 Å can be absorbed by an Fe atom in the right state (that is, line absorption) or by an H^- ion, or by an H atom in $n > 3$ (Paschen or higher continuum), or scattered by an electron. To apply the transfer equation (4.12) to line formation, we express the absorption coefficient k_ν as the sum of continuous absorption κ_ν and line or selective absorption l_ν,

$$k_\nu = l_\nu + \kappa_\nu = \kappa_\nu(1 + \eta_\nu). \tag{6.49}$$

We can use as independent variable the optical depth τ_κ in the continuum. The transfer equation becomes:

$$cos\theta \frac{dI_\nu}{d\tau_\kappa} = (1 + \eta_\nu) I_\nu - S_0 - \eta_\nu S_\nu, \tag{6.50}$$

where S_0 is the source function in the continuum and S_ν that in the line. This form has the virtue that the source function in the continuum often has a wavelength dependence similar to the Planck function, even in non-LTE cases. In the case of coherent scattering, the emission equals the absorbed energy:

$$4\pi j_\nu = k_\nu \int_0^{4\pi} I_\nu(\theta,\phi)d\Omega, \tag{6.51}$$

so that the source function in the line is

$$S_\nu = \frac{1}{4\pi} \int_0^{4\pi} I_\nu \, d\Omega = J_\nu, \tag{6.52}$$

which is just the average intensity.

In the case of pure absorption, the source function is the local ratio of emission to absorption, j_ν/k_ν, which, in LTE, would be the Planck function. In each absorption process, let the probability be $1 - \epsilon$ that the photon is scattered coherently, while all the other photons are returned to the radiation field according to the source function (except for an extinction factor $1 - \delta$ which accounts for other photon losses). Then

$$\cos\theta\frac{dI_\nu}{d\tau} = (1 + \eta_\nu)I_\nu - S_0 - \epsilon(1 - \delta)S_\nu - (1 - \epsilon)\eta_\nu J_\nu, \qquad (6.53)$$

with J_ν defined by Eq. (6.52). One may now allow for noncoherent scattering by redefining J_ν as

$$\overline{J_\nu} = \frac{\int J_\nu \eta_\nu \, d\nu}{\int \eta_\nu \, d\nu}, \qquad (6.54)$$

where η_ν gives the frequency distribution in the line.

Eq. (6.54) is merely a formal statement of the transfer problem for a homogeneous atmosphere, with all possible effects included, which one can adapt as needed by the appropriate choice of ϵ, δ, and so on. It is clear from examination of high-resolution spectrograms that straightforward solution of this equation has no hope of producing the structure of the faint or strong absorption line, but it does provide a framework for calculation. Recent work in this field involves massive computer codes which evaluate

6.22. Flux and brightness temperature in the VAL model compared with values (AL) estimated by Ayres *et al.* (1986) from infrared molecular lines combined with a photospheric model by Ayres and Linsky (1976).

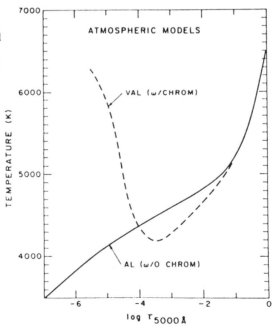

the non-LTE atomic equilibrium at each point in the atmosphere and add up the contributions of each level. Beckers and Milkey (1975) have defined a response function which measures the instantaneous response to local changes of the contribution from an element along the line of sight.

Lites (1973) was successful in fitting the profiles of the Fraunhofer iron lines by calculating the non-LTE equilibrium of Fe with 15 levels. He obtained an Fe abundance of 1.2×10^{-5} relative to H, using the line wings which are formed in deeper layers where LTE holds. By fitting parameters such as microturbulence and uncertain atomic constants he matched the profiles and intensities. But Evans and Testerman (1975) show that lines of different intensity require different dependence of microturbulence on height and question the whole procedure. With enough parameters we can fit many things, but we should be cautious in inverting these results.

6.12. Curves of Growth

The analysis of Fraunhofer absorption lines by the curve of growth method is based on the reasonable assumption that the strength of a line should depend in a regular way on the number of absorbing atoms; the only difficulty lies in the determination of that dependence. The curve of growth measures the growth of integrated line intensities, or equivalent widths, as a function of the number of absorbing atoms. It can be used to determine the relative abundances of the elements, as well as the excitation temperature of the lines. It depends completely on the assumption of LTE. We know that LTE does not exist because so many photons escape, but its use simplifies matters and makes possible direct comparison between atomic parameters and the observed lines. The use of integrated profiles reduces errors that might be introduced by poor understanding of line profiles. On the other hand, the curve of growth assumes all parts of the line are formed at the same temperature and abundance ratio. We know that the temperature at which the wings of a line are formed differs from that of the core, and new data suggest that abundances may be changing rapidly with height. So the results of this technique should not be regarded as constants of nature.

In the Milne–Eddington model the central depth of a line depends on the line/continuum absorption ratio η_ν, multiplied by the number of absorbing atoms/gram N/ρ

$$\eta_0 = \frac{l_0}{k_0}\frac{N}{\rho} = \frac{\pi\epsilon^2}{mc}f\frac{N}{\Delta\nu_D\sqrt{\pi}k_0\rho}. \tag{6.55}$$

The variation of absorption through a line with Doppler and natural broadening is obtained from Eq. (5.27):

$$\eta = \eta_0\left(e^{-v^2} + \frac{a}{\sqrt{\pi}}\frac{1}{v^2}\right), \tag{6.56}$$

where v is the distance from line center divided by the Doppler width.

Fig. 6.21 shows a tracing of the solar spectrum near the Mgb lines. The weak lines show a Gaussian form because only Doppler broadening is important. The strong lines develop triangular wings because of natural broadening. The weak lines simply get deeper and wider as the number of atoms increases, but the saturated triangular shapes of the strong lines grow much more slowly. The deep central cores of the strong lines are formed at the top of the atmosphere in a lower density region of low source function; pure scattering makes them quite dark. The transition of shape from weak to strong lines is the key to curve of growth analysis.

We define the equivalent width W as the integral of the line depth r:

$$W = \int_{-\infty}^{\infty} r \, d\lambda = \Delta\lambda_0 \int_{-\infty}^{\infty} r \, dv. \tag{6.57}$$

For emission lines an analogous expression can be formulated. In a weak line the wings are unimportant, and the equivalent width is obtained by simply integrating the first part of Eq. (6.53):

$$W = A\eta_0 \Delta\lambda_D \int_{-\infty}^{\infty} e^{-v^2} \, dv = A\Delta\lambda_D \sqrt{\pi}\eta_0$$
$$\propto N f / k_0, \tag{6.58}$$

which means that for weak lines the equivalent width grows directly with the number of atoms and the f-value.

In a strong line the Doppler core is relatively unimportant, and the equivalent width is determined by the shape of the wings, given by the second part of Eq. (6.54). Since the dependence of r on η is no longer clear, we write W as an integral over the line depth r:

$$W = 2\Delta\lambda_D \int_{r_0}^{0} v(r) \, dr, \tag{6.59}$$

using the intensity as a variable. If we drop the Doppler term in Eq. (6.56), we find for v:

$$v = \left(\frac{a}{\sqrt{\pi}} \frac{\eta_0}{\eta} \right), \tag{6.60}$$

so that the equivalent width W is now

$$W = 2\Delta\lambda_D \left(\frac{a\eta_0}{\sqrt{\pi}} \right)^{1/2} \int_{r_0}^{0} \frac{dr}{\sqrt{\eta}}. \tag{6.61}$$

The central depth r_0 of strong lines is always nearly 1, so the integral is a constant and W is proportional to $\sqrt{\eta_0}$, thus by Eq. (6.55) to the square root of Nf. We see that the equivalent width first increases as the number of atoms and then as the square root. In between there is an important flat transition zone in which the core is being saturated. If $\log W/\lambda$ is now plotted against $\log C$, which contains the dependence on N, η, and other variable parameters, we obtain the curve of growth. Fig. 6.23 shows curves of growth computed by Aller (1953) for the K line.

Although it is possible to calculate a theoretical curve of growth for any line as a function of N, the comparison with observational data is limited, because there is only one value of N for a single line in a single star, giving only a single point on the $\log W : \log N$ plot. But we can measure the equivalent widths of a number of absorption lines arising from the same lower level, and since the abscissa of the curve of growth is proportional to $\log Nf$, each line of different f-value will give a different point. Since this procedure still gives a limited number of points, we add other lines of the same element originating from other levels. Using the Boltzmann formula, we have for the population of the nth level of the ith ion,

$$\frac{N_{in}}{N_i} = \frac{g_{in}}{u_i(T)}e^{-\chi_{in}/kT} = \frac{g_{in}}{u_i(T)}10^{-\Theta\chi_{in}(\text{eV})}, \qquad (6.62)$$

where the terms are defined in Eq. (4.31).

6.23. Absorption profiles (top) and curves of growth for the CaII K-line computed by Aller (1953). The central optical depths in the top plot run from 1 to 100 000. The two curves in the lower plot are for different ratios of natural to Doppler broadening.

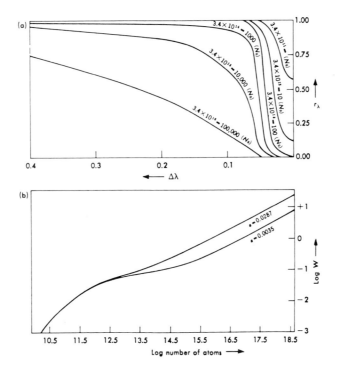

The empirical curve of growth may now be plotted as $\log W/\lambda$ *vs.* $\log X$, a quantity containing all the relevant factors:

$$\log X = \log N + \log gf\lambda - \Theta\chi - \log k_\lambda - \log \Delta\lambda_D - \log u(T) + \log \frac{\pi e^2}{mc}. \quad (6.63)$$

The only quantities on the right that vary from line to line are the second, third, and fourth; we bundle the others together in a constant C and write

$$\log X = \log C(i, T) + \log gf\lambda - \Theta\chi - \log k_\lambda. \quad (6.64)$$

To evaluate the right-hand side, it is necessary to know the value of the excitation temperature Θ . Since $\Theta\chi$ is different for each multiplet, we can fit each multiplet to the theoretical curve of growth by assuming Θ. If the constant C is independent of χ, we can assume that the excitation temperature of all the levels is the same. This is possible even if only relative f-values are known, but assumes that the curve of growth is independent of excitation potential, which it probably isn't. If we have good absolute f-values, we can determine the Doppler width $\Delta\lambda_D$, and specify the macroscopic velocities (oscillations, flows, *etc.*). Then the only remaining unknown is the relative abundance of each element. Because the curves for

6.24. The change in the solar constant measured by ACRIM, a radiometer on the SMM space-craft (Hudson *et al.* 1982). The large sunspots (Figs. 1.1–2) produced a drop in the Sun's luminosity corresponding to the area they darkened. Since this should have no effect on the Sun's energy production millions of years ago, it is hard to understand.

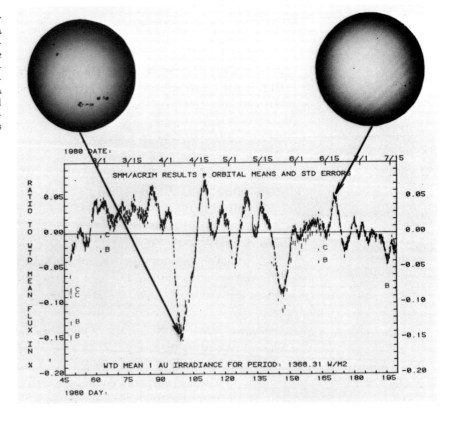

each element are not straight, we slide them horizontally till they match, and the distance the curves are moved gives the abundance.

Most of the accepted values for abundances in the Sun and stars have been obtained with the curve of growth method, notably the work of Goldberg *et al.* (1961). This work gave reasonable abundances for a number of chemical elements with the exception of iron, for which the values were about ten times less than those obtained from the solar corona. The discrepancy was removed when it was found that the f-values used for iron and nickel were in error by a factor of ten. By contrast, the coronal f-values are known exactly from Eq. (5.17). The solar abundances of many less frequent elements were studied by Lambert and co-workers using the Oxford spectrometer (Lambert *et al.* 1969), and the general solar-stellar abundances are summarized by Meyer (1985b).

The curve of growth method has been criticized by proponents of non-LTE analysis. Since the ionization and excitation vary rapidly through the region we consider, it is inaccurate to use the Milne–Eddington approximation that the ratio of line to continuous absorption is constant. Furthermore, we know that the temperature is different at different points in the line, and the triangular line profiles observed cannot be fitted by present theory. Although the equilibrium equations are satisfactory in a condition of detailed balance, at the surface half the photons emitted escape into space. But for an atom at $\tau = 1$, where the source function at disk center is established, only 1/6 the atoms escape, and the upward looking atom sees radiation not much below the local flux. For this reason LTE is not a bad approximation in weaker photospheric lines at disk center, and only gets much worse as we go to lines emitted at the top of the photosphere, either strong lines or those at the limb. So non-LTE effects are a problem for the chromosphere. In favor of the curve of growth is that it is an integral method, which adds up so many factors that errors in some of them may be fatal. The problem of photospheric non-LTE effects is discussed thoroughly by Mihalas (1978).

Recently the question of relative abundances in the photosphere and above has been reopened by the determination (Breneman and Stone 1985; Geiss and Bochsler 1984) that atoms of high first ionization potential are deficient in the solar atmosphere, corona, solar wind and solar cosmic rays. If the separation occurs in the photosphere, then the curves of growth are measuring a moving target, variable abundances, and cannot be depended on.

Another difficulty which brings into question our understanding of the photosphere is the discovery that (Fig. 6.24) when sunspots appear on the surface they actually block out the light that was generated many years ago in the core. How this energy could be stored is a mystery. Foukal (1987) has shown that simultaneous measurements from two spacecraft tracked each other perfectly, so we know that the Sun really varied. Foukal and

Lean (1986) have evaluated the facular component and found that while this additional emission does not balance the spots at any one time, the long lifetime of the faculae may enable them to balance the missing energy over a longer period.

References

Athay, R. G. 1976. *The Solar Chromosphere and Corona*. Reidel: Dordrecht.

Bray, R. J., R. E.Loughhead and C. J. Durrant 1984. *The Solar Granulation*. Cambridge: Cambridge Univ. Press.

Goldberg, L., and A. K. Pierce 1959. "The Photosphere of the Sun," in *Hdb. d. Phys.*, v.52. Berlin: Springer.

de Jager, C., ed. 1965. *The Solar Spectrum*. Dordrecht: Reidel.

Jefferies, J. T. 1968. *Spectral Line Formation*. Waltham: Blaisdell.

Kourganoff, V. 1963. *Basic Methods in Transfer Problems*. New York: Dover.

Lüst, R., ed. 1965. *I.A.U. Symposium No.22, Stellar and Solar Magnetic Fields*. Amsterdam: North-Holland.

Mihalas, D. 1978. *Stellar Atmospheres*. San Francisco: W. H. Freeman.

Pierce, A. K., and J. Waddell 1961. *Mem.Roy.Astr.Soc.* **68**, 89.

Schwarzschild, M. 1958. *Structure and Evolution of the Stars*. Princeton: Princeton Univ. Press.

Thomas, R. N., and R. G.Athay 1961. *Physics of the Solar Chromosphere*. New York: Interscience.

Unsold, A. 1957. *Physik der Sternatmospharen*. Berlin: Springer.

White, O. K. 1977. *The Solar Output and Its Variations*. Boulder: University of Colorado.

7

The Chromosphere

7.1. Structure

The chromosphere is the region between the temperature minimum and the corona. The atmosphere becomes transparent in the continuum, so the chromospheric continuum can only be observed at the limb during total eclipse, when the Moon blocks out the overwhelming scattered photospheric light. Sir Norman Lockyer named the chromosphere for its appearance as a pink flash (because of the dominant red Hα line) during a total eclipse. Although invisible against the disk in the visible continuum, the chromosphere is opaque in radio wavelengths longer than 100μ and can be studied there. In the UV $\tau = 1$ occurs near the temperature minimum, so ionization continua formed higher up are seen, even if they are not optically thick. The chromosphere is optically thick in strong resonance lines such as Hα and CaII K, where it may be observed with monochromators, as in most of the illustrations in this chapter. The weaker lines of the chromosphere may only be observed at the limb in eclipse, or in the UV, where the photospheric background is weak. Because these observations are difficult and offer low spatial resolution, most of our knowledge of chromospheric structure is based on observations in Hα and CaII K.

Since each scale height of a hydrostatic atmosphere contains as much material as all the overlying layers, the chromosphere contains about 10^5 more material than the corona. Since all the material of the solar wind must pass through the chromospheric state, it is completely replaced about once a year.

The magnetic forces due to the field concentrations in the photosphere fall in strength in a height equal to their lateral scale, thousands of km, while the gas pressure decreases with the barometric scale height, ≈ 100 km. Thus in the chromosphere the magnetic field dominates the gas motions. Compared to the chaos of the granulation, the chromosphere shows a well-ordered structure governed by the magnetic pattern of the photosphere, principally network and active regions. Velocities and oscillations in the chromosphere are of greater magnitude than in the photosphere, but controlled by the field. In the quiet magnetic network the chromosphere is

7.1. A pair of photos in the K line (top) and Hα of the edge of the disk. The numbers are to assist comparison of features. The chromospheric network, the locus of field concentrations, is marked by bright elements in both. The upper frame is taken with 0.6 Å bandpass so spicules are not seen, but in the lower we can see the dark spicules above the bright elements (we would see them in a narrower bandpass K picture, too). Some bright spicules also appear (pt 6). In Hα the inside of network cells is covered by spicules and fibrils. These structures of the general chromosphere are visible below 6 and 7. Note the elongated bright spicules near 6. (BBSO)

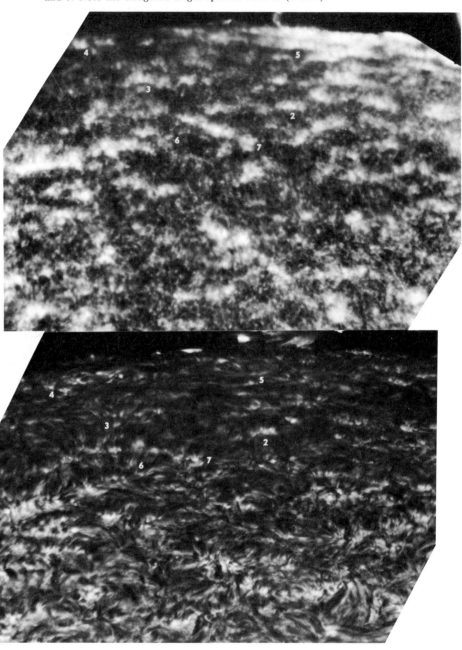

7.2. A scan of Hα in the same region (Sept 23, 1971) as Fig. 7.1. (a) Hα − 0.7 Å:
(b) Hα; (c) Hα + 0.7 Å; (d) Hα + 1.0 Å. Spicules are clearly seen to come out of elements
of the bright network. The filigree is visible in (a) and (d). Because of the projection, only
a few spicules pointed toward us can be seen. There is little correspondence between blue
and red wings, showing that we are really seeing shifted, rather than broadened material.
The centerline frame shows the bright bases of spicules in the network. Numerous double
spicules can be seen. (BBSO)

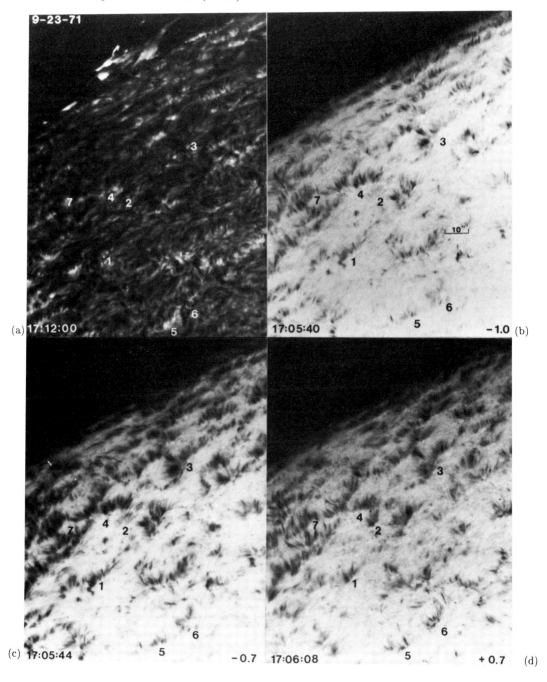

more chaotic. The structure varies strikingly from one magnetic regime to another. While the granulation is more or less similar everywhere outside of sunspots, the chromosphere exhibits several completely different structural forms. Velocities are so high that material may pass completely through the chromosphere before it reaches equilibrium. All this structure makes the chromosphere more interesting than the photosphere.

While fitting the black-body emission curve enables us reliably to set the temperature of the photosphere at 6000°, and the ionization equilibrium tells us that the corona is about $1\,000\,000°$, we can only tell that the chromosphere is somewhere in between. This is because the temperature gradient through it varies according to the magnetic structure. At the limb, where we have height resolution, the different regions are mixed up along the line of sight, while on the disk we have little height discrimination. We know that the temperature rises more rapidly in regions above the magnetic network, but it gets up to coronal values above other regions, too, since the corona is everywhere. The mystery of what produces the temperature reversal and the description of the variations of this gradient from one chromospheric region to another is the engrossing challenge of chromospheric studies.

One approach to modelling the chromosphere was to extrapolate photospheric models. Photospheric models like VAL (Table 6.6) are based on solid data up to $\tau_{5000} = 10^{-2}$, where $T \approx 4000°$. Above this point there is insufficient evidence to extend them higher. Data listed in Table 6.6 above the temperature minimum is *pro forma*. It was long assumed that above the last measurement the temperature suddenly rose toward coronal values. But introduction of high resolution observations shows that this principally occurs where the magnetic field is strong. Since the temperature must eventually increase everywhere, the picture is confused. None of the present chromospheric models is any use at all, except for an occasional Ph.D. thesis.

There is considerable evidence that in fact the chromospheric (or high photospheric) temperature continues to drop. When the temperature minimum was thought to be above 4500°, measurements of the limb darkening at 5–22μ by Lena (1970) indicated that the brightness temperature is 4200° and still decreasing at $\tau_{5000} = 10^{-4} - 10^{-5}$. Ayres and Testerman (1981) found that it was even lower. They measured the intensity in the CO bands at 2.1μ and 4.7μ near the limb, and compared the brightness temperature with the continuum at disk center, where the intensity could be tied to known values at lower heights. They found that the intensity of the absorption line cores near the limb fitted a brightness temperature of only 3500°. This result depends on the assumption of LTE, and the curve in Fig. 6.22 is a lower limit. Comparison of observations in these bands by Ayres *et al.* (1986) with data for the K line showed that the temperature at those heights ($\tau_{5000} = 10^{-4}$) was about 500° higher in a plage. As we go up from the photosphere the temperature drops less rapidly in the regions of enhanced magnetic field, leaving them brighter than their surroundings (Fig. 6.11).

The overall temperature reversal can be observed in the 12μ lines and radio waves. The $812\,\mathrm{cm}^{-1}$ (12.32μ) MgI line observed by Brault and Noyes (1983) displays strong double reversal, meaning that the temperature is still dropping at the heights at which the line wings are formed. The emission core indicates that the temperature gradient reverses at the height we see to there. Since the continuum opacity at 12μ is at least 6 times that at 5μ and the absorption line, about 20% deep, the temperature may fall as low as $4000°$ at $\tau_{5000} = 10^{-6}$. Possibly the temperature reverses at a lower height, but this would require an even higher temperature in the line core, and we should see the core broaden near the limb, which it doesn't.

To study these matters further, we (Popp and Zirin 1987, in preparation)

7.3. Spicules in the red wing of Hα. We cannot say much about the time histories because we do not have long enough sequences with such good conditions. (BBSO)

made further measurements of the surface distribution of the 812 cm^{-1} MgI line. We found that the reversal occurs everywhere, but the spatial resolution was low. At the same resolution the K line would also show resolution everywhere, under high resolution we see that this only occurs in discrete points. But the 812 cm^{-1} line shows stronger limb brightening, increasing from 1.1 to 1.4 times the continuum. The increase is due to both the temperature rise and increased path length.

Brault and Noyes found that the nearby H and HeII lines corresponding to the MgI and SiI transitions were only in emission above the limb. Apparently the temperature is still too low to excite much hydrogen. Boreiko and Clark (1986), observing from a balloon in the 110 to 170 μ range, found that the higher Mg and Si lines were weak but the H lines $16 \to 15$, $15 \to 14$ and $14 \to 13$ now appear in emission against the disk. At intermediate wavelengths they found both metallic and hydrogen lines in emission. This can be explained by the increasing temperature level of $\tau = 1$ at these long wavelengths. All these levels lie so close to the continuum that we may use LTE formulae. This is because the cross-section of the outer orbits is very large and the energy differences are small, so frequent collisions establish LTE.

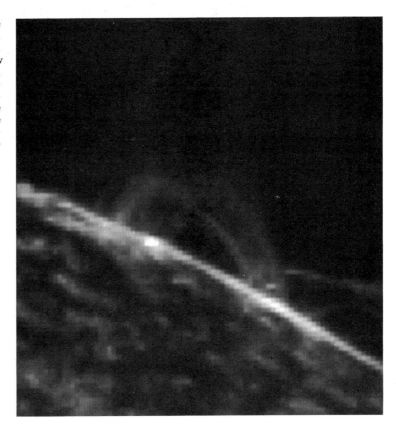

7.4. The chromosphere at the limb in the 1550 Å CIV line, with some loop-like prominences. A sharp limb-brightening only a few arc-sec thick appears, and on the disk we see that the emission comes mainly from the network. The network elements are not high enough to merge at the limb, so the emission is probably not from spicules. (SMM – UVSP)

Surprisingly, Brault and Noyes found the 812 cm^{-1} line emission is weakened over plages, where the K line and all the UV lines are enhanced. It is hard to explain this observation, but we will see below that spicules and other chromospheric phenomena are absent over plages, where the temperature rise to the corona may be steep. So the population of these levels must drop off rapidly as the temperature rises. The FWHM of the 12 μ lines is about twice the thermal width, indicating mean motions of 3 Km/sec. Aside from the pioneering work of Brault, Noyes and Clark, the extraordinary potential of these infrared lines for investigation of the transition zone has been virtually ignored by the solar community. The lines have also been observed in stars, but only in absorption so far.

When the continuum becomes transparent at the limb, the chromospheric lines are still relatively optically thick, so an emission spectrum appears, Hα being visible up to 8000 km. C. A. Young of Princeton predicted

7.5. Spicules at various points in Hα. In line center (0 Å) we see the spicule forest extending up to 7000 km, masked at lower heights by the general chromosphere. Beyond $\pm 1/2$ Å, the narrow emission of the general chromosphere disappears, and only the broad-line spicules remain. (R. B. Dunn, SPO)

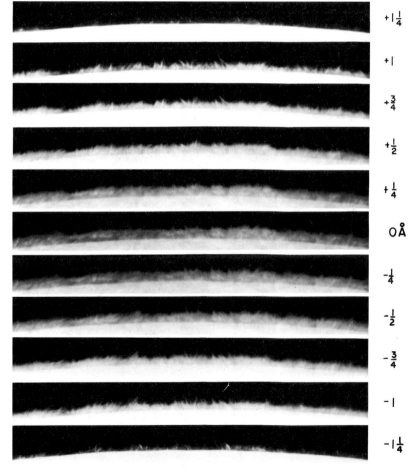

$+1\frac{1}{4}$

$+1$

$+\frac{3}{4}$

$+\frac{1}{2}$

$+\frac{1}{4}$

0 Å

$-\frac{1}{4}$

$-\frac{1}{2}$

$-\frac{3}{4}$

-1

$-1\frac{1}{4}$

this and observed the "flash spectrum" (Fig. 7.27) at the 1870 eclipse. Janssen found intense, hitherto unidentified lines of an unknown element in prominence spectra at the 1868 eclipse; Lockyer made the same discovery outside of eclipse. The element was named "helium" for its source and later identified by Ramsay and Soddy in radium decay products on Earth. The high excitation energy of helium suggested a high temperature for the chromosphere and led astronomers on a wild goose chase for years until the helium excitation by coronal ultraviolet was understood.

The flash spectrum reflects the temperature transition between photosphere and corona. In the first 500 km above the limb, low excitation lines of rare earths dominate; these quickly disappear leaving lines of singly-ionized atoms as well as hydrogen and helium. In the UV, lines of much higher excitation are observed from the same region. So the solar spectrum looks like a G star in the photosphere, a K star in the low chromosphere, and a B star in the upper chromosphere. But these changes happen differently in different places, and we have height resolution only at the limb, where they are all mixed up. Our task is to understand the three-dimensional structure of the hot and cold regions relative to the underlying photospheric magnetic fields.

The superposition problem can be understood by calculating the length L along a line of sight to the limb from the tangent point to a height h. From Pythagoras we have:

$$L^2 + R_\odot^2 = R_\odot^2 + 2hR_\odot + h^2. \qquad (7.1)$$

If we neglect h^2, the double path (in front and back of the limb) is

$$L/H = 2\sqrt{2R/H}. \qquad (7.2)$$

Thus the tangential path through a layer of scale height $H = 150$ km is $200\,H$. Chromospheric structures 1000 km high are projected against the limb if they are 37 000 km in front or behind. An optically thin line would be strengthened at the limb relative to the center disk by this ratio. The fact that only a few lines like helium increase by that much shows us that the atmosphere is not stratified. Optically deep resonance lines, like Hα or Ca K show no change in intensity as we step across the limb, but weaken higher up. The spicules, which have average height 7 000 km, (7.2) cover a 200 000 km path, so these thinly distributed jets appear as a thick forest at the limb. The convergence toward the limb is a bit unexpected. One thinks that 70° is only two-thirds of the way to the limb, but in projection the last 20° take up only a tenth of a radius, and observations beyond 70° are quite difficult.

Figs. 7.1–3 and 6.14 show the appearance of the chromosphere in Hα and K lines. The images are dominated by the magnetic network, from the boundaries of which the spicules project. In CaII K the emission from

7.6(a) An active region in Hα.
(b) A VMG of same, white is
preceding polarity. (1) field transition arches (FTA), connecting
the p and f spots at left. (2) a
filament, separating white from
dark polarities; (3) similar, but
much more sheared. (4) an element of opposite polarity connected to the spot by FTA, and
(5) an element of network polarity. Outside the main spots
and plages the structure quickly
breaks up into network.

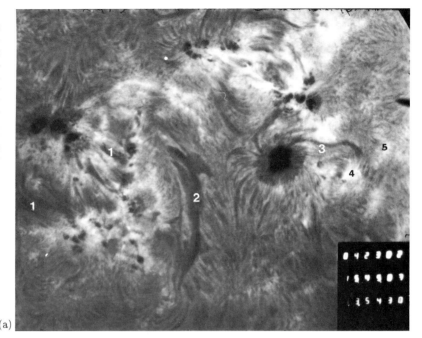

(a)

(b)

enhanced magnetic fields is more prominent because we are further out on the exponential tail of the Planck function. In the wing of Hα the spicule absorption dominates because the Doppler broadening is greater; the emission is reduced because the temperature contrast is lower at the lower heights to which we see in the line wing. In Fig. 7.4 we see that in the strong UV line of CIV the chromosphere is rather thin, low, and limited to network elements. It does not show the extended limb spicule structure of Fig. 7.5.

Monochromatic images of the chromosphere reflect the patterns of magnetic field outlined in Sec. 6.8. This is especially useful for understanding the transverse field, which cannot easily be detected with conventional magnetographs. Figs. 7.6 and 7.7 show this correspondence, which can be set down in a few easy rules (Zirin 1972).

1. All regions bright in Hα centerline (except flares) correspond to peaks of longitudinal (vertical) magnetic field.
2. Dark fibrils in Hα (centerline or off band) mark horizontal lines of force connecting opposite magnetic polarity. When the opposite field is far away, all we see is a spicule.
3. Dark filaments (seen as prominences at the limb) separate opposite magnetic polarities, the fibril structure running parallel to the boundary.

These rules apply to the K line as well, except that the dark fibrils and prominences are harder to see. In Fig. 7.7 we see the correspondence of field and off-band Hα for the quiet Sun, with the spicule elements marking the network.

The magnetic network is strongly enhanced near active regions, but suppressed in filament channels. The dark arches crossing the magnetic inversion line in active regions are called field transition arches; they are perpendicular to the magnetic inversion line, while filaments are parallel. In either case the absorption is due to the low excitation temperature in this material suspended by the magnetic field. In every case a magnetic pole must be connected by force lines to some other pole, and we can follow these in the Hα structure.

The clumps of magnetic network are the site of spicules and chromospheric heating; any wavelength that shows the chromosphere is bright there. The intranetwork (IN) areas are an unstructured oscillating mess. The fields there may be too weak to affect the chromosphere. Some of the small IN emission points seen in K2v (dark in the Hα wing) correspond to IN magnetic elements, but others are transient oscillating features. The appearance of these bright points on the violet edge of the Ca II emission is thought to be due to upward propagation of energy, as in the granulation.

In Fig. 7.8 we see the profile of double reversals of the CaII K in quiet and active regions. The emission of the chromosphere is primarily from the line core. There are two small emission peaks called K2v and K2r on either

7.7.(top) magnetogram and (below) Hα−0.7 Å in a quiet region. The numbers are to facilitate comparison of features. (1) and (2) are just emerged ephemeral regions. The evolution of these fields is shown in Fig. 6.13.

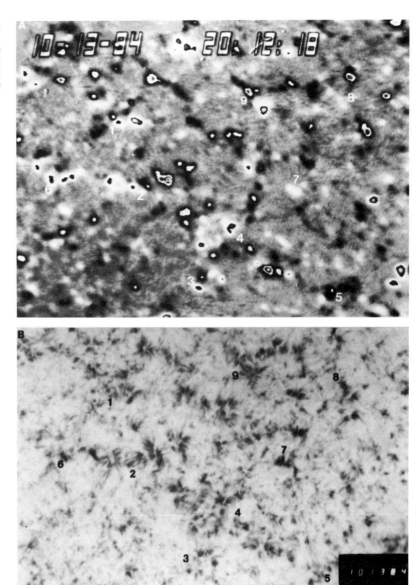

side of the line center, and a central dark absorption, K3, between them. This double reversal appears in all strong resonance lines such as those of MgII (near 2700 Å) and Lyα, mapping out the temperature structure through the temperature minimum. But at $\tau = 1$ in the line center the density is low and the excitation temperature is lower, so the central K3 reversal occurs, even though the temperature is still rising. Contributing to K3 reversal is the fact that, like the comic strip character who always had a cloud above him, every bright network element has a little cloud of spicule jets above it. Plages, where there are no spicules, show little K3 reversal.

All Fraunhofer lines broaden markedly near the limb, for reasons that are still unclear. Exploding granules must broaden the weaker lines, and there is considerable horizontal motion in the chromospheric oscillation. The overlap of vertical chromospheric structures probably also contributes.

Fig. 7.9 shows how the appearance of the chromosphere changes as we sweep across the Sun and vary the position in the K line at the same time. While the overall impression is of a doubly reversed line, we see that the effect is produced by discrete structures. At any point in the quiet chromosphere, there is emission only in either K2v or K2r – but rarely both – and

7.8. Profiles of the K line (White and Livingston 1981): (top) the reversal in the average quiet Sun at sunspot minimum in 1976 and sunspot maximum in 1979; (bottom) an active region and a quiet region. (NS0 – KPNO)

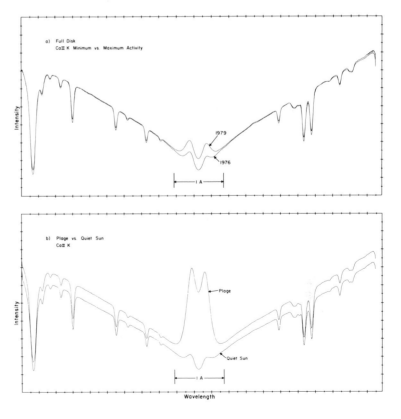

7.9 Spectroheliogram made by R. B. Leighton by sweeping the second slit across the CaK line while the first slit scanned the Sun. The dark lines at left and right are at ±1 Å; from left we approach K2r and see the bright network; K3 appears as a dark band at the center; the network is slightly weaker there. Further right, K2v shows the network filled with IN bright points similar to those seen in the CIV spectroheliogram Fig. 7.4. Note that the emission and absorption are broader near the limb.

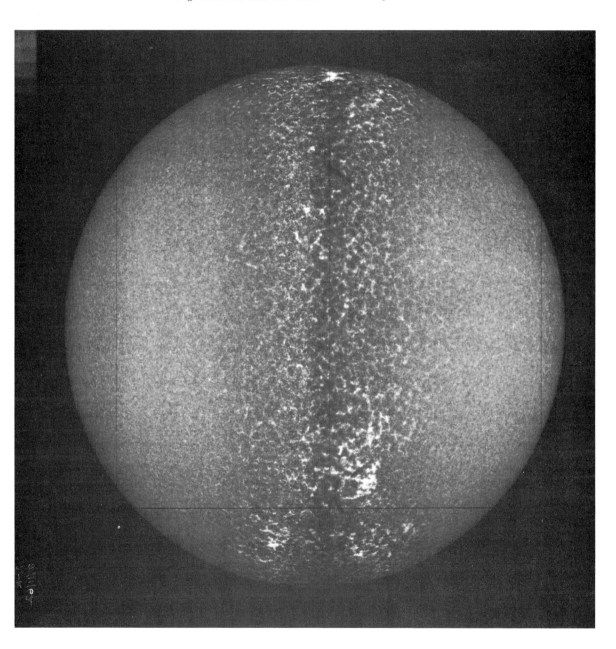

sometimes there is no reversal at all. K2v and K2r are simultaneously bright only in plages and the network elements. The K3 central absorption is predominantly found only in the network, and is rather weaker in plages, where there are no spicules.

Because the central emission core of the K line is so strong, we can identify K-line spectra with corresponding magnetic features. In Hα, by comparison the reversal is so marginal that we cannot ascribe particular spectral characteristics to different features, although they are clear in monochromatic images. The K line rules are:

1. Both plages and elements of the chromospheric network have symmetric emission in K2 with a dip (but still emission) in K3. They also are brighter in K1.
2. The intranetwork region shows emission almost exclusively in K2v and small dark clouds in K2–3. All these structures oscillate strongly.
3. Sunspots show narrow K3 emission superposed on K1.
4. The emission features in K3 and K1 are obscured at the limb, but those in K2 extend right to the limb.
5. Bright spicules are much more evident in K3 than Hα (Zirin, 1974).
6. There is a general high-lying absorption in K3 which seems associated with spicules and may account for the sharp increase in the separation of the K2 peaks near the limb.

The chromosphere is quite uniform in its response to magnetic fields. Magnetic elements are *always* bright in K and the UV. Filaments *always* separate opposite fields.

Since the chromosphere is so dynamic, it is best understood from Hα or K-line movies, which are dominated by the five-minute oscillation except in active regions, where it is suppressed by the magnetic field. Bhatnagar and Tanaka (1972) found the chromospheric brightness oscillation period to vary according to the structure. In Hα they found a dominant 300 sec period in plages and the network, but only 170 sec inside the cells. Possibly the network is tied to the photosphere by magnetic fields while the intranetwork material oscillates on its own. Because of a dip in the B–V frequency (Fig. 6.4), there is a possibility of a *p*-mode resonance in the chromosphere.

When we study the chromosphere in Hα, we often see small transient features. These include surges and eruptions, small brightenings called microflares, and transient small filaments separating small bipolar elements. The variety is limited only by the currently available resolution.

7.2. Spicules

If the chromosphere had the same scale height as the photosphere its density would drop to coronal values in 12 scale heights or 2000 km, and solar astronomers would have been deprived of the fun of many eclipse expeditions. But the first eclipse observations revealed an extended, ragged structure (Fig. 7.5). It was called a "sierra" by Airy; others called it a "burning prairie". Roberts (1945) introduced the term "spicule" for the jets which dominate the chromosphere seen at the limb, shooting up to 10 000 km. The photosphere, which is in hydrostatic equilibrium with a 150 km scale height, changes abruptly to a extended region where the density gradient is determined by the trajectories of rapidly moving jets and arches. Much of the apparent extent of the chromosphere is a matter of projection: the high spicules, which only cover about 10% of the surface, project against other regions at the limb. The true chromospheric scale height is between hydrostatic equilibrium and the spicule trajectories.

Although spicules appear to form a thick forest in Hα pictures at the limb, Cragg, Howard and Zirin (1963) recognized that they occur only at the edges of the magnetic network and subsequent pictures confirmed this fact. In the center of the disk (Figs. 6.14, 7.7), the spicules are seen to protrude in all directions from the network elements, tracing the connecting flux loops. The lower velocity of the IN structure produces a fairly narrow Doppler profile. Thus this widespread structure, which may be thought of as the general chromosphere, is not seen further from Hα line center than ±0.6 Å. At the limb (Fig. 7.5) spicule and IN chromosphere are all mixed up. In centerline Hα there is a fairly continuous band about 7000 km high: since this disappears at ±1/2 Å, we identify it with the IN chromosphere. Outside this wavelength only spicules are seen on disk or limb. Although long exposures with a coronagraph will show spicules extending up to 10 000 km, the fact that most spicules are tilted means that they are no higher than the IN chromosphere. Below 1500 km there is little structure visible in centerline, but at Hα±0.3 Å a dark band, due to a minimum in the Doppler broadening, can be detected. The chromospheric heating and acceleration of spicules seem to start near the top of this band.

What do we know about the spicules themselves? Measurements of limb spicules were made by R. B. Dunn (1960) at Sacramento Peak Observatory in 1958. Dunn found that the spicules are rapidly ascending jets that rise with velocities of about 30 km/sec to heights above 6000 km and then fade rapidly once they have reached this height. Dunn also found them to be extremely thin, finer than the 500-km resolving power of his telescope. Spicule lifetimes are not well-determined, but the common picture of $v = 30$ km/sec, $H = 10\,000$ km gives an up-and-down lifetime of 600 sec. It is possible that we are seeing a jet, through which material moves much faster than the rate at which the visible jet rises.

While the limb observations show the vertical scale of spicules, they emphasize the higher ones and are confused by line of sight effects. Swept frequency disk filtergrams in Hα (Fig. 7.2, 7.5) should reveal the real structure, but so far we have not been successful with these. One requires good seeing during the entire spicule lifetime, and the spicule evolution has to be documented on both wings of the line. The spicules do not come down where they went up. The spicule structure is distinct on the blue wing and diffuse on the red, as though a little surge moves up, then dissipates and comes down as diffuse material following the force lines. Tanaka (1974) has pointed out that doubling of spicules is often seen (Fig. 7.12). This may be due to emission in the center of a spicule, changing to absorption in the edges. It is not seen at the limb.

Mosher and Pope (1977) measured the tilt of over 5000 spicules; after

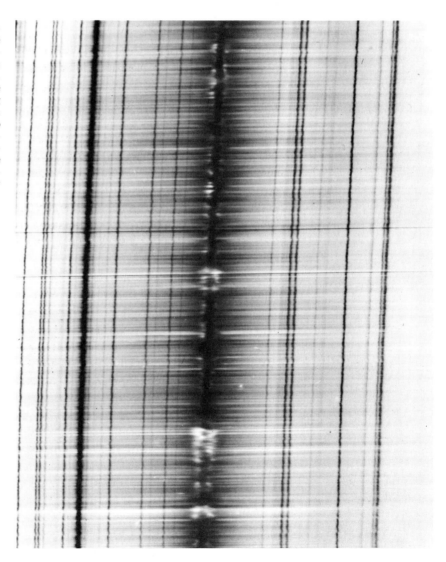

7.10. A superb Sac Peak spectrogram in CaII K. We see the strong and symmetric K2 reversal in network elements, with bright streaks in the continuum. The bright points inside of the cells show only K2v brightening and do not match the continuum brightenings. Recently Suemoto has shown that many bright elements near the K line do not coincide with bright granules. The weak lines nearby show the "wiggly line" structure produced by solar oscillations. (SPO)

correcting for projection, they found 30° to be the most popular inclination to the vertical. Almost none were tilted more than 60°. They found no dependence on the general solar field.

All spicules follow the local magnetic field direction. This is particularly evident in Fig. 10.6, where we see the spicules pointing outward from an active sunspot. In this case coronal streamers showed that the fields from the active regions actually led in that direction (Webb and Zirin 1981). Those spicules associated with the enhanced network follow stronger and more organized fields and generally all run the same way. In the quiet Sun the network elements are apparently not connected in any regular way and the spicules fan out in all directions; these formations were called "rosettes" by Beckers (1963) and may be seen in Fig. 7.5. The rosette form tells us that each magnetic element consists of fields connected to various other poles. Since our VMG movies show merging of small fields into network elements, as well as the decay of the latter, it is an interesting question at what point a magnetic element is decorated with a spicule rosette. The data is not yet available, but at any point in time all network elements except ephemeral regions have spicules.

Near active region filaments spicules flatten and run parallel to the filament, because the magnetic fields are horizontal. Under the filament a "channel," free of network or spicules, appears. But under quiet Sun filaments, where the magnetic field is weak, the network is normal. Because spicules are elongated, we see few of those pointed toward us. There are no spicules over plages (Figs. 7.13, 7.29), probably because of a sharp transition to coronal temperatures. This can be seen at the limb as a hole in the spicule forest, but the lower chromosphere is higher and brighter above plages.

Foukal (1971) discussed the relation between spicules, fibrils (long, sometimes curved spicules coming from the edges of plages) and threads. The last appear to be a thin arch of dark material connecting two spicules arising in opposite magnetic polarities. Foukal argued that all were the same physical phenomenon, material injected into force lines. The only difference I can see is Doppler broadening; the line profiles of fibrils are narrower in Hα and involve less motion. Thus these features represent material injected into different flux loops in different ways.

Some spicules are bright against the disk in centerline, but these, too, are dark in the wing because the background (now outside the absorption line) is brighter. In the K line the story is the same. There has been some controversy about whether the spicules at the limb are bright or dark against the disk in Hα; unfortunately the limb is so confused in centerline that one cannot resolve the question. Probably the bright spicules are more visible above the limb, but they are not frequent. Bright spicules must be hotter than the chromospheric background, but dark spicules are not necessarily cool, because the low density above the surface would result

7.11 An emerging flux region (see Sec. 10.4) and nearby quiet Sun spicules observed on Sept 5 1971 in various parts of the Hα line: (a) −1Å; (b) −0.75Å; (c) −0.5Å; (d) CL; (e) 0.75Å; (f) 0.5Å. In (a) we see the many Ellerman bombs and one blue-shifted arch filament.   In the other frames we see the arch filaments, their tops blue shifted because of the upward motion of the flux loops, their ends red shifted because of the downward draining material. Most of the arch filaments terminate in growing spots. The left (W) part of the arches is prominent in (e) because the region was at 30° W longitude, showing limbward as well as downward flow. The average length of the spicules here is 10 000 km, and the average tilt is around 45°.   Many off-band spicules are double.   In the upper part of the frame we see enhanced network associated with the active region Mt. Wilson 18547 (Fig. 10.6). Spicules coming from this network turn parallel to the filament F, and spicules on the other side turn in the opposite direction, showing that the lines of force run parallel to the filament as pointed out by (Foukal 1971). (BBSO)

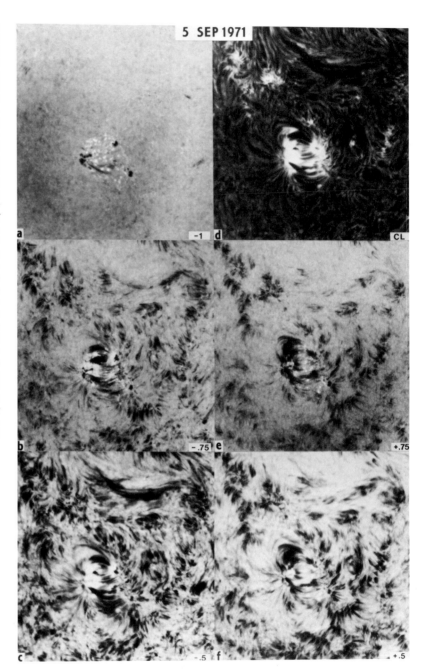

in a low excitation temperature even for hot spicules.

For some obscure reason there has been considerable interest in the total number of spicules on the Sun. In the limb pictures in the wing, individual spicules are easily visible. The line of sight $2L$ is 100 000 km and we see only about one spicule along it. Since there is about one spicule per arc second, there are about 5000 around the limb in a 100 000 km-deep band and therefore about 70 000 on the whole Sun at any time. Athay (1976) gets about 60 000 by a different method, so these numbers are not too bad.

A new kind of spicule, a long jet following the polar plumes, was discovered in Skylab HeII 304 overlappogram images (Fig. 7.20). Moore *et al.* (1977) showed that these "macrospicules" are connected with tiny Hα limb flares in ephemeral active regions. They argue persuasively that there is a continuum of flare-like events going down to the smallest sizes; the smallest are the brightenings at the base of spicules. The macrospicule events are never simple jets, but spray-like. This idea fits well with the concept of spicule heating of chromosphere and corona. LaBonte (1979) found macrospicules showing helium D3 emission in the quiet Sun near the pole; such emission occurs only under high excitation conditions. He inferred a rate of 1400 per day on the solar disk. At any time about 5–10 macrospicules are visible around the limb; if we allow for the extra visibility of these events there is only one for several thousand ordinary spicules.

There appears to be a resemblance between spicules and surges (Chap. 9). Both are linear ejecta from flare-like brightenings. While the base of a surge is a real flare, the Hα brightening sometimes visible at the base of spicules is small and confused by the presence of the bright network. But (Chap. 11) flares occur only on neutral lines, and the network elements where we see spicules are unipolar. It is possible that there is hidden bipolar structure in these elements, left over from magnetic merging, but invisible features are a poor explanation for anything.

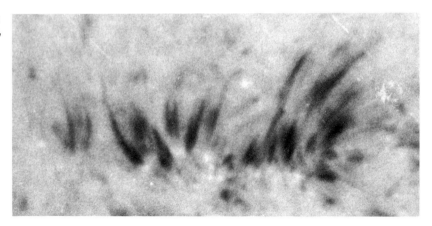

7.12. An enlargement of a spicule bush from Fig. 7.11(b), blue wing, showing double spicules (Tanaka 1974). The twinning only occurs in the blue wing, a characteristic of the spicule injection process. (BBSO)

7.13. The relation between Hα and magnetic structure in active regions and enhanced network: (a) HαCL; (b) VMG; (c) Hα − 0.7Å; (d) Hα + 0.5Å. (1) is an area of FTA, (2) is a sheared neutral line (note that the sunspot to the right of (2) is covered by fibrils in the CL frame and therefore flare-prone by Sec. 11.9), (3) is a filament and channel, and (4) is an included region of white polarity surrounded by FTA. (BBSO)

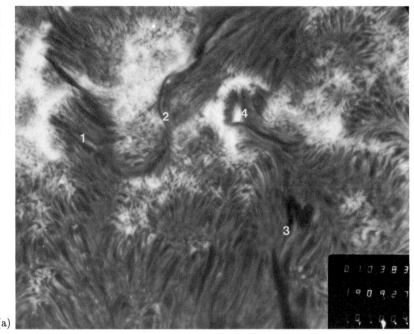

(a)

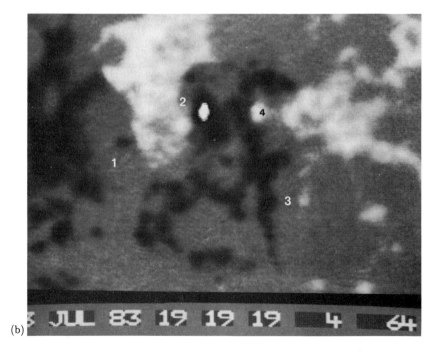

(b)

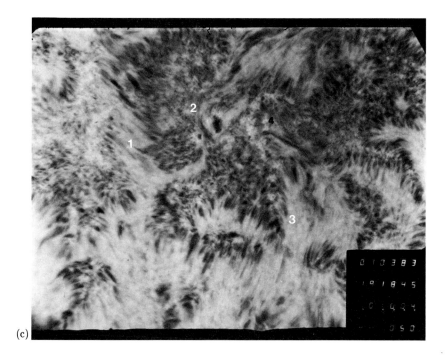

(c)

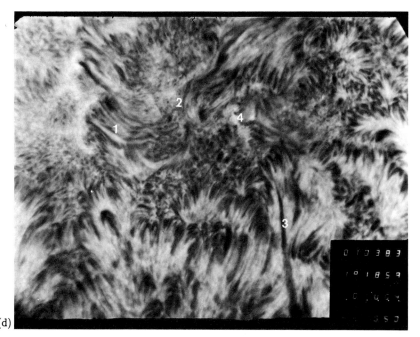

(d)

What is the temperature of spicules? I really don't know, and neither does anyone else. There are some observations which imply that spicules produce absorption near the limb, both in radio and UV wavelengths. Because the spicule is an injection of relatively dense material into the thin, hot corona, it is the locus of some UV emission, but that does not mean that it is hot relative to its surroundings; it could just be a place where coronal electrons are easily converted into photons. Still, there is no question that the flash spectrum shows the excitation to increase upwards as the importance of spicules increases; metallic lines disappear and H and He lines dominate the spectrum. Since spicules are a dynamic phenomenon which travel rapidly from the cool base to the hotter top of the chromosphere, it is reasonable to consider that they are hotter than their surroundings near the base and cooler near their top.

7.3. Radio observations

Radio observations are a powerful tool for chromospheric studies because the atmosphere is opaque in the continuum at wavelengths > 1 mm. Because collisions maintain a Maxwellian distribution, there are no non-LTE effects. The thermal emission is given by the simple Rayleigh–Jeans law, Eq. (4.27a), $viz.$ $S_\nu = B_\nu = 2kT/\lambda^2$. The wavelength dependence and disk distribution of radio brightness is a powerful tool to explore the chromosphere. But study of the fine structure is limited by the low resolution of radio dishes and the weakness of the chromospheric sources. Eclipse observations help improve the resolution, because we can subtract successive observations as the Moon covers or uncovers slivers of the surface. It is most important in this work to avoid active regions, which are strong sources of radio emission. Only recently have decent radio maps of the chromosphere been obtained.

It is convenient to use the brightness temperature, which is that value of T which gives the observed emission when substituted in the Rayleigh–Jeans law after allowing for the angular size of the Sun. The opacity at radio wavelengths depends on different ways in which the electrons can convert the electromagnetic energy into kinetic energy. For free-free absorption (Eq. 4.56),

$$k_{\text{ff}} = 2 \times 10^{-23} Z^2 g \lambda^2 N_e N_i T^{-3/2}, \qquad (7.3)$$

where λ is in cm and the Gaunt factor g is, at radio wavelengths:

$$g = 1.27\left(2.78 + logT - \frac{1}{3}\log N_e\right). \tag{7.4}$$

In the presence of a magnetic field electrons spiral around the field lines in resonance with the electromagnetic field and re-radiate the energy. This gyroresonance absorption occurs at harmonics of the gyrofrequency, 2.8 MHz/gauss (Eq. 3.14). The opacity is great but decreases at higher harmonics. Except in active regions, the magnetic field is usually too weak for this process to be important.

In the absence of a magnetic field, the electrons oscillate in the electrostatic field of the protons at the plasma frequency ν_p (Eq. 3.32). From Eq. (3.33) we know that below this frequency the index of refraction becomes imaginary and the radiation is absorbed. Because the plasma frequency is proportional to $\sqrt{N_e}$ while free-free opacity depends on N_e^2, the former predominates in the corona and the latter, in the chromosphere. The free-free opacity from the Earth to a particular height in the solar corona can be estimated by multiplying the absorption at that height by the scale height $H = 5 \times 10^9$ cm:

$$\begin{aligned} \tau_{ff}(\text{cor}) &= k_{ff}\lambda^2 H N_e^2 \\ &= 7.5 \times 10^{-22}\lambda^2 N_e^2. \end{aligned} \tag{7.5}$$

If we evaluate Eq. (7.5) with coronal densities of 10^8 we find the free-free optical depth of the corona to be less than unity for all wavelengths below a few meters, but the plasma frequency for that density is 90 MHz ($\lambda = 3$ m). At frequencies above 90MHz we will see through the corona to the point of $\tau_{ff} = 1$.

7.14 Scans of the disk at various position angles (PA) (Horne *et al.* 1981) at $\lambda = 1$ mm. These profiles represent the remainder after removing the beam (bottom) convolved with a rectangular solar profile. The smooth curves represent the appearance of the Sun without limb brightening, convolved with the beam. The limb brightening is about 20%.

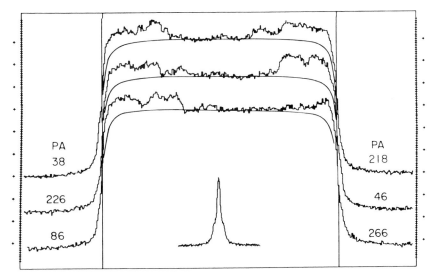

For the chromosphere, we can substitute $T = 6000°$ and $H = 1.5 \times 10^7$ cm in Eq. (7.3) to get the optical depth,

$$\tau_{ff} \ (chrom) = 2.25 \times 10^{-21} \lambda^2 N_e^2. \qquad , \qquad (7.6)$$

Thus $\tau > 1$ for wavelengths in the cm range for the typical chromospheric densities 10^{12} cm^{-3}. Changing the wavelength gives $\tau = 1$ at different densities, so we probe different levels of the chromosphere and map the dependence of temperature on height.

We can estimate the consequent emission from chromosphere and corona if we assume that the corona has constant temperature and low optical depth. Then from Eq. (4.21).

$$I_\nu = B_\nu(T_{cor})(1 - e^{-\tau_{cor}}) + B_\nu T_{chrom} e^{-\tau_{cor}}, \qquad (7.7)$$

where T_{cor} and τ_{cor} are the temperature and optical depth for the corona at the wavelength used and the same convention is used for the chromosphere. This formula simply adds the contributions of the chromosphere filtered by the corona to the coronal emission.

Using the brightness temperature, we have:

$$T_{br} = T_{cor}\tau_{cor} + T_{chrom} e^{-\tau_{cor}}. \qquad (7.8)$$

This assumes that the chromosphere is opaque at all wavelengths of interest and the Sun fills our antenna (not always the case at low frequencies). At long wavelengths τ_{cor} is not zero and, because T_{cor} is large, there will be a small coronal contribution. At short wavelengths τ_{cor} is small, and we get the brightness temperature of the chromosphere. This model assumes only two distinct temperatures and nice spherically symmetric and homogeneous layers. Results using $T_{chrom} = 6\,000°$, coronal base density $N_e = 5 \times 10^8$ and $T_{cor} = 10^6$ deg are given in Table 7.1. Of course we don't believe the chromosphere is 6000° throughout. In the "adopted" column we give a run of chromospheric temperature which fits the observed data, such as it is. This goes from 6000° where the density is 2×10^{11} at about 1800 km to a temperature of about 11 000° at $N_e = 2 \times 10^9$ (about 3500 km). More detailed, but less exact radio models of the chromosphere have been made by Beckman *et al.* (1973) and Linsky (1973a).

The "observed" values in Table 7.1 are based on a plot in Krüger (1979) and values from Simon and Zirin (1969), Linsky (1973a), Horne *et al.* (1981), and Marsh *et al.* (1981). The temperatures quoted are only approximate because of calibration and resolution problems. The only source against which the Sun may be calibrated is the Moon, which is imperfectly known (Linsky 1973b) but at least is the right size. Calibration against noise diodes is difficult because we don't know the antenna filling factor. One searches in vain for data more recent than 1971.

As in the visible spectrum we can also try to map the height variation with temperature by looking at center-limb variation. For a symmetric layered atmosphere

$$\tau_{\text{ff}} \propto \lambda^2 \sec\theta \qquad (7.9)$$

where θ is the angular distance from the Sun's center. For any value of $\lambda^2 \sec\theta$, T_{br} must be the same. Since the observed temperature increases for increasing λ, the same should occur for increasing $\sec\theta$, *i.e.* we expect to see limb brightening. Wrong. Limb brightening is almost undetectable.

Using current pictures of a hot chromosphere and spherically symmetric corona, Smerd (1950) produced straightforward models which showed big limb-brightening at most wavelengths. Early measurements confirmed his picture; since the antennas did not track well observers let the Sun drift through the antenna beam from east to west. The active regions are spread along the equator east-west, and their projected density increases steeply toward the limb, so a spurious limb brightening was observed. Although newer antennas showed no limb brightening, Smerd's plots, in total disagreement with observations, were dutifully reproduced in all the books, including mine, for years.

High resolution arrays which could scan the Sun north-south soon appeared, and they showed no limb brightening. Simon and Zirin (1969) pointed out this discrepancy and the fact that it disagreed with all extant models of the solar atmosphere (it is much easier to show existing models wrong than to produce one that fits the data). Only in the last few years has limb brightening of about 25% in the last 30 arc sec ($.03R_\odot$) from the limb been observed at 6 cm (Furst *et al.* 1979; Marsh *et al.* 1981) and about 10% within 20 arc sec from the limb at 1 mm (Horne *et al.* 1981). These data show that the problem is only with the older models that used hot chromospheres; for example Smerd used a $30\,000°$ layer which we now know doesn't exist.

Inside the limb, the optical depth of lines should increase with $\sec\theta$,

Table 7.1. Central radio temperatures from a simple chromosphere corona model.

λ(cm)	τ_{cor}	$\tau_{\text{cor}}T_{\text{cor}}$	T_{br} Calc	T_{br} Observed	T_{chrom} Adopted	$N(\tau=1)$
100	1.88	10^6	10^6	10^6	–	2×10^8
10	.019	$19\,000$	$25\,000$	$30\,000$	$11\,000$	2.1×10^9
3	.0019	$1\,900$	$7\,900$	$11\,500$	$9\,600$	7×10^9
1	.00019	190	$6\,190$	$8\,900$	$8\,700$	2.1×10^{10}
.3	.00002	19	$6\,019$	$7\,000$	$7\,000$	7×10^{10}
.1	0	0	$6\,000$	$6\,000$	$6\,000$	2.1×10^{11}

where θ is the angle from Sun center. This holds for optically thick clumps with vertical extent as well as optically thin structures of any shape because the clumps crowd together toward the limb. But it does not hold for optically thick platelets, which cover a constant fraction everywhere.

If we use the right chromospheric model, such as Table 7.1, we see that the problem of missing limb brightening is minor so long as we don't have an extensive hot layer. For example, the brightness temperature at $\theta = 0$ and $\lambda = 1$ cm (9000°) should match that at $\theta = 85°$ and $\lambda = 3$ mm; when we scan the Sun in 3 mm the brightness should rise from 7000° at the disk center to 9000° at $\theta = 85°$, only 3 arc sec from the limb. But close to the limb the antenna beam includes a certain amount of sky, and limb brightening is not directly observed at that point. So the fact that we see no limb brightening at these frequencies is due partly to the low resolution of radio telescopes and partly to the fact that the limb brightening is actually low.

The simple model of Table 7.1 fits the modern data reasonably well, and the complicated rough structure invoked by Simon and Zirin to explain the failure of Smerd's model is probably unnecessary, at least for the data presently available. Even at 6 cm, Marsh *et al.* (1981) found a reasonable fit to a symmetric corona. In fact the raw data from all these observations show no limb brightening; it appears only after deconvolution of the instrument profile.

7.15. (a) A rocket photo of the Sun in the continuum near 1600 Å. (b) The same in Lyα. (LPSP – LPARL) (c) Hα from BBSO. A small flare is ending in the region at top center. The filament extending from the region is much brighter in Lyα than Hα. The network is much more prominent in the UV images. Both hydrogen images exhibit more diffuse structure than the 1600 Å continuum, probably because they are formed higher. The filaments are not prominent in Lyα, possibly because the filter is wide. The rings in (b) are filter defects.

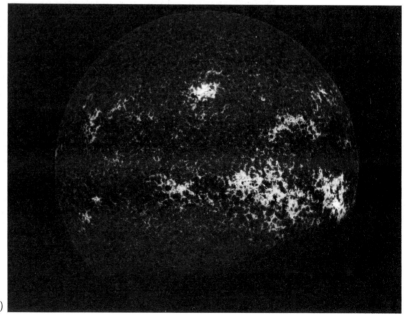

(a)

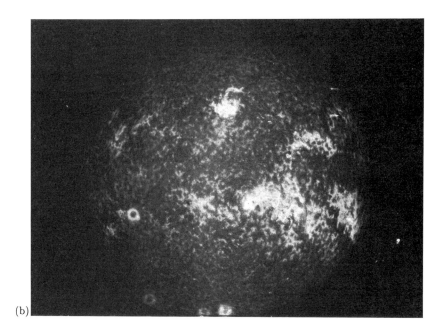

(b)

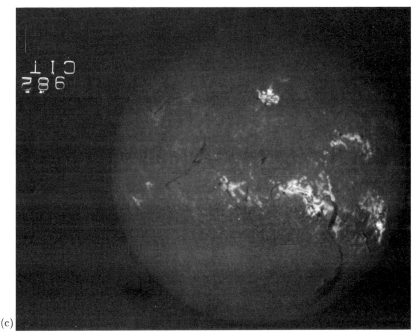

(c)

Fig. 7.14 shows the results of Horne *et al.* (1981) who used the 24 arc sec beam of Leighton's mm dish to measure the brightness distribution at 1 mm. They found a limb-brightening coefficient β between 0.1 and 0.2, where

$$T_{\mathrm{br}}(\theta) = T_0(1 + \beta - \beta \cos\theta). \qquad (7.10)$$

Above 10 cm there is a sharp increase in brightness temperature which should be reflected in limb brightening if the corona were uniform. But the clumpy structure of the corona and the fact that transition-zone regions are buried in the low chromosphere seems to suppress the limb brightening.

For what wavelengths will H^- opacity be important? Athay (1976) gives $k_\lambda = 1.17 \times 10^{-31} T^{0.15} \lambda^2 N_e N_H$ for the H^- free-free; if we compare with Eq. 7.3 for normal free-free absorption with $Z^2 = 1.5$ and $T = 6000°$, we find that H^- will be greater if $N_H/N_p > 1000$, roughly. This is about true in the photosphere. At greater λ, we see higher heights, N_H falls and ordinary free-free dominates beyond the ionization edge at 1.6μ.

There are other difficulties in the brightness distribution at the shortest wavelengths. Wannier *et al.* (1983) found limb darkening at 2.6 mm with a 6 arc sec beam; this result was after subtracting a spicule band extending 5–8000 km. They decided the spicule temperature was 6100° compared to 7200° for the disk. On the other hand Lindsey *et al.* (1982) find that the limb extends to 1000 km and Lindsey *et al.* (1984) found limb brightening at 300 microns. Since the temperature minimum occurs around 4000°, limb brightening at the short wavelengths is reasonable, with the cooler spicules absorbing at 3 mm but transparent at Lindsey's wavelength. The data are still inadequate.

Most surprising is the discovery by Kundu (1963), Babin *et al.* (1974) and Wefer and Bleiweiss (1980) that coronal holes are bright at 1 cm. These crude results were confirmed by maps made with the Nobeyama 40-m dish by Kosugi *et al.* (1986), who found an increase above the general level of 3–7% (250°–550°) at 36 GHz, but none at 98 GHz. Equatorial coronal holes are also enhanced, and long term observations by Efanov *et al.* (1980) show that polar brightenings have occurred near sunspot minimum, when we expect polar coronal holes to occur. To what can we attribute this remarkable result, which contradicts our concept that coronal holes are everywhere cooler than the rest of the solar atmosphere? Because $\tau T \approx T^{-1/2}$ for an optically thin medium, a decrease in the coronal temperature might do it, but Table 7.1 shows that the coronal contribution to T_{br} is only 190° at this wavelength. Polar faculae are prominent near minimum, but this would not explain the brightening in equatorial coronal holes. Because the effect occurs just at the transition from chromosphere to corona, it seems likely that it is due to increased opacity in some transition structures otherwise thin in radio waves. Feldman *et al.* (1977) found that transition zone lines extended higher in coronal holes. The XUV line intensity is higher in coronal holes above 5000 km or so. This implies increased density at that height, and it is possible that this increases the temperature at $\tau = 1$ to produce the observed brightening.

The radio brightness of the Sun of course includes contributions from active and other magnetic regions. High-resolution radio maps of the quiet Sun look like a K-line image, with bright network. At 6 cm, the radio temperature of network elements is about 30 000°, about twice the background. But the area of the network elements is so small that the total contribution is only about 800°. The match between radio and K-line images is perfect except that tiny active regions are relatively much brighter than the normal network elements, while ephemeral active regions are not.

7.4. The ultraviolet spectrum

The radio spectrum picks out the invisible transition zone because of the absorption of low frequencies at low densities; UV emission is selectively produced at higher temperatures. Because the photosphere radiates very little in the UV, we see only chromosphere and corona when we look at the Sun in those wavelengths. The tremendous temperature range makes a whole range of ionization stages visible, and the lines are resonance lines, the most important lines in the spectrum, rather than the subordinate lines we must be content with in the visible. But a 6000° black body does not put out much UV, and the entire emission below 1500 Å is 1/20th of that in a single angstrom at 5000 Å.

The absorption increases as we go into the UV until we reach the wavelength (about 1800 Å) where we see the temperature minimum. At shorter wavelengths only emission lines are seen. Because the continuum emission at this point is from the temperature minimum, the color temperature in the near UV is 4000°–4500°. The lines come from higher, hotter regions, and the exponential increase of excitation with temperature overweighs the falloff in density, resulting in emission lines. Absorption lines may also be present but have not been detected below 1500 Å.

The intensity of the UV lines is determined by excitation conditions, abundances, and atomic peculiarities. The resonance lines of H and He, the most abundant elements, are the strongest. Lyα of hydrogen at 1216 Å is as strong as all the other UV lines put together, and Lyα of HeII at 304 Å is as strong as all the UV lines below 500 Å. Because of the low density, collisional ionizations are not balanced by their counterpart, triple collisions. As a result, the most common ions have ionization potential five or ten times the thermal energy. The strongest lines observed come from those ions with low-lying, easily excited levels. These are typically the "screening doublets", transitions with n constant, $viz.$ $2s - 2p$, $2s^2 - 2s2p$, $3s - 3p$. Because only the screening changes in these ions the excitation

7.16. Comparison of a VMG (a) with a fragment (b) of Fig. 7.15a. Every element of flux corresponds to a bright feature in the UV continuum, showing that at this low height heating takes place where the field is strong. The VMG was aimed at the active region and could not reveal fields matching the bright IN grains; we do not know what magnetic phenomena, if any, the bright IN grains correspond to.

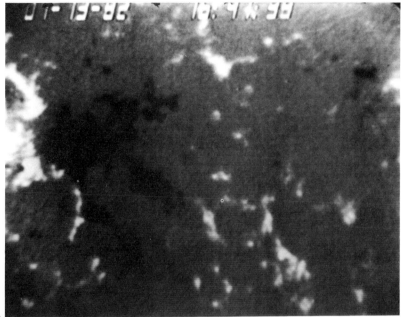

(a)

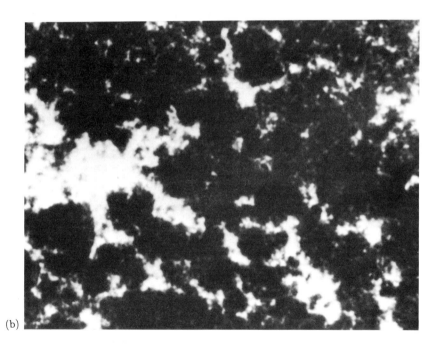

(b)

energy is low. Most of the ions in the corona are ionized up to a He-like or Ne-like state which is hard to ionize further. Those He- or Ne-like states are hard to excite, but they produce many Li-like ions by recombination. As a result, the strongest UV lines (excepting H and He) come from the Li-like ions CIV, OVI, NeVIII, MgX and SiXII. Note that we see the effects of nucleogenesis; the elements of even Z are most abundant (except for H and N). Another group of strong lines are those of Fe, for which easily excited coronal ions are abundant. But the sharpness of the transition to the corona is accentuated by the virtual absence of lines of elements with ionization potential 30–200 eV, except for the Li- and Be-like ions of the most abundant species.

Observations of the height distribution of UV lines above the limb ought to tell us where the temperature begins its sharp rise to coronal values. Feldman *et al.* (1979) measured the height distribution of many transition zone lines on Skylab spectra in the 1100 to 1900 Å, including ions such as CI, II, III, IV, SiIII, etc. They performed an Abelian inversion to determine the true surface brightness from that along the line of sight and found the peak emission (Table 7.2) to occur surprisingly low in the chromosphere; even the lines of ions common at 200 000° peak well below the top of the Hα band. The transition to coronal temperatures takes place within the Hα chromosphere, in the region where the radio brightness is still below 10 000°. Simple arithmetic tells us that the 200 000° material occupies less than 5% of the area; UV images tell us that this is at the chromospheric network edges. The UV line emission from inside the network cells is five or ten times less than at the edges.

The distribution of UV emission tells us the location of high temperature and density regions in the chromosphere. Figs. 7.23–25 show overlappograms with images in chromospheric and coronal lines. The distribution of line emission in transition-zone lines roughly corresponds to a K-line spectroheliogram. Both HI λ1216 and HeII λ304 show complete disks (Fig. 2.16) like optically deep lines, with no particular limb brightening. The brighter emission comes from the same sources as the K line, the magnetic network. Weaker chromospheric lines such as OIV show emission from the network with somewhat more contrast. There is definite but

Table 7.2. Heights of peak emission for UV lines (after Feldman *et al.* 1979).

Temperature Range	Height
$3\,000° < T < 6\,000°$	2 100 km
$6\,000° < T < 12\,000°$	2 900 km
$10\,000° < T < 20\,000°$	3 600 km

slight limb brightening in OIV and NVII; this can occur only if the vertical extent of emitting elements is sufficient to overlap the horizontal separation between elements. In the CIV image from SMM (Fig. 7.4) we see this happens a few arc seconds inside the limb. The images are consistent with a vertical scale of EUV brightening similar to that of Feldman *et al.* (1979).

The detailed distribution of UV limb brightening (or lack thereof) should tell us more about the structure of the emitting elements. If we have a flat distribution of bright and dark optically thick elements, the average brightness will be almost constant as we approach the limb, since the same proportion of either will be seen. This is observed in $\lambda 304$ and Lyα. Deviation from flatness will depend on whether the cooler or hotter elements have vertical extent. In the Lyman continuum it appears that vertical cool elements (spicules?) cover the bright features and suppress the geometric limb brightening. In other lines such as CIV we see increased coverage by bright elements near the limb, indicating that they are three-dimensional.

7.17. A high resolution image made with the HRTS rocket experiment of the continuum at 1600 Å. Its similarity to the K line shows that the lower network is brightened by temperature enhancement. (NRL)

Even if roughness reduces the limb brightening, optically thin lines jump by a factor of two just above the limb, because the layer on the far side

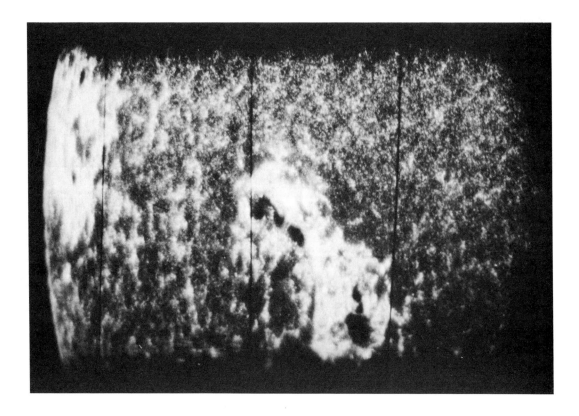

7.18. High resolution ultraviolet spectra with the HRTS rocket instrument from Lyα to 1700 Å. The position of the slit is at upper left. This and other HRTS spectra is a negative, the lines are in emission. The chromospheric lines are broad and strong in the network elements and weak and narrow, but not absent, inside the cells. The plages near the spot produce increased continuum and broader lines. Among the strong lines here are 1216 Å, Lyα; 1336 Å, CII; 1406 Å, SiIV; 1550 Å, CIV; 1660 Å, OIII. (NRL)

of the Sun becomes visible. This can be checked in Table 7.2; Doschek *et al.* (1976) and Feldman *et al.* (1977) find considerable limb brightening for all optically thin lines. The peak emission occurs between 1000 and 2000 km and falls off above that point with a scale height around 1000 km. This matches the height distribution of spicules counted by Lynch *et al.* (1973). But counting spicules, although objective, tends to emphasize outstanding individual structure. If one simply measures the typical projected height of spicule bushes inside the limb, as well as the centerline limb band, one finds most of the spicules do not exceed 5000 km. So the height distribution of UV line emission more or less matches that of the spicules.

There are interesting effects associated with coronal holes. Doschek *et al.* (1976) find the line intensities in coronal holes to peak at the same height as the quiet Sun, but the high temperature lines like OV are much

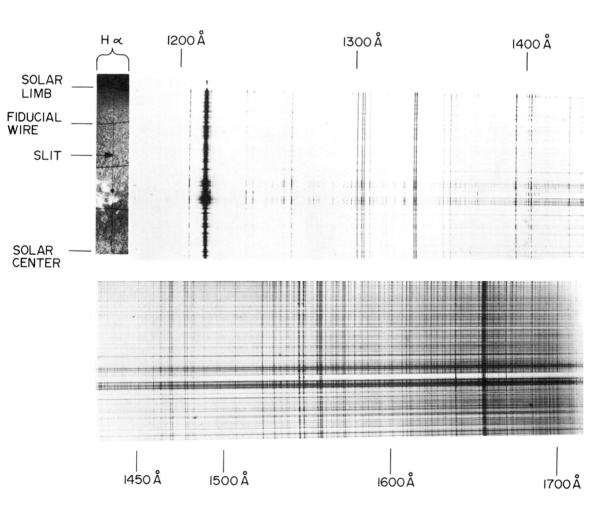

weaker. That kills the old idea that the chromosphere is produced by back-conduction from the corona; turning off the corona has little effect on the chromosphere. The coronal lines, of course, are weak in holes, as are the "subcoronal" lines, like NeVII. Surprisingly, this is true of the helium lines as well. Both λ 584 and λ 304, the resonance lines of HeI and HeII, are dramatically weakened in holes, as are all other He lines. This is because photoionization by coronal ultraviolet plays an important role in the excitation of these lines. In the He lines the normal chromospheric network is weakened in coronal holes, showing the effect of the more diffuse coronal photoionization. Does this mean that NeVII also is excited by coronal back-radiation? Quite possibly. Although the NeVII images are dark under the holes, a bright limb ring is still visible at the poles. The limb ring is probably due to downward conduction from the corona. Some weakening of hydrogen Lyα in holes also appears.

The HRTS spectrograph (Bartoe and Brueckner 1975) has produced high spatial resolution spectra and images of the chromosphere which shed some light on the distribution of UV emission. Cook *et al.* (1983) studied continuum images around 1600 Å, finding a background temperature of 4300° with the bright points a few hundred degrees higher. Because the contin-

7.19. Spicules photographed in CI by HRTS, compared with Hα (SPO). There is excellent correspondence between CI emission and spicules. (NRL)

EUV SPICULES

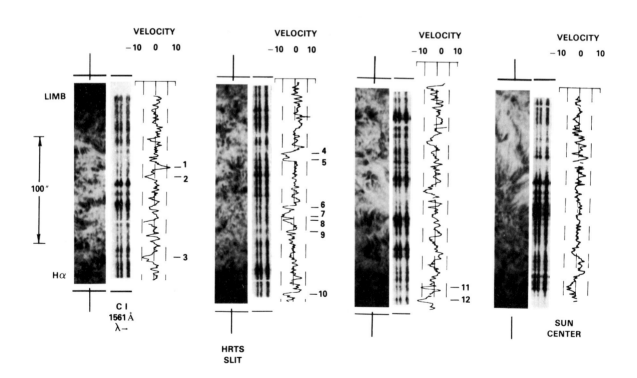

uum is probably close to LTE, that is the temperature of the layer we are seeing. Fig. 7.16 shows the perfect correspondence between the magnetic network and a high resolution continuum image at 1600 Å from Bonnet (1982). All the magnetic elements correspond to continuum brightening. Unfortunately the magnetograms on that date were not deep enough to test the correspondence between IN bright points and magnetic fields.

The emission lines are not formed in LTE. Although their brightness temperature is not far above 4500°, they must be produced at a much higher temperature, a plasma hot enough to produce the particular ion. The intensity is reduced by the low filling factor and optical depth. Dere *et al.* (1983) examined the relation (Fig. 7.19) between the CI lines and an Hα filtergram. Both atoms have similar ionization potential, and there is a good coincidence between the CI emission and Hα spicules. But CI events with large velocity shifts come from inside the network, where there are no optical counterparts. Although the authors conclude that these are spicules, they probably are not, since spicules come from the network. The comparison between CIV and CI found by Dere *et al.* is only fair. CIV is bright over more extensive areas than CI, and may simply extend higher along a canopy. This is supported by the limb brightening data of Doschek *et al.* (1976) which show CI peaking at the limb, and CIV, about 2000 km higher. These results show that UV images diverge from K-line images once we can resolve the network elements. While everyone claims high resolution for UV images, five arc seconds is rarely exceeded. To judge for yourself, use a ruler to divide the solar disk into 1800 one arc second parts. You will see nothing that small.

Brueckner and Bartoe (1983) discuss "turbulent events" and jets observed in CIV. The former are marked by strong broadening (up to 200 km/sec), the latter by Doppler shifts of 400 km/sec. They feel these may play a role in the heating of the corona and in the source of the solar wind. But the features they show are unrelated to bright points, and therefore are probably not associated with coronal heating. Similar features – big bushy spicules – are observed at Hα − 1 Å.

Line broadening indicates high velocities in the transition zone. Hα widths correspond to 20–40 km/sec, and Doschek and Feldman (1978) find a mean velocity of 26 km/sec for NV in the quiet Sun. These velocities apply to the spicules rather than the general chromosphere, which is transparent at Hα±0.5Å. The atoms cross the transition layer in tens of seconds, sufficient for collisional equilibrium, but not always for ionization. So there is no equilibrium. The transition zone lines also show persistent downflow. The temperature of material changes rapidly; gas is heated as it moves up in the spicule and cools again as it falls. Kanno (1979) and others have noted the importance of continuum absorption in the ultraviolet beyond the Lyman limit; if the layer is convoluted, there can be cool material in front of hot, especially at the limb.

Although the UV emission inside the network cells is weak, at some point there must be a transition to coronal conditions, because there is some corona everywhere above the surface, even in coronal holes. There is a "*Ratskeller*" model, in which arches of magnetic field are supposed to overlie the cell interiors. But an arch has to have two feet, and the enhanced network is unipolar. Another variant is that the field lines diverge from the network in arch form, but are uniformly radial in the corona.

With the possible exception of H and He, no UV line is so optically thick that the UV photons created do not escape. The photons emitted are the sum of all the collisional excitations. This makes quantitative analysis possible. That we are far from LTE is obvious: a line such as CIV is produced typically at $70\,000°$, but its brightness temperature is only $5000°$. The brightness of the line is down by about 10^8 from the black body value. This is because the low density results in a high level of ionization for the temperature. Only occasionally (every 10 sec or so) does an excitation take place, and the great A_{21} causes de-excitation in less than 10^{-8} sec. Since the ionization and excitation "temperatures" may be quite different, it is important to evaluate them separately.

One simple approach is to find certain supermultiplets where the different multiplets have upper terms whose energies differ considerably. Because only the angular momentum coupling changes the relative transition probabilities are known. A good example is the $1s^2 2s^2 2p$ ground configuration of OIV, NIII, and CII (Fig. 7.21), whose lone $2p$ electron gives a term ^2P (spin 1/2, angular momentum 1). The lowest excited configuration in these ions is $1s^2 2s 2p^2$, where one of the s electrons has been excited to a p state.

7.20. Polar sections of an over-exposed HeII 304 Å image showing macrospicules at both poles. (NRL)

This configuration has three doublet terms (a polyad) ^2S, ^2P, and ^2D, all of which can combine with the ground state because the parity (the sum of the electron angular momenta) differs. Furthermore, the cross-section for exciting each of these terms must be simply proportional to the line strengths from Eq. (5.9) and Eq. (5.10), because in each case an electron is excited from a $2s$ to a $2p$ orbit. All photons escape, and they can be excited only by collisions because the photospheric radiation is so weak at this wavelength. The downward transition probabilities are so great that all atoms emit the line once they are excited; the intensity is equal to the total rate of collisional excitation. That rate is given by the integral over the Maxwellian electron velocity distribution Eq. (3.4) of the cross-section σ, times the velocity v, where the integral is limited to those electrons that have enough energy to excite the level we want:

$$
\begin{aligned}
C_{1n} &= 4\pi \left(\frac{m}{2\pi kT} \right)^{3/2} \int_{\chi_n = mv^2/2}^{\infty} \sigma(v) e^{-(mv^2/2kT)} v^3 \, dv \\
&= \frac{8\pi}{m^2} \left(\frac{m}{2\pi kT} \right)^{3/2} \int_{\chi_n}^{\infty} \sigma(E) e^{-E/kT} E \, dE.
\end{aligned}
\tag{7.11}
$$

If we take the ratio of the excitation rates for two doublets of the same supermultiplet, the constant terms will cancel, the cross-sections will be in the ratio of the line strengths S and statistical weights g, and, if the velocity dependence of the cross-sections is the same (which it ought to be), the ratio of the integrals will be just $\exp[-(\chi_1 - \chi_2)/kT]$. Thus the ratio of photon numbers in the two doublets will be:

$$
\frac{I_1}{I_2} = \frac{g_1 S_1}{g_2 S_2} e^{-(\chi_1 - \chi_2)/kT};
\tag{7.12}
$$

7.21. Term diagram of OIV. The upper levels result from a $2s$ electron excited to $2p$.

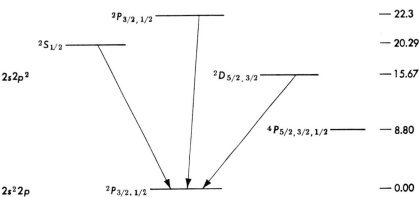

and we get the temperature by taking the natural logarithm:

$$T = \frac{\chi_2 - \chi_1}{k \ln(I_1 S_2 g_2 / I_2 S_1 g_1)}. \tag{7.13}$$

The nice aspect of this means of determining the temperature is that T is proportional to the logarithm of the measured quantity and hence insensitive to errors of measurement, mistakes in our assumptions about the cross-sections, and so on. In other words, the number of exciting collisions varies exponentially with the temperature, so a small change in the temperature produces a large and easily detectable change in the line ratios.

For the lines from $2s2p^2$ ^2D and ^2P downward to $2s^22p$ ^2P the ratio $S_2 g_2 / S_1 g_1$ is unity. Inserting the observed intensities for three ions in Eq. (7.13), we have

$$
\begin{array}{lll}
\text{OIV}: & 1.4 = 10^{5040 \times 6.61/T} & T_1 \approx 216\,000° \\
\text{NIII}: & 5 = 10^{5040 \times 5.55/T} & T_2 \approx 40\,000° \\
\text{CII}: & 56 = 10^{5040 \times 4.41/T} & T_3 \approx 12\,700°
\end{array} \tag{7.14}
$$

Examples of derived temperatures are given in Table 7.3. The values agree roughly with the temperature at which each ion is most abundant, according to Jordan (1969). We thus have a handle on the kind of temperatures that exist in the chromosphere. We can use this method whenever we have two lines of known line strengths S from the same ion and are confident that the cross-sections have the same energy dependence, so we can apply Eq. (7.11) and Eq. (7.14). Unfortunately, higher excitation lines come from higher temperatures and we deduce temperatures that are too high. This is the weakness of the use of temperature and density diagnostic ratios; in an inhomogeneous plasma, lines which radiate preferably from high temperature regions cannot simply be compared to lines of lower excitation which overweigh lower temperatures. Line ratios are rarely what we expect them to be (Zirin's rule).

We can determine the extent of the transition zone by summing the number of exciting collisions C_{1n} given by Eq. (7.11) over a model layer of temperature T, scale height H, and density N_e. Working backward from the fluxes observed at the Earth we determine the product of emission measure and element abundance for each ion. The abundances, cross-sections, and geometry are not well understood, and it is difficult to sort

Table 7.3. Temperatures derived from the excitation of XUV lines.

Ion	CI	CII	CIII	CIV	NII	NIII	OIII	OIV
Temperature	9 000	12 700	39 000	< 80 000	< 40 000	40 000	137 000	216 000

out the relative contributions of the quiet Sun and active regions. But important upper and lower limits on all these quantities may be found.

Let us calculate the photon flux at the Earth in CIII from (Eq. 7.10):

$$\phi(i,\chi) = 3.7 \times 10^{-12} \times 10^{-6.25 \times 10^7/\lambda T} \beta_i N_e N_a H, \qquad (7.15)$$

where λ is in Å and β_i is the fraction of the ions in the ith stage within scale height H.

For $\phi = 4 \times 10^9$ ph/cm^2/sec at the Earth,

$$N_e N_a H \approx 4 \times 10^{22}. \qquad (7.16)$$

So if $N_a/N_H \approx 10^{-4}$, and $N_e \approx 10^{10}$, H is 40 km for a layer which is all CIII; four or five such layers, about 200 km, would comprise the entire transition from neutral to fully ionized carbon. These results, which we first obtained in 1963, have been further extended to show that the transition layer emission measure is smaller yet. Pottasch (1963) and others have calculated such integrals for many ions (finding $N_e N_a N_H < 10^{23}$ for all), derived an average curve for the integral of N_e at each temperature, and used this standard curve to derive abundances.

Eq. (7.13) is quite sensitive to temperature; we have used an observed effective temperature derived from Eq. (7.11). Pottasch used 75% of the temperature at which each line produced peak emission, but it is unlikely that the emission is so efficient; the sharp density gradient weights the lower temperatures. In any event, it is always better to use empirical temperatures like those of Table 7.3. The emission measures obtained from the UV are orders of magnitude below those of the high-temperature chromosphere models; together with the radio data they force a more realistic view of the chromosphere.

While these results show that the chromospheric transition is only tens of kilometers thick, the limb measurements of Doschek *et al.* (1976) show that it extends over thousands of kilometers. This points to a thin, transient shell surrounding the cooler structures extending into the corona. Further, Zirin *et al.* (1963) found that all the UV lines emitted in the 100 000° range are missing or relatively weak. The calculations by Jordan (1964) and Pottasch (1964) of elemental abundances from UV intensities also show a deep minimum in emission measure at this temperature. Since spicules occur along flux loops which must restrict the conductivity across the axis they may be thermally isolated from their surroundings, and the transition zone emission may come from a thin jacket around them.

Recent work on solar energetic particles (SEP) has revealed (Meyer 1985a,b; Breneman and Stone 1985), that atoms of high first ionization potential (FIP) such as A, Ne, or O are systematically underabundant by about a factor four in the solar wind and in energetic particles from flares.

At some point the elements of high FIP are filtered out of the solar material. Geiss (1982) and Geiss and Bochsler (1984) suggest that this might occur in spicule acceleration. The abundances deduced by these authors from SEP agree with those derived from flare X-ray lines studied by Veck and Parkinson (1981). But they also roughly agree with results of Mariska (1980) for the N and O abundances in the chromospheric UV lines measured by Doschek *et al.* (1976). The C abundance given by Mariska cannot be used because it also fixes the density. Mariska finds $O/Si \approx 14$, while Breneman and Stone give 6 in SEP and Allen (1973) gives 20. The studies of abundances carried out by Pottasch, Jordan, Withbroe and other authors also show underabundances of high-FIP elements like C, N, O and Ne relative to Mg and Si, but the error bars are large. Further, measurements of argon coronal lines in solar flares (Zirin 1964) show that high FIP element is underabundant relative to calcium.

One would think that we could watch this abundance change take place as we move upward through the atmosphere, and find out what makes it happen. But the UV-line height gradients measured by Doschek *et al.* show no evidence of different gradients for the different elements, particularly the singly ionized atoms that should be most strongly affected. As we shall see in Sec. 7.7, the flash spectrum, which should give accurate height dependence, gives no indication of differing gradients. One should also remember that the high-FIP elements are most resistant to photospheric abundance determinations; the noble gases are not observed, and C, N, and O have few good lines. So the intriguing mystery of where and how the high FIP elements are filtered out is not yet settled, and more focussed studies are in order.

7.5. Non-LTE effects

Because the chromosphere is mostly optically thin, non-LTE effects are quite important. For example, in the visible and UV the dominant radiation field is that of the photosphere, typically a 5770° black-body with dilution factor 1/2 (the temperature quoted is the effective temperature, limb-darkening included). Even if the kinetic temperature is also 5770°, we do not have LTE because the radiation field is dilute and the radiation temperature is lower in the UV and IR. In general, if the element of plasma of electron temperature T_e can see light of a different radiation temperature, and if that light plays an important role in excitation and ionization, we will not have detailed balance and require a non-LTE approach.

As noted in Sec. 4.4, pseudo-LTE can exist if the main transition rates between the states in question are in detailed balance, even though other factors are not. If all transitions are produced by collisions, we will have pseudo-Boltzmann equilibrium at the local kinetic temperature, and if all transitions are due to radiation, the equilibrium will be at the local radiation temperature. If we plot the intensities of lines of triplet neutral helium in the flash spectrum or in a prominence, we find they are well fitted by a Boltzmann formula with $T = 5700°$ and dilution factor $1/2$. This is because radiative excitation by the photosphere balances spontaneous emission, so the relative population is independent of the local temperature. This is true of almost all transitions in the visible that can absorb the photospheric radiation. If we look at the population relative to the ground state, we see it is impossible to excite triplet helium at $5770°$, and the amount of triplet helium will depend on other factors. But within the triplet levels, these processes regulate the equilibrium.

Properties of continuum emission are determined by the free-electron velocity distribution, which is maintained in a Maxwellian distribution by the frequent elastic collisions. The spectral distribution in the continuum, *i.e.* radio measurements or recombination continua, will be determined by T_e weighted by the optical depth. If the temperature changes in a distance shorter than the mean free path, then non-Maxwellian effects occur, and the same is normally true of the electron distribution in transient events such as flares.

The emissivity of optically thin chromospheric lines in the visible will be determined by the scattered photospheric light. An emission line will appear only if local collisional excitation dominates over the radiative excitation in the photosphere. This requires that the local temperature be greater than the photospheric radiation temperature at that wavelength, and the density sufficient for collisions to dominate. For lines like D3 or Hα this requires $N_e > 10^{12}$ cm^{-3}. In the UV the photospheric emission is so low that lower densities produce emission lines. The intensity emerging from the chromosphere will be the scattered intensity plus all photons added along the path (by collisional excitation, recombination, and cyclical collisional excitation through other levels) minus those photons lost to the kinetic energy field by collisions of the second kind or continuous absorption. This intensity cannot build up indefinitely; eventually gain and loss from the kinetic-energy field balance and we saturate at the Planck function. For subordinate lines there are complicating factors, such as breakup into intermediate transitions or jumps to other levels.

The non-LTE picture gets complicated if the line in question is so thick that not all photons escape, or if it is formed in a place (such as the solar surface) where physical parameters change rapidly and the local radiation field is important.

If we are dealing with an isolated gas cloud where the external radiation field is unimportant (such as the chromospheric UV lines), we can track the creation and destruction of photons by transfer of energy to and from the kinetic energy field. Each photon will take a random walk through the gas till it escapes or is destroyed. The length of each step is $\tau = 1$, and the progress in any direction is the square root of the number of steps, or $\sqrt{\tau}$. If the radial optical depth is τ, it takes τ^2 steps to escape. But danger awaits the photon on this journey: every time it is absorbed, it may either be re-emitted (probability A_{21}) or lose its energy by a collision of the second kind (probability $N_e C_{21}$). The latter returns the photon energy to the kinetic-energy field. The loss probability is

$$\epsilon = \frac{N_e C_{21}}{N_e C_{21} + A_{21}}. \tag{7.17}$$

Because of Doppler broadening of the line, many photons are re-emitted in the wings where $\tau < 1$ and escape immediately. When a photon travelling in a given direction is absorbed and re-emitted in the same direction, its frequency will not change and neither will τ. But the atomic velocity in the orthogonal directions is independent of that along the original direction of propagation, with the familiar Doppler distribution

$$f(\Delta\lambda) = \frac{1}{\sqrt{\pi}} e^{-(\Delta\lambda/\Delta\lambda_D)^2}, \tag{7.18}$$

where $\Delta\lambda_D$ is the Doppler width. The optical depth of the chromosphere at any wavelength is given by a similar formula:

$$\tau(\Delta\lambda) = \tau_0 e^{-(\Delta\lambda/\Delta\lambda_D)^2}. \tag{7.19}$$

For some wavelength shift

$$S = \Delta\lambda_D \sqrt{\ln \tau_0}, \tag{7.20}$$

the optical depth is unity. At each redistribution in direction of propagation, every quantum emitted at a wavelength shift of S or greater has an excellent chance to escape directly, because the optical depth is less than unity; this probability of radiative escape, which we designate q, is the sum of all re-emissions farther from the line center than S:

$$\begin{aligned} q &\approx \frac{2}{\sqrt{\pi}} \int_S^\infty e^{-(\Delta\lambda/\Delta\lambda_D)^2} \, d\Delta\lambda = 2\mathrm{Erf}\,(S/\Delta\lambda_D) \\ &\approx \frac{0.53}{\tau\sqrt{\ln \tau}}. \end{aligned} \tag{7.21}$$

If continuous absorption is significant, photon energy is converted to the energy of an electron which is quickly thermalized, so the photon is lost to the kinetic-energy field. If the ratio of continuous to line opacity is

$$R_\nu = \frac{k_\nu}{l_\nu} \tag{7.22}$$

then the effective photon loss ratio becomes

$$\epsilon' = \epsilon + R_\nu - q. \tag{7.23}$$

What are the magnitudes of these numbers? The collisional excitation rate at $10\,000°$ can be obtained from Eq. (4.64). For Lyα, $P \approx 0.02$, $f = 0.8$, and multiplying by a density 10^{12} we obtain the rate $N_e C_{12} = 0.14$/sec. The rate of de-exciting collisions C_{21} is readily obtained by using Eq. (4.41). The statistical weight of the ground state is two and that of the ^2P term, six; so we find

$$N_e C_{21} = \frac{g_1}{g_2} e^{-\chi/kT} C_{12} = \frac{2}{6} \times \frac{0.14}{7.23 \times 10^{-6}} = 6450 \,/\text{sec}. \qquad (7.24)$$

The transition probability for Lyα is $A_{21} = 6.25 \times 10^8$ /sec, so the photon escape probability ϵ at $N_e = 10^{12}$ is:

$$\epsilon = \frac{N_e C_{21}}{A_{21}} = 1 \times 10^{-5} \quad \text{or} \quad 10^{-17} N_e. \qquad (7.25)$$

Let us evaluate the Lyα emission from a model spicule with thickness 10^8 cm, $N_H = 10^{12}$, $N_e = 10^{12}$, and $N(\text{HI}) = 10^{11}$, *i.e.* 90% ionized. From Eq. (5.25) we find the absorption coefficient at the center of Lyα at $10\,000°$ to be:

$$l_0 = \frac{g_2}{g_1} \lambda^3 \frac{A_{21}}{16\pi^2} \sqrt{\frac{m_H}{kT}} = 2.35 \times 10^{-14} \qquad (7.26)$$

we have

$$\tau_0 = l_0 N(\text{HI}) L = 2.35 \times 10^5 \quad \text{and} \quad q = 6 \times 10^{-7} \qquad (7.27)$$

from Eq. (7.21). We see that the escape probability $q \ll \epsilon$ and most of the photons will be lost to equilibrium processes. Depending on the surface conditions, the emergent Lyα intensity will be close to the Planck function. The continuum absorption under these conditions is small and need not be considered. For densities ten times lower we can calculate the Lyα flux by adding up the photons created by collisions. The Planck emittance in the wavelength scale is, from Eq.(4.30)

$$F_\lambda = \pi B_\lambda = 1.01 \times 10^{15} \text{erg/cm}^2/\text{sec/ster/cm}. \qquad (7.28)$$

Using the Doppler width for $10\,000°$ from (5.29), which comes out $5 \times 10^{-5}\lambda$, the emittance in the line is 5×10^5 erg/cm^2/sec/ster.

The intensity in a spectrum line will fall somewhere between the sum of the collisional excitations and the Planck function. If we could count all the physical processes and understand the geometry we would probably get the right answer. This calls for iterative calculation of the radiation field, which depends on the equilibrium all through the gas. No matter how big our computer or how sophisticated our codes, the calculation is useless unless we have the right model and the right data. So most of the effort should go to understanding the physics.

We often use an approximation derived by Thomas (1957):

$$S_{ij} = \frac{\int J_\nu \phi_\nu d\nu + \epsilon_{ij} B_\nu(T_e) + I_{ij}}{1 + \epsilon_{ij} + \eta_{ij}}, \qquad (7.29)$$

where the first term in the numerator is the scattering of the incident radiation field, the second is the rate of collisional photon creation, and the third represents cyclic excitation via the continuum. The denominator includes our familiar term for the ratio of photon loss to emission and a term for photon loss via the continuum.

7.6. Helium

Helium, the solar element, presents fascinating astrophysical problems and to this day is a subject of dispute. The limb-disk contrast in helium is remarkable. The strong $\lambda 5876$ line (called D3 because of its proximity in the spectrum to the D1 and D2 lines of NaI) is about 0.2 times the disk intensity at the limb, nearly as bright as Hα. Yet on the disk it has a central depth less than 1%. The strongest He I line, 10830 Å (see Fig. 5.5 for the energy level diagram), has a depth of a few per cent on the disk but is as strong as Hα at the limb. Because the He lines have excitation potential of about 20 volts, they are not easily excited; so it was soon decided the chromosphere must be hot. This idea was reinforced by the high scale height and broad line profiles in the chromosphere, as well as the knowledge that the million degree corona lay above. But quantitative analysis of the radio and UV spectra showed that if the chromosphere were hot enough to excite the observed He lines, the emission in those wavelengths would be orders of magnitude greater than observed (Zirin and Dietz 1963).

We can summarize the characteristics of the solar He spectrum as follows:

1. The neutral He lines in the visible are enhanced by a factor of 50 at the limb. Flash spectrum measurements show that He emission increases above the photosphere to a peak at about 1200km. While other chromospheric lines come mainly from spicules, D3 comes from a low thin layer (Fig. 7.24) far below the spicule tops.
2. All helium lines are sharply weaker in coronal holes.
3. The resonance lines of HeI and HeII, $\lambda 304$ and $\lambda 584$, are optically deep and have brightness temperatures about 5000°, even though HeII only exists above 15 000°.
4. The He lines are weak in filament channels, but the filaments themselves show absorption.
5. The lines are enhanced near active regions and over unipolar regions, but the structure is fairly diffuse.

I believed for years that chromospheric He was locally excited, even after we discovered (Zirin and Dietz 1963) that there could not be an extensive

7.22 (above) Spectroheliogram of the Sun in λ10830 compared with (below) Hα filtergram. The λ10830 absorption is enhanced in active regions and old field (marked by spread-out plage). The filament channels are weak in this line, but the filaments themselves show strong absorption. A coronal hole winds across the right part of the disk. Harvey uses λ10830 to detect coronal holes on the disk. (KPNO).

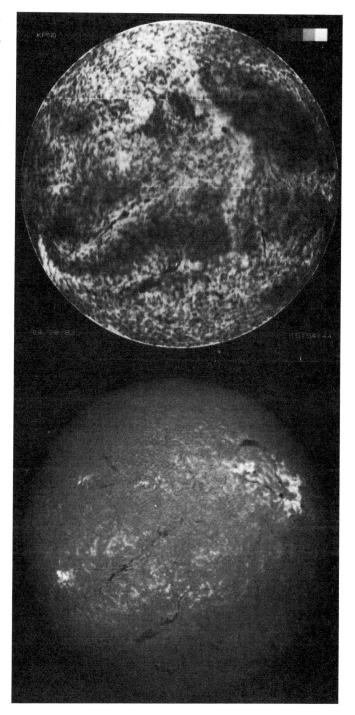

hot chromosphere. In 1974 I looked at the solar limb in D3 for the first time, using our new universal Lyot filter. What I saw convinced me that the authors (Goldberg 1939, Hirayama 1971 and Nikolskaya 1966) who had proposed that the He lines are excited by coronal UV had been correct. I saw a narrow, uniform bright band between 1000 and 2500 km above the limb, with faint spicules above and a clear dark band below (the effect cannot be seen in λ10830 because the resolution is poor with existing detectors for that wavelength). The line was missing in the polar cap coronal holes, and spicules were faint.

It was already known that the surface brightness of the helium lines in the visible peaks at a relatively low height, about 1000 km (Thomas and Athay 1961; Belkina and Dyatel 1972). This is below the peak found by Doschek *et al.* (1976) for comparable XUV lines and far below the top of the Hα chromosphere. The height of D3 emission is thus determined by the optical depth to vertically incident coronal UV emission which ionizes He. Chapman (1931) showed that such a layer formed in the terrestrial ionosphere, as the solar UV ionizes various components. In a Chapman layer, the ions formed increase with the local density, but the ionizing radiation decreases exponentially downward as it is absorbed. A sharp peak results, below which no excitation takes place. This is exactly what we see in the chromosphere. In coronal holes the Sun tests the hypothesis for us; we turn off the corona and the He emission goes away.

While one may argue with details of the model, all the observational data support it. While almost all UV lines are concentrated in clumps, with limb brightening of the order of 5 or 10, the weak D3 line, virtually undetectable on the disk, is one of the strongest lines in the flash spectrum. It is enhanced 20-fold or more at the limb and shows a sharp spike. It must arise in a thin uniform layer, with much less vertical structure. The relation of He excitation to coronal UV is confirmed by the excellent correlation of λ10830 absorption and SXR emission in stars (Zarro and Zirin 1986) and the fact that the line is absent from stars where only low temperature chromospheric emission lines appear (Baliunas *et al.* 1984). The only contrary evidence on the P–R model was produced by Feldman *et al.* (1974), who measured the ratio of components of the $3 \rightarrow 2$ λ1640 HeII. Since recombination favors high l states, one would expect the $3d \rightarrow 2p$ component to dominate, while they found the components corresponding to collisional excitation of $3s$ and $3p$ dominant. This is a strong argument for collisional excitation, but there is some confusion by a blend with FeII which may affect the measurement. More important, Feldman *et al.* also found the intensity of λ1640 to increase 15 times within 10 000 km at the limb, while the other UV lines simply double. That, as we have noted, can only happen in a flat, radiatively excited layer. The ratios of subordinate HeII lines relative to λ304 also fit a recombination spectrum and differ by a factor 10 from collisional excitation ratios (Zirin 1975).

7.23. Mosaic of the Sun in Hα (BBSO – Tel Aviv) compared with a Skylab λ304 HeII image (NRL). In general there is good correspondence, but prominences are not so dark in HeII and the emission from active regions, somewhat greater. Note the extra emission from the active limb prominence (lower left); the upper image was taken earlier and does not show it. λ304 really matches the K line better than Hα. The biggest difference is the coronal holes at the poles, where the general helium emission disappears.

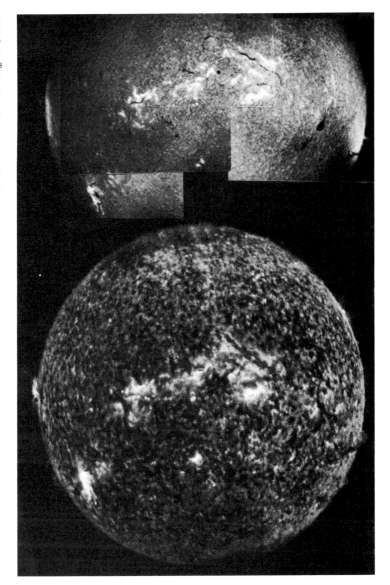

Although other explanations of the coronal hole effect have been offered, none have attempted to explain the complete observational data. Most depend on transition zone phenomena, which occur much higher than the He band. The EUV lines show a broad peak starting well inside the limb, doubling at the limb, while D3 and $\lambda1640$ increase more than 20 times. The He lines must come from a layer no more than 2000 km thick.

Calculations of the photoionization–recombination (P–R) model are simple (Zirin 1975; Avrett *et al.* 1976). We calculate the photoionization as a function of height (the UV being absorbed by HeII, HeI and HI) and equate it to the recombination:

$$\frac{N(\text{HeII})}{N(\text{HeIII})} = \frac{5 \times 10^{-8}}{E_1(\tau)} N_e \qquad (7.30)$$

where $E_1(\tau)$ is the first exponential integral, appropriate for a localized coronal XUV source. If the source is a layer, the second integral $E_2(\tau)$ must be used. This produces a layer of ionized helium down to the point where $\tau = 1$ in the HeII continuum. If the opacity is due to helium alone,

$$\tau = 1.8 \times 10^{-18} \, H \left[N(\text{HeI}) + N(\text{HeII}) \right] \qquad (7.31)$$

because HeI and HeII contribute in equal amounts to the opacity. If we take $H = 1000$ km, $\tau = 1$ occurs at $N(\text{He}) = 5 \times 10^9$ or $N_\text{H} = 5 \times 10^{10}$. There is absorption by other ions such as hydrogen, but this is much reduced at 228 Å, the HeII limit. At this point HeI and HeII are about equal in population if the ionizing flux has a filling factor of 10, *i.e.* the XUV sources are concentrated in the network with that factor.

The brightest line in the UV spectrum is HeII $\lambda304$, which also contributes to the ionization of the HeI below it. This is confirmed by flash spectrum data given by Thomas and Athay (1961), who show HeII $\lambda4686$

7.24 (a) The chromosphere in D3, a faint narrow band above the limb with a dark band underlying, produced by absorption of the coronal SXR emission. To the right is the beginning of a coronal hole. The height of the peak emission is about 1500 km. (BBSO)

peaking at 1600 km while HeI λ7065 peaks 300 km lower. The HeI triplets are populated by recombination from HeII to HeI 2^3S and then excited by photospheric emission. Livshits (1976) presents other aspects of the P–R model.

It should be emphasized that the UV photons do not produce the strong emission of the helium triplets in the visible; they only get atoms up to the triplet levels, where they radiate by scattering photospheric emission. Athay and Menzel (1956) showed that the excitation temperature of the He triplets is 6000°, as we would expect if they are excited by the photosphere.

Hirayama and Irie (1984) measured the Doppler widths of HeI 4713 and HeII 4686 in the flash spectrum and found narrow profiles. They estimate the source temperatures to be 7000° and 20 000° respectively. This measurement sets a definite upper limit on the temperature of the He source and can only be explained by the P–R mechanism.

Most authors currently agree that the He lines are produced by the P–R

7.25. Excerpt from an overlap-pogram covering the range from NeVI 402 Å to NeVII 465 Å with coronal lines of MgVIII in between. The lower, slightly shifted spectrum is that of a flare, where the strongest line is the FeXV λ417 intersystem transition. The polar coronal holes are clear in NeVII, but with no break in the bright limb. (NRL)

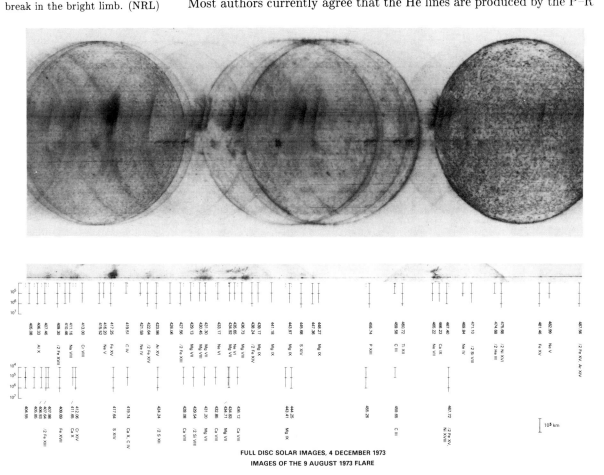

FULL DISC SOLAR IMAGES, 4 DECEMBER 1973
IMAGES OF THE 9 AUGUST 1973 FLARE

FROM AN ATLAS, IN PREPARATION, OF DATA
RECORDED BY THE NRL SPECTROHELIOGRAPH
ABOARD SKYLAB

mechanism, but there are difficulties with the strength of the λ304 line. While the P–R mechanism produces enough HeII, the λ304 intensity is too great to explain by recombination alone. The observed HeII continuum is much weaker than the λ304 flux; it should be roughly equal if recombination preceded each excitation. But the line is not very intense, with brightness temperature of about 5000°, and it can be excited by collisions once the ions are produced by photoionization.

Jordan (1979) suggests that λ304 is excited by diffusion of He atoms through the thin transition layer into the corona, which would bombard them with higher energy electrons and produce anomalously high He emission. The lines would be weaker in coronal holes because the temperature gradient is lower and the mechanism less effective. We do not know exactly where the transition zone is, but we should probably not see the thin, homogeneous layer that is observed. There is no evidence that the temperature gradient in coronal holes is much different than the normal corona. Still, when we have a better picture of chromospheric structure we may have to use such models that recognize the dynamics of excitation.

Milkey (1975) pointed out that the P–R mechanism would lead to a self-reversed core in the He UV resonance lines, because there would be little internal emission from the gas. Imagine my pleasure when Phillips *et al.* (1982) and Wu and Ogawa (1985) found, using rocket observations with a resonance cell, that both λ584 and λ537 are indeed self-reversed. Using Milkey's argument we may conclude that HeI λ584 is excited by recombination (and scattering of λ304). It is most important to resolve the λ304 line.

7.7. The flash spectrum

The spectrum of the chromosphere is very rich. By observing at the limb with the high dispersion KPNO spectrograph, Pierce (1968) measured and identified 11,500 lines, about one-third of the Fraunhofer lines, giving wavelengths accurate to 0.01 Å. Identified lines range from the common elements we have discussed to Terbium, Holmium and Lutetium. He comments that the rest of the Fraunhofer lines might be observed with a darker sky. This can be done at an eclipse; the flash spectrum is the last bright light that we see as the Moon covers the Sun and the first that appears as the Moon moves off. Not only can we see faint lines, but we can get excellent height resolution. To obtain spectra like those shown in Fig. 7.27, astronomers

7.26. (opposite page) The entire overlappogram range, with a flare in progress. The filled disks are chromospheric zone lines. Limb brightening is most prominent in transition lines like NeVII. The coronal holes are prominent in coronal and He lines, but not seen in OV and OIV. The higher ionization coronal lines come predominantly from above the limb and active regions. (NRL)

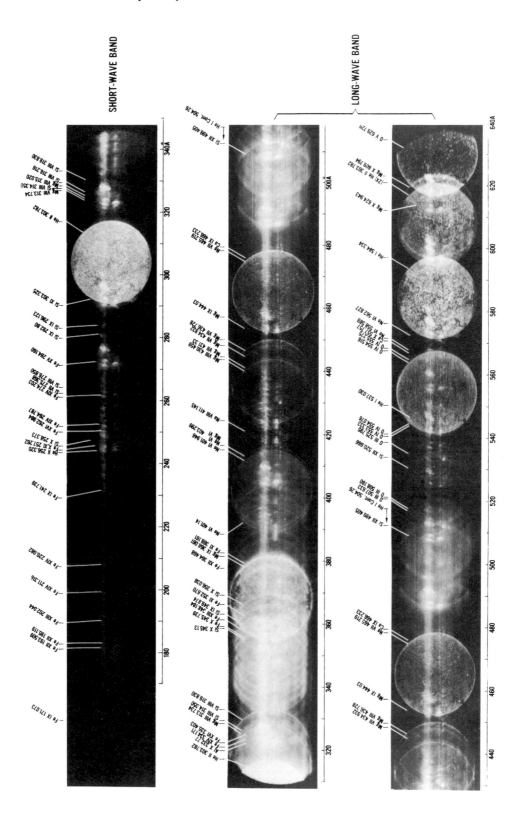

have ventured to the far corners of the globe, for total solar eclipses occur rarely (no more than two a year over the whole Earth), and the fraction of the Earth where they are readily accessible is not large. The astronomer is often disappointed by clouds or rain on the day of totality, but if the sky is clear and his instruments properly adjusted, he reaps a rich harvest of material that can be gotten in no other way. In any event, the main point of an eclipse expedition is the fun, the challenge, and the exotic location, and most of the eclipse data are never analyzed or published.

The flash spectrum affords the high spatial resolution and photon flux available in the visible region combined with the height resolution of the eclipse. The most common means of observation is the "jumping film" method. A slitless spectrograph is used, to project the spectrum on film, which is rapidly transported between exposures, so that a sequence of spectra at closely spaced intervals is obtained. Each spectrum shows the chromosphere above the edge of the Moon, which moves across the chromosphere at a rate (translated into chromospheric heights) of about 310 km/sec. Typically about two spectra per sec may be obtained, giving a height resolution of 150 km. The intensity of each spectrum line is the integral along the line of sight of the integrated intensity in a line λ above a certain height h:

$$E(h,\lambda) = \int_h^\infty \int_{-\infty}^\infty E(y,z)dy\,dz \qquad (7.32)$$

where y is the position in the plane of the sky and z the position along the line of sight. By taking the difference between successive pictures at different base heights h and fitting the geometry we may find the height variation of the emission.

There are three obvious defects in this system. First, it is always dangerous to difference data; second, no line profiles are obtained, and finally, there is structure along a line of sight. One must also correct for the form of the lunar limb. The general height variation is obtained, but probably not to 150 km resolution. The whole problem of deducing chromospheric structure from the flash spectrum is treated in detail by Athay (1976) and by Thomas and Athay (1961) and we will discuss only the simpler aspects.

In Table 7.4 I list a number of scale heights derived from gradients given by Athay (1976) for chromospheric lines. Many lines show a flat brightness distribution at low heights; others fall off steadily with height. He is the only element which increases upward in the first 1000 km. The scale heights listed are the asymptotic values at greater heights. The data are somewhat contradictory, which probably explains why people quit trying to analyze the flash spectrum some years ago. Weak lines of neutral metals drop off sharply with height as the atoms are ionized. Strong lines are saturated at lower heights, then drop off at greater heights with a scale height influenced by the chromospheric density scale height there and the effects of further ionization. HI, HeI, HeII and CaII reflect the spicule scale height.

Emission scale heights in the flash spectrum jump abruptly from 150 km to 500–1000 km as we cross the 500 km height level; the XUV lines give the same result. Is this due to a transition from uniform chromosphere to spicules? Only partly. Fig. 7.5 shows that the spicule forest is not thick at Hα − 0.5 Å, and the material seen at centerline doesn't really look like spicules. It would appear that the IN chromosphere plays a role up to 4 or 5 thousand km.

Interpretation of the flash spectrum is so complex that in recent years there has been little further study of it. The elaborate expedition in 1962 (Dunn *et al.* 1968) resulted in fine spectra which were measured with enormous effort but received only limited analysis. The integrated intensity data are published, however, and the reader in search of truth can double differentiate these data if he wishes. The data for different lines are contradictory, but there is important information not available elsewhere which should not be ignored. I suspect the problem with the old flash spectrum data was the recording on photographic film, which does not afford sufficiently accurate photometry to permit differentiation. With modern

7.27. The flash spectrum at the 1966 Peru eclipse photographed by the Tokyo Astronomical Observatory. Left, CaII H and hydrogen Hϵ; right, the K line. The continuum shows the spectrum of the last bit of photosphere. Many rare earth lines appear in emission.

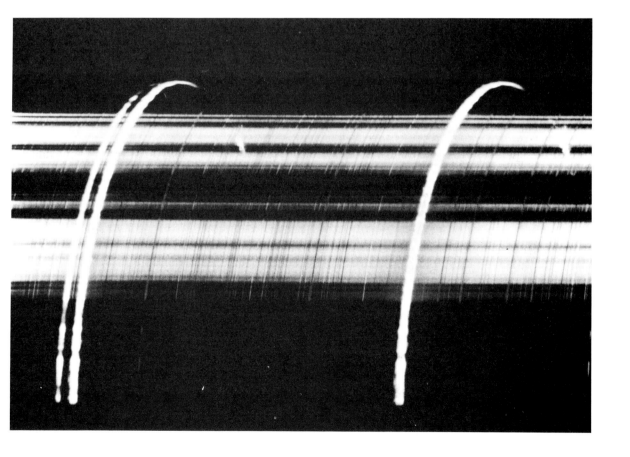

CCD's one could get really accurate photometry and once more use the flash spectrum to probe the chromosphere.

Continuum-intensity measurements give the most accurate picture of height gradients of density. The most accurate measurements of the continuum gradients were carried out by Weart (1970). As is usual with eclipse data, the results were never published, but they were summarized by Athay (1976). We display them along with data from other observers in Fig. 7.28. These modern results are in good agreement. The intensity falls sharply with scale height 90 km up to 500 km above the limb, whereupon it flattens out and falls with scale height 760 km. The H^- intensity should fall off faster, roughly as $N_e N_H$. The scale height above 500 km is much greater than we would expect from hydrostatic equilibrium, but agrees with a recent result of Hermans and Lindsey (1986) from the sub-millimeter data. Although the low gradient is partly an artifact of rising ionization rather than the true density falloff, we have to accept the fact that even the general chromosphere is not in hydrostatic equilibrium.

So long as the negative hydrogen emission is negligible, a good clue to physical conditions in the chromosphere is provided by the ratio of the intensity on either side of the Balmer limit. On the red side we have electron scattering, which is proportional to N_e; on the blue side the dominant contribution is from recombination, which is proportional to $N_e^2 T^{-3/2}$. The ratio of these two quantities should give us $N_e T^{-3/2}$ in the emitting region, independent of the various photometric problems one gets into in trying to measure the absolute brightness of anything. Athay (1976) finds $N_e N_p T^{-3/2} = 2.0 \times 10^{16}$ at 1500 km. If $T = 10\,000°$, $N_e = 1.4 \times 10^{11}$.

At low heights the flash spectrum is rich in rare-earth lines which fall off rapidly with height and disappear by 500 km. They are strong because their spectra are rich and they are easily excited at low temperature; they are

Table 7.4. Gradients of selected chromospheric lines (Athay, 1976).

Line	Log E_0	H(km)	Line	Log E_0	H(km)
Metals			*Resonance lines*		
FeI λ3735	14.6	550	CaII λ3969 (H)	16.1	910
FeI λ3800	13.7	200	CaII λ8498	15.3	600
FeII λ4542	13.8	140	SrII λ4078	14.6	730
CaII λ3706	14.2	330			
Hydrogen			*Helium*		
Hβ	15.8	625	HeI λ7065	14.8	910
Hγ	15.7	625	HeII λ4686	12.6	900

not overabundant. The most abundant atoms are less prominent because of high excitation and ionization potentials.

The hydrogen lines are the dominant factor in the chromospheric spectrum, and their crescents extend much farther than other lines. Many members of the Balmer series are visible; the list given by Thomas and Athay goes up to H28 ($n = 28 \rightarrow 2$). One would think that the Inglis–Teller formula Eq. (5.31) could tell us something about the electron density where the highest lines are formed, but unfortunately the lines on slitless spectra have an intrinsic width due to the finite extent of the chromosphere, so we can only guess where the hydrogen lines merge. Slit spectra might help if taken in such a way that we could tell at which height they had been obtained.

The spectral variation of the recombination continua depends almost exclusively on the temperature, since it is determined by the energy distribution of the free electrons. If the continuum is optically thick, however, the slope is distorted. Noyes and Kalkofen (1970) analyzed the Lyman continuum slope from spacecraft data and found the temperature derived from the slope (about 9000°) much higher than that from the observed absolute intensity (about 6400°). Perhaps the curve is distorted by the contribution of hotter elements. There are similar problems with the Balmer continuum, which may be observed in the flash spectrum. One way or another, these data give another upper limit on chromospheric temperature.

7.28 Chromospheric continuum brightnesses measured by different authors. One is tempted to identify the sharp break at 500 Km with the temperature minimum.

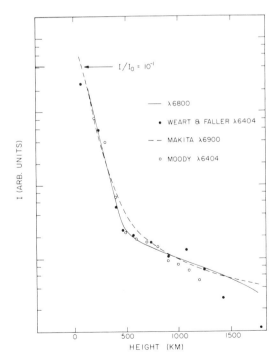

7.8. Heating the chromosphere

The anomalous and mysterious process that heats the chromosphere and corona has been responsible for many Ph.D. theses and has justified many experiments. Stellar astronomers are interested in it because other stars have chromospheres, too, and we can best understand these by studying the Sun. Unfortunately much of the work has ignored the actual evidence on the temperature inversion, and some of the best work was done before details of the solar atmosphere were known.

The key observational fact is that heating of the chromosphere–corona takes place only over magnetically enhanced regions. As we have seen, the network elements are brighter than their surroundings as low as 300 km above the photosphere. Higher in the atmosphere the role of trapping increases the contrast between regions of strong magnetic fields and the rest of the Sun. The discovery of coronal holes showed that coronal material will escape into the solar wind when it is not trapped by closed field lines. Thus the peaks in X-rays observed above closed dipoles are as much a measure of trapping as heating. This effect produces increased emission measure at the loop tops but cannot be connected with increased brightening at footpoints.

Several additional points in the heating process need to be explained:

1. In the first few hundred km above the photosphere the temperature drops from 6000° down to 3500°, but falls off much more slowly where there is vertical magnetic field.
2. The field must be vertical. Regions of horizontal field are not heated.
3. Every area of increased vertical field is bright; thus the heating cannot depend on episodic effects such as reconnection of network elements.
4. Regions where reconnection is taking place (usually emerging flux reconnecting with existing flux) are always sites of additional and strong heating. There is some evidence that this is more effective at greater heights; the 1600 Å continuum picture does not show much extra brightness at low heights, nor does white-light plage show special intensity in such regions.
5. Except in reconnection sites, heating is proportional to magnetic field intensity, as evidenced by the LaBonte (1986) recursion relation

$$HK = 0.21 + 0.0060|B| \tag{7.33}$$

where HK is an index of the intensity of 1 Å bands centered on H and K. Thus we need a mechanism that starts just above the photosphere and produces heating proportional to field intensity along vertical field lines.

The second law of thermodynamics does not permit us to heat the chromosphere with the thermal energy of the cooler photosphere. But if non-thermal energy is present in the photosphere, it might somehow be focussed on the tiny amount of material above so as to heat it. Non-thermal energy is available in the photosphere in the form of magnetic fields and convective motions.

Biermann (1948), Schwarzschild (1948), and Schatzman (1949) suggested that small-amplitude acoustic (longitudinal compression) waves originating in the hustle and bustle of the convection zone would transport energy from that source to the upper layers. The energy transport in these sound waves is the product of the kinetic energy per cm^3 of the moving material, ρv^2, and the sound velocity V_s:

$$F = \rho w^2 V_s = 2 \times 10^8 \text{ergs/sec/cm}^2 \tag{7.34}$$

for a mean material velocity 1/2 km/sec and $\rho = 10^{-7}$ gm/cm^3, the value at the top of the convective zone. The acoustic waves propagate upward with little dissipation so long as the velocity is below the speed of sound.

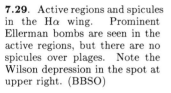

7.29. Active regions and spicules in the Hα wing. Prominent Ellerman bombs are seen in the active regions, but there are no spicules over plages. Note the Wilson depression in the spot at upper right. (BBSO)

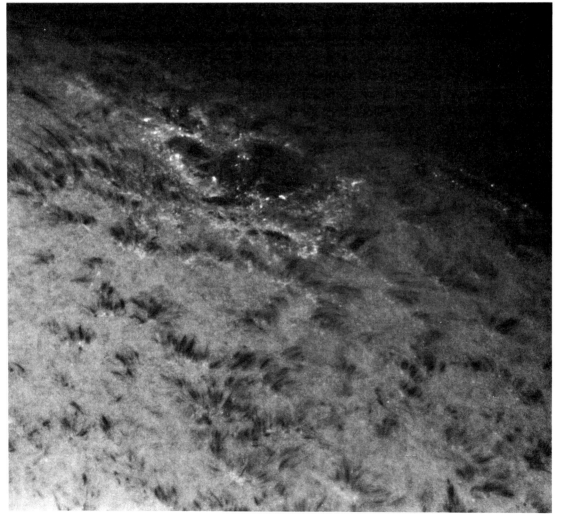

The temperature in the photosphere and low chromosphere varies slowly, but the density drops sharply with scale height 100 km or so. The sound velocity at the temperature minimum is about 5 km/sec. Since there is no dissipation, ρV is constant, so that we reach sound velocity at $\rho = 5 \times 10^{-10}$. At this point the waves steepen into shocks and the energy is rapidly dissipated. This mechanism works much like the crack of a whip, in which the energy of motion of the entire whip is concentrated and dissipated in a small amount of material at the end.

This picture, which we shall refer to as the BSS model, was the main direction for studies of chromospheric heating for many years, even though it bore almost no relation to reality. The BSS model predicts uniform heating over the entire Sun, with the exception of regions of magnetic field, where the motions and consequently the heating would be suppressed. Instead the chromosphere and corona show heating concentrated in regions where the magnetic field is strongest. So the BSS hypothesis, although physically reasonable, fails because it did not try to explain the observational facts (which were not all available at that time, but also ignored after they became known). Does this mean that acoustic heating does not work at all? We cannot say, because while there is a general upturn in temperature outside the magnetic regions, it may be due to conduction along field lines spreading from the magnetic knots. Also, the increased sound speed in heated regions tends to refract away the sound waves.

The connection between magnetic fields and chromospheric and coronal heating was first studied by Osterbrock (1960), who studied the generation of Alfvén waves in the convective zone and their propagation upward. He concluded that the convective zone generates predominantly fast-mode Alfvén waves, finding that both sound and slow-mode Alfvén waves are strongly absorbed in the photosphere, so that only the fast mode escapes. As the waves run outward, the amplitudes increase and shock waves result, with consequent high dissipation. Osterbrock found, however, that the rapid increase of the B–V and the acoustic cutoff frequencies (Fig. 6.4) prevent propagation above the temperature minimum.

Osterbrock's model was built before the magnetic network was understood and hence could not address the problems raised above. The problem of refraction away from high-velocity zones would equally affect propagation of waves into network elements. The vertical magnetic fields of the network should form a natural pathway for energy to flow from below to above the surface. While waves could not enter the network elements, slow mode Alfvén waves can follow the lines of force. In fact, propagation across the field lines would be in the fast mode, and the wave would be rapidly refracted into the vertical direction. This idea is supported by the lack of heating in horizontal field regions.

The chromosphere is vigorously heated above plages and even more so above emerging flux regions. The amount of heating is proportional to the rate of magnetic change.   While the magnetic path outlined above is important for stable plage, there is no doubt that field reconnection is the important mechanism in the brightest plages. Our observations of the weaker fields shows reconnection occurring all over the Sun; this may account for much of the chromospheric and coronal heating.

One way or another, the high temperature of the Sun's atmosphere must be a consequence of the rapid falloff in density at the surface, coupled with the magnetic elements playing the role of "whip" in tying the subsurface to the atmospheric regions. In this picture, only stars with convective zones would have chromospheres and coronas, and that has been found to be the case.

7.9. Chromospheres in other stars

The presence of this activity in other stars may be detected by

1. self-reversal (emission cores) in the CaII K line
2. absorption or emission in HeI $\lambda 10830$
3. emission lines in UV spectra
4. soft X-ray emission

Only the first of these is a direct measure of chromospheres; the last three depend on coronal emission measure, but all are reasonably well correlated. These effects would be only marginally detectable on the Sun if it were at stellar distances, so there is a tendency to study stars more active than the Sun. This could mean that the chromosphere as a whole is hotter and denser, or that active regions are more frequent or extensive than on the Sun. The evidence favors the latter.

The level of activity, presumably magnetic, has been shown to depend on the age of the star. Activity is also high in double stars, and higher still in close double stars. There is particularly strong activity in RS CVn stars, evolved binaries where one star has overflowed its Roche lobe in the course of evolution. These stars, with their atmospheres stripped away, appear to reveal even more intense activity in their inner parts. But they are highly evolved and cannot be compared to the Sun.

The K line in a star of high activity shows the symmetric double peaked profile characteristic of solar plages. In Figs. 7.8–7.10 we see that the chromospheric K-line profile is dominated by emission in the blue K2v component. One therefore concludes that increased activity in stars is connected with greater plage intensity or area, rather than a major increase in chromospheric emission measure.

Wilson and Bappu (1957) studied the double peaks of K-line emission in cool stars of spectral class G, K and M emission. They found a direct relation between the absolute luminosity of these stars and the separation of the two emission peaks. This is probably the only real *chromospheric* function generated by solar-stellar studies so far. Wilson (1963) later found the K-line intensity of a star to be strongly connected with youth and high rotation rate, a result subsequently supported by the work of many authors. The Wilson–Bappu effect suggests some connection between the scale height of the atmosphere (luminosity is directly connected with extended atmospheres and low gravity) and the breadth of the absorption. It is tempting to speculate that chromospheric velocities are greater under low-gravity conditions, thus producing the Wilson–Bappu effect.

Because HeI is excited by coronal UV emission, the $\lambda 10830$ line can be used to sample coronal emission measure. Zirin (1982) measured the line in 455 stars, finding the Sun to be about average. Coronal activity is most frequent in binaries and in F and M dwarf stars and strongest in G and K giants and supergiants. In G and K stars there is a good correlation between the $\lambda 10830$ and K-line equivalent widths, but in M stars there is none, because there is little He absorption in M stars (with the possible exception of dwarfs). The K line appears relatively more intense in M stars because the peak of the Planck curve is out around 9000 Å, so the K line behaves like the UV lines on the Sun. Since the K line measures the extent and brightness of plages on the stars and $\lambda 10830$ measures the corona, this means that the two are correlated on stars as they are on the Sun.

$\lambda 10830$ has a broad profile in supergiants and a narrow one in dwarfs. Thus the chromospheric turbulent velocity depends on scale height, and the absorption by this moving material probably produces the Wilson–Bappu effect. Kraft *et al.* (1964) showed that Hα line width in stars is also correlated with the absolute magnitude, no doubt because of the same effect.

Although the K line in the more active stars tends to resemble that of plages, it is important to understand how much of the peak is due to enhanced chromospheric emission inside the network. Skumanich *et al.* (1984) tried this for the K-line data for the integrated Sun obtained by White and Livingston (1981). They somehow end up with a Sun 80% covered with K emission, which simply does not agree with reality. It would be more useful directly to measure spectroheliograms and filtergrams; superficially I see no variation. The network K-line spectrum is indistinguishable from that in plages, except for the brightest, while the IN elements show the characteristic excess of K2v over K2r. If there were a relation between the quiet network and the cycle, it would mean that local field eruption increases near spot maximum.

In stars earlier than spectral class F, it is thought that the atmosphere is fully ionized and no convective zone exists. On the other hand, it is in the A stars, which are even hotter, that the greatest stellar magnetic fields are observed; fields as high as 5000 gauss have been found. In almost all cases the field appears to alternate between north and south polarity, in or out of phase with various spectrum lines. It is as though these large stars are completely covered with enormous spots 1000 times larger than the biggest sunspot. A substantial fraction (10%) of A stars show peculiar spectra, but not all are magnetic variables. XUV and other observations have confirmed the absence of chromosphere in these stars, so the magnetic fields must be due to something other than the interaction of convection and rotation.

Following Wilson's (1978) survey of temporal changes in activity in the CaII lines, Vaughan *et al.* (1978) developed a special photometer for measurement of H and K line intensities with the 60-inch reflector on Mt. Wilson. For the first time, detailed long-term synoptic observation of physical properties of stars was carried out. The surprising result of these observations (Vaughan *et al.* 1981) was that the HK intensity distribution was so non-uniform that rotational modulation was observable and accurate stellar rotation rates could be measured. Further, the phase of the modulation remained unchanged for long periods. By contrast, the chromospheric emission on the Sun is so uniformly distributed that rotational modulation is barely detectable and the phase changes rapidly as activity complexes come and go. And certainly whatever modulation is observed has to do with active regions rather than chromospheres, which would not show a rotation-modulated signal.

Noyes *et al.* (1984) tabulate the most recent determinations of stellar rotation periods and HK intensities from the Mt. Wilson survey. They propose that the level of chromospheric activity should be proportional to the Rossby number, the ratio of the rotation rate of a star to the convective overturn rate. In the terrestrial atmosphere this may be thought of empirically as the number of eddies in a hemisphere. Because the convective overturn rate depends on the ratio α of mixing length to scale height, they plot the HK emission index as a function of Rossby number for different values of α and find an excellent fit for $\alpha = 2$, a slightly worse fit for $\alpha = 3$ and a poor fit for $\alpha = 1$. It is interesting that $\alpha = 2$ corresponds to current values for the depth of the convective zone derived from helioseismology. On the other hand Noyes *et al.* give a plot of the HK index as a function of rotation period alone which looks as good as the correlation with Rossby number, so the matter is not completely settled. In any event the HK index in these stars may be more a measure of magnetic activity than chromosphere.

Extensive studies of the EUV emission from stellar chromospheres have been carried out with the IUE explorer spacecraft (Chapman 1981; Mead *et al.* 1984). The ultraviolet lines show that in some stars the final rise to temperatures above $100\,000°$ does not take place, even though chromospheric (up to $20\,000°$) lines are observed. These stars, which also do not show $\lambda 10830$ absorption, are usually M giants and supergiants. In some stars, mostly giants and supergiants, $\lambda 10830$ emission is observed. One might think that this is due to a superhot and dense chromosphere, but these stars do not show hydrogen emission. It is thought that the emission arises in a shell, but the absence of hydrogen emission still must be explained.

Summary. The temperature of the solar atmosphere drops to a minimum of 3500°–4000° at a few hundred km above the photosphere. Above this point the atmosphere is no longer in hydrostatic equilibrium. The temperature above magnetic elements of the network rises steeply to coronal values, but the temperature increase in the regions in between is more moderate. The chromospheric network is probably the result of the supergranulation flows.

The magnetic elements are also the site of spicules. Coronal loops extend outward from the spicule bushes and form the "quiet" corona. The spicule jets apparently establish the interface between chromosphere and corona; the interface inside the network is not known. There are no spicules over plages, where the chromospheric emission measure is particularly high.

References

Athay, R. G. 1976. *The Solar Chromosphere and Corona: Quiet Sun.* Dordrecht: Reidel.

Athay, R. G. (ed.) 1973. *Chromospheric Fine Structure.* IAU Symp. No. 56.

Bray, R. J., and R. E. Loughhead 1974. *The Solar Chromosphere.* London: Chapman and Hall.

Chapman, R. D. 1981. *The Universe at Ultraviolet Wavelengths.* NASA Conference Publ. No. 2171.

Leighton, R., R. W. Noyes, and G. W. Simon 1962. *Ap.J.* **135**, 474.

Mead, J. M. *et al.* 1984. *Future of Ultraviolet Astronomy Based on Six Years of IUE Research..* NASA Conference Publ. No. 2349.

Pagel, B. E. J. 1964. *Ann. Rev. Astron. Astrophys.* **2**, 267.

Thomas, R. N., and R. G. Athay 1961. *Physics of the Solar Chromosphere.* New York: Interscience.

Zirin, H., and R. D. Dietz 1963. *Ap.J.* **138**, 664.

8

The Solar Corona

8.1. Structure

With the possible exception of the sunspot cycle, the biggest enigma in our study of the Sun is the existence of a million-degree atmosphere. The mystery of the solar corona is enhanced by the fact that until recently it could only be detected at eclipses or with coronagraphs, and then only at the limb. Figs. 8.1 and 8.2 give examples of the marvelous appearance of the corona at eclipses.

Newkirk and Altschuler (1970) compared the eclipse corona with the potential field distribution obtained by using the photospheric magnetic fields as a source (Fig. 8.2). These images show clearly how the coronal structure is shaped by magnetic fields. The coronal holes appear where the field lines are open, while the streamers occur when two unipolar regions are magnetically connected. At great distances from the Sun, the outer magnetic flux loops, instead of connecting on the Sun, extend outward into the solar system, forming what we call a helmet streamer, after the Kaiser's World War I helmets. A filament is often found at the bottom, separating the unipolar regions. Most of the features that do not match are due to the fact that the streamers are very high and are seen in the plane of the sky even though they may be in front of or behind the limb.

Extension of astronomical detectors to new wavelengths has made it possible to study the corona against the disk. Because of its high temperature the corona radiates strongly in the ultraviolet and soft X-ray range, which may be studied from spacecraft using Wolter-type telescopes (Sec. 2.1). Excellent UV coronal images down to 200 Å are obtained with the "overlappograph" (Fig. 7.25). The corona is optically deep in radio waves longer than 1 meter and may be mapped from the ground with radio arrays. Radar pulses at longer wavelengths are reflected by the corona and also used to analyze its properties. Eclipse photos permitted observations only to $2R_\odot$; externally occulted coronagraphs on spacecraft now permit observations to $5R_\odot$.

We use the following model parameters for the corona; they probably are within a factor two of truth:

$$T = 10^6 \deg \quad v_{\text{therm}} = 5.5 \times 10^8 \quad \text{cm/sec} \quad \nabla \mathbf{A} = 0$$
$$H = 5 \times 10^9 \text{cm} \quad N_0 = 5 \times 10^8 \quad \text{particles/cm}^3 \quad \mathbf{j} = 0 \tag{8.1}$$

Figs. 1.3, 8.3 and 8.4 show soft X-ray (SXR) images of the corona. The emission depends on $N_e^2 \exp(-h\nu/kT)$, so the images enhance the actual temperature and density contrast. Almost all the emission is in bright magnetic loops. The brightening in active regions (AR) is due to the combined effects of heating and trapping. Thus, instead of pressure equilibrium, the corona has the peculiar characteristic that density and temperature usually increase together, and the pressure disequilibrium is contained by the dominant magnetic field. The SXR pictures show many bright points, which correspond to a range of very small AR's. All of the bright points show some Hα brightening in the chromosphere. The brightening above the limb is due to the path doubling; there is little brightening inside the limb because the structure has the same scale as the slant path.

8.1. The solar corona photographed at Palem, India 16 Feb 1980 by a team from High Altitude Observatory and Southwestern at Memphis, using a special camera developed by G. A. Newkirk, Jr. The camera has a radial density gradient to preserve the coronal morphology. Note the helmet streamers over limb prominences. Since this is near sunspot maximum there are no polar coronal holes, the corona is seen all around the Sun. (HAO)

Fig. 8.4 compares the SXR structure of the corona with the underlying Hα structure; magnetograms show that coronal loops run from one magnetic pole to the opposite (what else should they do?), just like Hα fibrils; the difference is only that the X-ray loops are visible over their entire trajectory, while Hα is not emitted from the corona. Why is one flux loop filled with bright material and the others, empty? The bright loops are associated with some magnetic activity at their base which feeds material into the corona (Webb and Zirin 1981). There is an EFR only twelve hours old at lower left in Fig. 8.4. In Fig. 8.4(b) we see it already had captured a coronal loop and was connected to the big active region at B. The points C and D mark the footpoints of loops.

Although large sunspots are always associated with enhanced X-ray emission, the bright loops never end in the spots themselves, either because no heating takes place there, or because the sunspot cools the corona along that magnetic path. The X-ray image of an active region is one of the few images that does not match the Hα or K-line image. The latter are bright

8.2. Superposition (Newkirk and Altschuler 1970) of the solar corona photographed Nov. 12 1966 on an Hα image of the disk. The lines of force corresponding to a current-free fit to the photospheric fields are indicated. We see that the polar coronal holes correspond to open field lines, while some of the bright streamers correspond to closed magnetic regions. Because of the great height of the streamers, some may be associated with features behind the limb. (HAO)

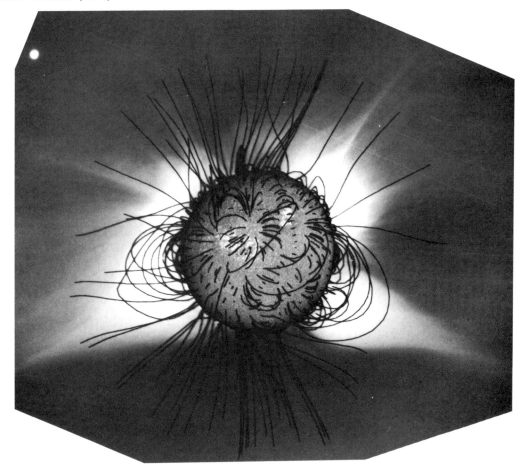

where the field is vertical, while the X-rays are bright over the neutral lines. Prominences in AR's will often have a coronal condensation above them. The Hα spicule structure at the base of coronal loops points along them; so the loops do mark the lines of force. Bright corona requires (1) an energy source, usually flux emergence and (2) a trap, closed loops or a filament. Thus the coronal emission usually is brightest over the neutral lines, where Hα is dark.

The peaking of X-ray intensity at the loop tops may be understood in terms of conduction, the tops being farthest from the cool surface and hence the hottest part of the loop. But in post-flare loops the density also peaks at the loop tops, and that violation of hydrostatic equilibrium is not understood. It must be a dynamic phenomenon associated with the condensation from the corona there, and is quickly corrected by the material pouring down. Loops may connect opposite polarities in different AR's: often they will cross the equator. Although one would think that

8.3. A large coronal hole photographed from a rocket flight 27 June 1974 with X-ray passbands 8 – 39 and 44 – 64 Å. Bright points are easily visible against the hole. The bright points are all tiny dipoles, spotless active regions, but their exact relation to ER's is still unclear. (AS&E)

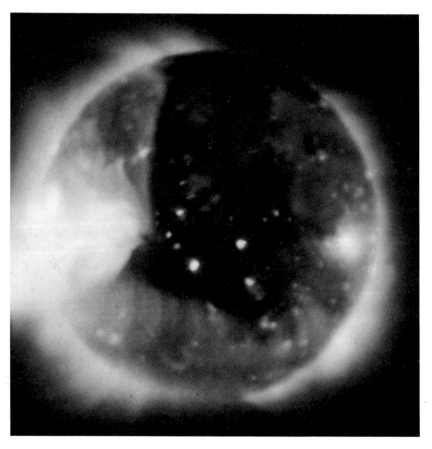

N

W

8.4. An active region complex July 6 1973 (see Figs. 10.9–12) in Hα (above) and X-rays (below). (a), (b) view of three interconnected regions. Arrows B, C, and D mark the termination of the coronal loops. Although the emerging region at the lower left of (a) and (b) is only 12 hrs old it is already connected to the others by a coronal loop ending at B. The arrows in (c) mark small pores covered by X-ray emission in (d). The brightest arches in (d) lie over AR filaments; A marks an arcade of X-ray loops over the largest filament, which erupted in Fig. 9.13. (Webb and Zirin 1981).

reconnection should be slow in the high conductivity of the corona, new connections form within hours of the emergence of new sunspot groups (Fig. 8.4). The new fields seem to capture existing lines of force going to more distant points.

Rosner, Tucker and Vaiana (1978) pointed out that a loop in hydrostatic equilibrium must have its temperature peak at the loop top if it is to be stable. They introduced the dimensionless RTV scaling law:

$$T \approx 1.4 \times 10^3 (pL)^{\frac{1}{3}} = 1.4 \times 10^3 (2n_e kTL)^{\frac{1}{3}} = 0.092 (n_e TL)^{\frac{1}{3}} \deg \quad (8.2)$$

where p is in dyn cm^{-2} and L, in cm (the factor 2 accounts for proton and electron pressure). For $n_e = 5 \times 10^8$ and $L = 10^{10}$ cm, this gives $T = 1.5 \times 10^6$. It is roughly valid for hot active-region loops as well, but gives values for post-flare loop temperatures that are too high. However

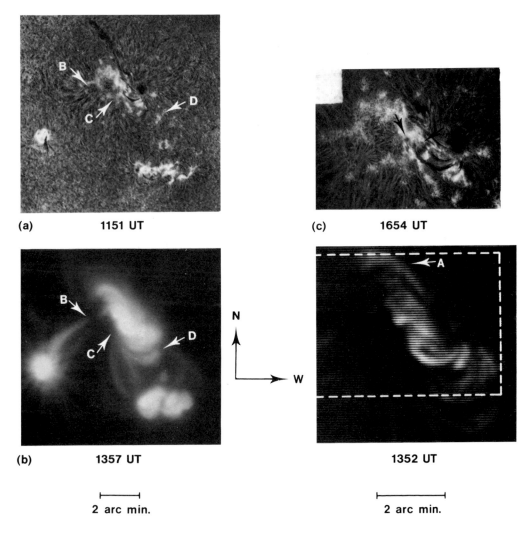

(a)   **1151 UT**

(c)   **1654 UT**

(b)   **1357 UT**

N

W

(d)   **1352 UT**

2 arc min.   2 arc min.

many loops of fixed geometry change temperature as the result of heating; whether the RTV scaling holds then is not established.

The corona evolves slowly except for flare-connected phenomena and new flux eruption. The loops change only as fast as the surface magnetic features that they connect, but the relative brightness of loops can change rapidly in response to activity. Flares and filament eruptions produce coronal changes within seconds, and waves of excitation have been seen to move from point to point (Rust and Webb 1977).

The most striking feature of the X-ray photos of the corona are the great dark areas called coronal holes (Figs. 1.3, 8.3, 8.17). In coronal holes the temperature is somewhat lower, and the density, a factor of ten lower than the values in the normal corona. Coronal holes occur in regions of open field lines where material can easily flow outward into the solar wind. Thus a high emission measure in the corona is only possible where closed magnetic fields keep the hot gas from flowing out, and the patterns of brightness we see result from the interplay of heating and trapping in the corona.

The larger scale coronal structure is revealed by eclipse pictures and ex-

8.5. A plot of the coronal continuum intensities *vs.* radial distance. Although the E corona is much fainter than the K, we can detect it by limiting the wavelength range. (After Van de Hulst 1953).

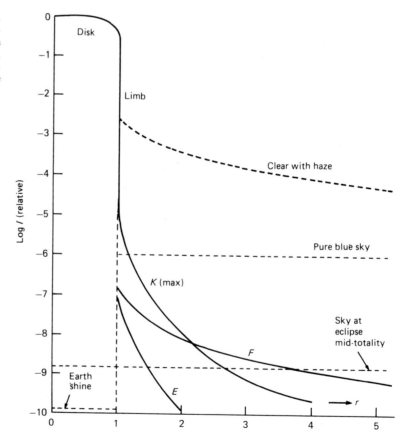

ternally occulted coronagraphs, which permit observation of the coronal continuum out to several solar radii over long periods. Long duration observations (Saito *et al.* 1977) show the K corona (Sec. 8.2) to be dominated by streamers and holes, although occasionally no structure of this sort was seen. One streamer was followed through five rotations by Poland (1978). The outer structure of the corona reflects the interplay of the solar wind above $5R_\odot$ with the magnetic field and the surface sources. Externally-occulted satellite coronagraphs permitted extended observations of the corona for the first time, revealing many coronal mass ejections (CME), great waves or bubbles of higher density (Kahler *et al.* 1978) propagating outward through the corona. These transients (Figs. 8.21–3) are mainly caused by flares and erupting prominence. They account for about half the total outflow from the Sun.

In the inner corona the magnetic field dominates the plasma, while in the outer corona the radial solar wind flow dominates everything. The coronal structure therefore reflects the structure of the general solar magnetic field. The shape of the corona varies with the phase of the 11-year sunspot cycle. At minimum the large-scale solar field is a dipole, and the open polar field lines produce coronal holes at the poles. Emission is limited to lower latitudes. Long streamers extend far out along the equator, marking the extension of surface fields by the solar wind. In the polar holes, fine thin streamers following the polar fields may be seen. At sunspot maximum the overall solar field is no longer poloidal (Section 12.3), the polar holes disappear and there is bright corona all around the limb. Active regions produce coronal condensations and their magnetic remnants give rise to helmet streamers.

8.2. The continuous spectrum

Because of its million-degree temperature the corona is a plasma of electrons and protons, with some heavier ions thrown in. The mean molecular weight is half that of a proton. The continuous emission at the low coronal densities is negligible compared to the photosphere except for the radio region. We therefore expect a continuous spectrum dominated by Thomson scattering of the bright photospheric emission by free electrons. Such scattering should fall off with a scale height corresponding to a million degree gas with mean molecular weight $m_p/2$, and that is what we observe in Figs. 8.1 and 8.2. Fig. 8.5 shows a plot due to van de Hulst of the radial dependence of various kinds of coronal emission in the visible. Note that only the K corona is visible against the pure blue sky. In fact, we never have such a dark blue sky, so none of these emissions is visible in integrated light. The corona close to the Sun has about the same intensity as the full Moon.

During a total eclipse the local sky is only illuminated by light from outside the eclipse path; its brightness drops to the level shown in Fig. 8.5 and the corona may be seen. The continuum observed consists of two components: the K corona (*Kontinuierlich*), which is polarized and free of absorption lines, and the F corona (*Fraunhofer*), which is unpolarized and shows all the Fraunhofer absorption lines of the photospheric spectrum. The K component is observed at all position angles and decreases much more rapidly with height than the F corona, which is concentrated toward the plane of the ecliptic. Emission lines are also observed, and referred to as the E corona. The integrated E corona, as shown in Fig. 8.5, is much weaker than the continua, but in the wavelengths of the lines its intensity is much greater than the continuum.

The K component is produced by Thomson scattering of photospheric light by coronal electrons. The thermal electron velocity is 1/60 the speed of light, so the broadening of Fraunhofer lines is 1/60 of their wavelength and they are all washed out. The F component, on the other hand, has nothing to do with the solar atmosphere; it is simply the inner zodiacal light, the glow produced by diffraction of sunlight by solid particles in the

8.6. An image of the corona taken from the lunar surface by the Surveyor spacecraft. The F corona (really the inner zodiacal light) could be traced out to 60 solar radii. The planet at the right is Venus. (Norton *et al.* 1967)

plane of the ecliptic (Fig. 8.6). These interplanetary solid particles can be anywhere between Sun and Earth. They are heavy and slow moving, and the diffracted absorption lines are not washed out. As we look at the Sun we have no idea where the scattering particles lie, so they appear to be part of the corona. The falloff in brightness is due to the angular properties of the diffraction. If the sky is dark we can see the zodiacal light visually just after sunset or before sunrise.

The components of the corona may be separated by polarimetry; the white light coronameters discussed in Chap. 2 also use this technique to separate the K corona from the sky as well. Then the integral equation for the scattering is solved for the density distribution that fits the data. In Figs. 8.1 and 8.2 we see that even at eclipse the corona is visible only to a few solar radii. This is because light from outside the eclipse path illuminates the sky. To see the outer corona we must go into space and use externally occulted coronagraphs (Sec. 2.2).

Simple arithmetic immediately gives us rough values for the coronal density. The K corona has intensity $10^{-5}I_{\odot}$ near the surface, produced by scattering of photospheric light by electrons along about one solar radius. Using Eq. (4.58):

$$I_{\text{scatt}} = \sigma_T \times N_e R_{\odot} I_{\odot}$$
$$10^{-5}I_{\odot} = 3.3 \times 10^{-25} N_e \times 7 \times 10^{10} I_{\odot} \tag{8.3}$$

which gives $N_e = 4 \times 10^8$ at the base of the corona. The run of density is given by the barometric equation Eq. (6.11). For constant temperature (maintained along a flux loop by the high conductivity), the run of density in the corona in hydrostatic equilibrium is given by

Table 8.1. Newkirk's (1961) model.

r	$N_e(10^8)$ Quiet	$N_e(10^8)$ Polar	$N_e(10^8)$ Active
1.000	900.*	430*	1630*
1.125	290.	140.	570.
1.250	120.	57.	240.
1.375	57.	27.	114.
1.500	33.	16.	61.
1.625	20.	9.5	35.
1.750	13.	8.2	21.3
1.875	8.8	4.2	13.2
2.000	8.2	2.9	13.

*Extrapolated

$$N = N_0 \exp - \left(\frac{h}{H} \frac{r^2}{R_\odot^2} \right), \qquad (8.4)$$

where we have adjusted the base scale height for the variation in gravity with radius. This leads to a finite density at infinity, which is one reason for the solar wind; but it is a good fit close to the Sun.

The electron density distribution has been solved in detail by Bogorodsky and Khinkulova (1950) and by Van de Hulst (1950), taking into account the polarization geometry. Newkirk's model (Table 8.1) based on radio and optical data is commonly used. Separate values are given for the average corona, the polar caps and active regions. Newkirk gives the approximate formula

$$N_Q = 4.2 \times 10^4 \times 10^{4.32/r}. \qquad (8.5)$$

This is close to the hydrostatic equilibrium distribution Eq. (8.4) with $T = 1.2$ million deg, for which Eq. (6.13) gives $H = 77\,000$ km.

8.3. Coronal emission lines

Superposed on the coronal continuum are many emission lines of intensity up to $2 \times 10^{-4} I_\odot$. In the far ultraviolet the coronal emission lines dominate the spectrum.

The identification of the coronal emission lines in the visible is a fascinating story. Most of them were observed at various eclipses, and their wavelengths accurately determined. By 1930 astronomers were in the embarrassing position of observing a number of well-defined emission lines of unknown origin, which were attributed for simplicity to "*coronium*".

A number of possible candidates were advanced: CaIII, doubly excited helium (which we know to be unstable), neutral oxygen, and so on; but none were plausible. Few guessed that the corona might be hot; it was assumed that the atmospheric temperature kept on dropping toward zero. Astronomers were aware of the great coronal scale height, but because they had no idea how or why the necessary high temperature might arise (we still have no good explanation), they decided that either there was an unknown source of support or that the corona must consist of an extremely light gas, perhaps just electrons, which would have the same density distribution as a heavier gas at a higher temperature. As we now know, a pure electron gas could not exist, for it would have a colossal static charge which would suck up ions from below until it was neutral. Besides, as S. A. Mitchell (1923) pointed out, the electrons would have to collide with and excite something in order to radiate the coronium lines.

Meanwhile, three important pieces of data were obtained which allowed the puzzle to fall into place when the right man came along. At Mt. Wilson

Observatory Adams and Joy (1933) discovered that the recurrent nova RS Ophiuchi (a binary where periodic accretion episodes on a white dwarf produce optical and X-ray bursts) showed some of the coronal lines. In Sweden, B. Edlén carried out a series of fundamental studies of the UV spectra of highly ionized atoms; and Bowen and Edlén (1939) showed that lines in the spectrum of the nova RR Pictoris agreed with wavelengths that Edlén had found for FeVII.

The evidence was put together by Walter Grotrian of Potsdam. Grotrian had long held the unpopular view that the corona was rather hot, basing his view on absence of lines from the K corona, which meant that the temperatures must be high enough to give big electron velocities. In a short note in *Die Naturwissenschaften* (Grotrian 1939), he pointed out that the existence of coronal lines in high-excitation objects as ordinary and recurrent novae implied that these lines arise under conditions of high temperature. Even better, he showed that there was an almost exact correspondence between the wavelength of the red coronal line at 6374 Å and that derived from Edlén's measurement of the separation of the levels of different J-values in the ground state of FeX. Edlén (1937) had not measured this splitting directly; the line is too weak to measure in the laboratory. Instead he had measured the wavelengths of four permitted lines of FeX produced in a hot spark. Each pair of permitted lines connected a common upper level with the two split lower levels, permitting the determination of the separation between the lower levels.

8.7. Term diagram for FeX. The four lines near 95 Å were observed by Edlén, and the difference (Table 8.2) gave the ground–state splitting which agreed with the wavelength of the "coronium" line λ6374.

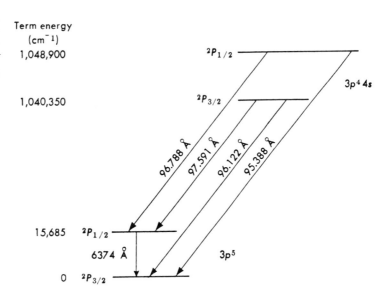

The term structure of FeX is sketched in Fig. 8.7, with the wavelengths attached; the data for the lines identified by Grotrian are given in Table 8.2. The first column gives the terms of the lines observed by Edlén in the laboratory; the second and third give the wavelength and inverse wavelength. The difference between the pairs of laboratory lines is given in the fourth column. For FeX we see that the splitting of $3s^2 3p^5$ ^3P is determined by taking the difference between lines from the upper $3s^2 3p^4 4s$ ^2P term in two ways, giving $15\,687$ cm^{-1}, which coincides almost exactly with the observed reciprocal wavelength of the red coronal line $\lambda 6374$. Levels of higher J lie higher, except when electron shells are more than half full, as in this case.

Grotrian also identified the strong line at 7892 Å with a transition in the ground term of FeXI, as shown in Table 8.2. These transitions were unlike any previously observed, in that they were magnetic dipole transitions between different J-levels of the same term of orbital angular momentum L

8.8. Emission from active regions at the limb in the FeXIV 5303 Å line, using a narrow-band filter and coronagraph. The loops at lower left are a post-flare phenomenon. (SPO)

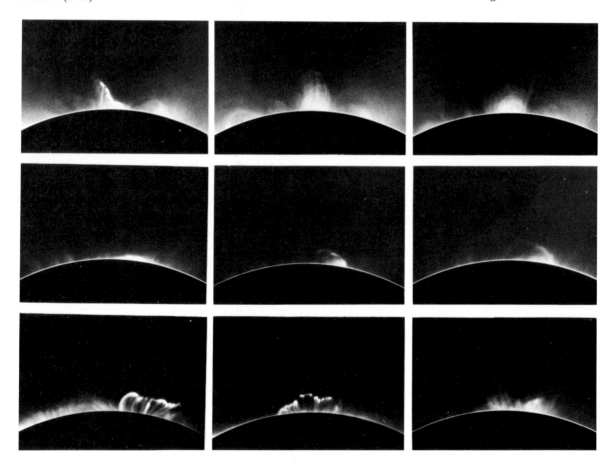

and spin S. In normal atoms these levels are so close that the transitions
between them are far in the infrared; but the spin–orbit interaction splitting them increases as Z_{eff}^4, so in highly ionized atoms the separation is
large enough to give rise to lines in the visible. The transition probabilities
are given in Table 8.3 and can be derived from Eq. (5.18); for lines in the
visible they range from 10 to 100 sec^{-1}.

Spurred by Grotrian's discovery, Edlén (1942) searched for further coin-
cidences and found them for the coronal lines $\lambda3328$ and $\lambda4086$, which he
could identify with transitions in CaXII and CaXIII. The data taken from
Edlén's (1942) basic paper are given in Table 8.2. Although Edlén had
direct experimental evidence only for the ions in this table, he was able to
identify most of the other strong lines by extrapolating wavelengths along
isoelectronic sequences. For example, the ArI isoelectronic sequence has
the configuration $1s^2 2s^2 2p^6 3s^2 3p$ and a ground term ^2P which splits into

Table 8.2. Identification of coronal lines (Edlén, 1942).

UV Transition	Wavelength		Level Splitting	Wave Number, Coronal Line
FeX $3s^2 3p^4 4s \rightarrow 3s^2 3p^5$	λ, Å	$\bar{\nu}$, cm^{-1}	cm^{-1}	cm^{-1}
$^2P_{1/2} \rightarrow {}^2P_{3/2}$	95.338	1 048 900		
$^2P_{1/2} \rightarrow {}^2P_{1/2}$	96.788	1 033 186	15 714	
$^2P_{3/2} \rightarrow {}^1P_{3/2}$	96.122	1 040 345		
$^2P_{3/2} \rightarrow {}^2P_{1/2}$	97.591	1 024 685	$\underline{15\ 660}$	
			15 687	15 683
FeXI $3s^2 3p^3 4s \rightarrow 3s^2 3p^4$				
$^3D_2 \rightarrow {}^3P_2$	87.025	1 149 095		
$^3D_2 \rightarrow {}^3P_2$	87.995	1 136 428	12 667	
$^3S_1 \rightarrow {}^3P_1$	89.185	1 121 265		
$^3S_1 \rightarrow {}^3P_1$	90.205	1 108 586	$\underline{12\ 679}$	
			12 673	12 668
CaXII $2s^2 2p^6 \rightarrow 2s^2 sp^5$				
$^2S_{1/2} \rightarrow {}^2P_{3/2}$	141.036	709 039		
$^2S_{1/2} \rightarrow {}^2P_{1/2}$	147.273	679 011	30 028	30 039
CaXIII $2s2p^5 \rightarrow 2s^2 sp^4$				
$^3P_2 \rightarrow {}^3P_2$	161.748	618 246		
$^3P_2 \rightarrow {}^3P_1$	168.412	593 782	24 464	24 465

$J = 3/2$ and $J = 1/2$. Edlén had experimental data on the members of the series through ScIX, and he knew that theoretically the splitting should increase as Z_{eff}^4. He therefore extrapolated the splitting of FeXIV and NiXVI, predicting lines at 5304 and 3602 Å. These wavelengths matched the green coronal line at 5303 Å (Fig. 8.8) and a weaker line at 3601 Å. In this way Edlén was able to identify many coronal lines, and since then the list has been further augmented. Since both Grotrian and Edlén published in German journals during the war, the information was carried out of Europe informally and not widely known until after World War II.

A compilation of coronal line identifications and brightnesses (Jefferies *et al.* 1968) is given in Table 8.3, with transition probabilities and the ionization potential necessary to ionize the next lower ion. The relative intensities are not really comparable; because high ionization lines appear only in active regions their intensity (in parentheses) is brighter because of high density and temperature in the active regions. Other lines have been observed but not conclusively identified; these exist only as barely visible specks on eclipse plates, and we have omitted them from the present list. Edlén (1969) has shown that the ArX line is probably at λ5533.

Table 8.3. Important forbidden coronal lines in the visible (Jefferies *et al.* 1971).

λ(Å)	Transition	A_{21} (sec^{-1})	Excit. Pot.	Ioniz. Pot.	Int*
3328	CaXII $2p^5$ $^2P_{1/2} \to$ $^2P_{3/2}$	488	3.72	589	(17)
3388	FeXIII $3p^2$ $^1D_2 \to$ $^3P_2$	87	5.96	325	37
3601	NiXVI $3p$ $^2P_{3/2} \to$ $^2P_{1/2}$	193	3.44	455	(18)
3642.9	NiXIII $3p^4$ $^1D_2 \to$ $^3P_1$	18	5.82	350	1.5
3986.9	FeXI $3p^4$ $^1D_2 \to$ $^3P_1$	9.5	4.68	261	
4086.3	CaXIII $2p^4$ $^3P_1 \to$ $^3P_2$	319	3.03	655	(22)
4231.4	NiXII $3p^5$ $^2P_{1/2} \to$ $^2P_{3/2}$	23	2.93	318	8
4412	ArXIV $2p$ $^2P_{3/2} \to$ $^2P_{1/2}$	112	2.84	682	16
5116.03	NiXIII $3p^4$ $^3P_1 \to$ $^3P_2$	157	2.42	350	2
5302.86	FeXIV $3p$ $^2P_{3/2} \to$ $^2P_{1/2}$	60	2.34	355	190
5445	CaXV $2p^2$ $^3P_2 \to$ $^3P_1$	83	4.45	814	(15)
5539	ArX $2p^5$ $^2P_{1/2} \to$ $^2P_{3/2}$	106	2.24	421	5
5694.42	CaXV $2p^2$ $^3P_1 \to$ $^3P_0$	95	2.18	814	(28)
6374.51	FeX $3p^5$ $^2P_{1/2} \to$ $^2P_{3/2}$	69	1.94	233	40
6701.83	NiXV $3p^2$ $^3P_1 \to$ $^3P_0$	57	1.85	422	(27)
7059.62	FeXV $3s3p$ $^3P_2 \to$ $^3P_1$	38	31.77	390	5
7891.94	FeXI $3p^4$ $^3P_1 \to$ $^3P_1$	44	1.57	261	50
8024.21	NiXV $3p^2$ $^3P_2 \to$ $^3P_1$	22	3.39	422	
10746.80	FeXIII $3p^2$ $^3P_1 \to$ $^3P_0$	14	1.15	324	100
10797.95	FeXIII $3p^2$ $^3P_2 \to$ $^3P_1$	9.7	2.30	325	50

Since the ultraviolet was opened up a number of additional coronal lines have been identified. These are particularly valuable because they permit imaging of emission from some of the highest stages of ionization, normally observable only in the region where grazing incidence imaging must be used. A particularly remarkable achievement was the flight of a rocket spectrograph into the eclipse path in March 1970 (Speer *et al.* 1970) which permitted the recording of 28 coronal lines between 977 Å and 2200 Å. Jordan (1971) identified 20 of these with forbidden transitions between common coronal ions. Table 8.4 gives a list of these and other coronal lines in this wavelength range compiled by Feldman and Doschek (1977). These lines differ from those in the visible in that they are transitions between different terms (like λ 3388 of FeXII) rather than between different J-levels of the same term. But the terms are perturbed by spin–orbit interaction, resulting in intermediate coupling (Sec. 5.5), and the transitions are still magnetic dipole.

Not all of the abundant elements give rise to forbidden coronal lines. H and He are completely ionized in the corona; C, N and O are ionized down to the electrons of the innermost shell, where LS splitting is small. The same is true of the possible ions of Ne, Mg, and Si. Coronal forbidden lines are typically found in the ground configuration is p, p^2, p^4 or p^5.

The permitted lines of coronal ions in the extreme UV fall in three classes: the screening transitions of the Li and Be isoelectronic sequences between 300 Å and 1000 Å; the permitted transitions of FeIX to XVI, all of the form $3s^2 3p^n \rightarrow 3s3p^{n+1}$ or $3s^2 3p^n \rightarrow 3s^2 3p^{n-1} 3d$, which lie in the 170 Å to 400 Å range; and many weaker lines shortward of 170 Å due to lines with change of n. These are observed all the way down to the SXR region, where one finds the resonance lines of He-like ions, such as MgXI $1s^2 \rightarrow 1s2p$ at 9.31 Å. The lines shortward of 70 Å are tabulated by Tucker and Koren (1971).† The He-like ions for $Z > 8$ appear in solar flares and will be discussed in Chap. 11. The SXR lines are normally not particularly strong; a 10 Å line requires about 1000 eV, while kT is only 100 eV. So the lines are sensitive indicators of temperature. Despite the fact that the UV lines are permitted, they are far less bright than the forbidden lines in the visible because the latter re-emit the strong visible photospheric flux.

The permitted UV lines have a wide range of ionization potentials and one can use them to trace the temperature structure. In the overlap-pograms (Fig. 7.25) the low-ionization lines like FeIX are found around the entire limb (except for the coronal holes) while a high-ionization line like FeXV at 283 Å is seen only in the AR's. Moreover, in the AR's (Fig. 8.12) there is a clear progression from the low-ionization lines prominent at the

---

† Note that Tucker and Koren mistakenly attribute the 2 ^3S level in the He-like ions to $1s2p$, but it is $1s2s$.

Table 8.4. Forbidden lines between 1100 Å and 2000 Å (Feldman and Doschek 1977)

Table 1 — Forbidden Lines Between 1100 and 2000 Å (Spectrum A)

Ion and Transition		λ solar (Å)	λ^{*} cal (Å)	Intensity at Sun erg cm^{-2}s^{-1}st^{-1}	W_O(Å)**	"T_i"(K)
Mg VII	$2p^2\ {}^3P_1$ $-2p^2\ {}^1S_0$	1189.84	1189.7	1.4	0.20	1.2(6)$^+$
Mg VI	$2p^3\ {}^4S_{3/2}-2p^3\ {}^2P_{3/2}$	1190.07	1190.1	2.4	0.20	1.2(6)
Mg VI	$2p^3\ {}^4S_{3/2}-2p^3\ {}^2P_{1/2}$	1191.64	1190.6	0.8	0.19	1.1(6)
S X	$2p^3\ {}^4S_{3/2}-2p^3\ {}^2D_{5/2}$	1196.26	1196.9	0.62	0.20	1.6(6)
S X	$2p^3\ {}^4S_{3/2}-2p^3\ {}^2D_{3/2}$	1213.00	1213.6	3.1	0.22	1.9(6)
Fe XIII	$3p^2\ {}^3P_1$ $-3p^2\ {}^1S_0$	1216.46	1217.2	(1.5)$^{++}$	(0.25)$^{++}$	(2.5(6))$^{++}$
Fe XII	$3p^3\ {}^4S_{3/2}-3p^3\ {}^2P_{3/2}$	1242.03	1242.05	13.7 (5.6)	0.17 (0.20)	1.8(6) (2.6(6))
Ni XIII	$3p^4\ {}^3P_1$ $-3p^4\ {}^1S_0$	1277.23	1277.0	0.42	0.18	2.0(6)
Mg V	$2p^4\ {}^3P_1$ $-2p^4\ {}^1S_0$	1324.45	1324.4	0.26	0.17	6.8(5)
Fe XII	$3p^3\ {}^4S_{3/2}-3p^3\ {}^2P_{1/2}$	1349.38	1349.51	8.4 (4.1)	0.23 (0.20)	3.0(6) (2.2(6))
Mn XI	$3p^3\ {}^4S_{3/2}-3p^3\ {}^2P_{3/2}$	1359.59	1359.5	0.17	0.17	1.4(6)
		1368.84		1.5?		
		1370.54		0.15	0.20	
Ca XV	$2p^2\ {}^3P_2$ $-2p^2\ {}^1D_2$	1375.98	1376.6	0.15?		
Ar XI	$2p^4\ {}^3P_2$ $-2p^4\ {}^1D_2$	1392.12	1390.67	0.29	0.20	1.3(6)
		1408.66		0.24	0.18	
		1409.47		0.45	0.17	
		1428.76		2.6 (0.56)	0.22 (0.21)	
Si VIII	$2p^3\ {}^4S_{3/2}-2p^3\ {}^2D_{5/2}$	1440.49	1440.6	0.42	0.20	9.6(5)
Si VIII	$2p^3\ {}^4S_{3/2}-2p^3\ {}^2D_{3/2}$	1445.76	1445.78	9.3 (1.3)	0.22 (0.23)	1.2(6) (1.3(6))
		1452.70				
Fe X	$3p^4 3d\ {}^4F_{9/2}-3p^4({}^1D)3d\ {}^2F_{7/2}$	1463.50		2.0 (0.44)	0.18	1.1(6)
Fe XI	$3p^4\ {}^3P_1$ $-3p^4\ {}^1S_0$	1467.08	1467.42	6.2 (1.7)	0.18 (0.18)	1.5(6) (1.5(6))
Cr X	$3p^3\ {}^4S_{3/2}-3p^3\ {}^2P_{3/2}$	1489.04	1488.9	0.27	0.19	1.5(6)
Cr X	$3p^3\ {}^4S_{3/2}-3p^3\ {}^2P_{1/2}$	1564.10	1564.1			
Fe X	$3p^4 3d\ {}^4F_{7/2}-3p^4({}^1D)3d\ {}^2F_{7/2}$	1582.60	1583.5	0.65	0.21	1.1(6)
Fe X	$3p^4 3d\ {}^4D_{7/2}-3p^4 3d\ {}^2G_{7/2}$	1603.31		2.2	0.30	
Fe X	$3p^4 3d\ {}^4D_{7/2}-3p^4 3d\ {}^2G_{9/2}$	1611.7		0.35	0.18	
S XI	$2p^2\ {}^3P_1$ $-2p^2\ {}^1D_2$	1614.51	1614.6			
O VII	$1s2s\ {}^3S_1$ $-1s2p\ {}^3P_2$	1623.68	1623.63			
O VII	$1s2s\ {}^3S_1$ $-1s2p\ {}^3P_0$	1639.80	1639.87			
		1666.94		(3.6) 0.30	(0.37) 0.16	(1.6(6))
S IX	$2p^4\ {}^3P_2$ $-2p^4\ {}^1D_2$	1715.45	1715.1	2.2	0.26	1.4(6)
		1717.42		0.56	0.20	
		1749.6?				
Mg VI	$2p^3\ {}^4S_{3/2}-2p^3\ {}^2D_{3/2}$	1805.97	1805.87			
S XI	$2p^2\ {}^3P_2$ $-2p^2\ {}^1D_2$	1826.23	1826.15	1.8	0.26	1.2(6)
Fe IX	$3p^5 3d\ {}^3P_1$ $-3p^5 3d\ {}^3D_2$	1841.55	1841.3	1.2	0.22	1.4(6)
		1847.25				
		1862.76		1.1	0.16	
Ni XIV	$3p^3\ {}^4S_{3/2}-3p^3\ {}^2D_{5/2}$	1866.75	1867	2.0	0.26	2.1(6)
Fe IX	$3p^5 3d\ {}^3P_2$ $-3p^5 3d\ {}^1F_3$	1917.21	1916.8	2.8	0.25	1.8(6)
Fe X	$3p^4 3d\ {}^4D_{7/2}-3p^4({}^3P)\ {}^2F_{7/2}$	1918.27		5.2	0.27	

bottom of loops to high-ionization lines near the tops.

In the SXR region we see lines of even higher ionization stages, where the spin–orbit interaction is so large that normally forbidden lines, such as intersystem transitions, become permitted. In those cases the ratio of permitted to forbidden transitions becomes a valuable indicator of density.

There is some evidence for underabundance of high FIP elements in the corona. The only lines observable in the visible are two forbidden coronal lines of Ar. Jefferies *et al.* (1971) give intensities for $\lambda 5539$ of ArX and $\lambda 4412$ of ArXIV. Both can be compared to lines from ions of the same IP. The ArX line is at least five times weaker than $\lambda 6702$ of NiXV; in fact it is doubtful that it has been observed at all. It was once listed at 5536 Å , then split into lines at 5533 Å and 5539 Å. Edlén (1969) showed that the proper wavelength at which it should be found is 5533 Å. Examination of the eclipse plates shows that these lines may not be present at all; I have searched the literature in vain for a solid observation of a coronal line at either 5533 Å or 5539 Å. One may conclude that ArX is five or ten times underabundant relative to Ni in the quiet corona. Although Ar cannot be observed in the photosphere, its abundance in meteorites is about three times greater than Ca or Ni.

The ArXIV line $\lambda 4412$, on the other hand, is really there; it is observed in post-flare coronal condensations and is similar to the $\lambda 4086$ CaXIII line, both in IP and spatial distribution. You can find a picture in the Ap. J. (Zirin, 1964). I found it weaker than $\lambda 4086$, concluding that Ar was half as abundant as Ca in these condensations, or six times below the cosmic values.

Lower ionization states of Ar are not absent in the corona; a number of permitted (Behring *et al.* 1976) and forbidden (Table 8.4) lines of Ar are observable in the UV. Although the UV intensities are not accurate, all the Ar lines appear relatively weak. It is particularly striking that only one Ar line appears in Table 8.4, while numerous strong lines of sulphur, which has a lower cosmic abundance, are found. This tends to support the Breneman and Stone results. The abundances from UV lines of Ne are not sufficiently accurate for estimates of coronal abundances, and there are no forbidden lines. We conclude that the high FIP elements are probably underabundant above the photosphere, but since we cannot measure Ar or Ne in the photosphere, it cannot be excluded that they are underabundant there, too. The behavior of He is, as usual, difficult to understand.

8.4. Excitation and ionization in the corona

The problem of excitation of the forbidden coronal lines is straightforward. The processes are as in Fig. 4.2, but the relative rates are different and all the rates are slow. The collisional rates are small because the density is low, and the radiative rates are slow because the f-value of the forbidden line is small. Under normal conditions, most of the electrons spend their time in the ground state, and for normal densities all excitations are followed by downward transitions. The rate of excitation F_{12} by absorption of photospheric photons is given by the product of the radiation density u_ν times the Einstein absorption coefficient B_{12} times a dilution factor $1/2$:

$$F_{12} = \frac{1}{2}u_\nu B_{12}. \tag{8.6}$$

Using Eqs. (4.25–8),

$$\begin{aligned}F_{12} &= \frac{1}{2}\frac{A_{21}}{B_{12}}\frac{g_2}{g_1}\frac{1}{e^{h\nu/kT_\lambda}-1}\times B_{12}\\ &= \frac{1}{2}\frac{g_2}{g_1}\frac{A_{21}}{e^{h\nu/kT_\lambda}-1}\quad \text{sec}^{-1},\end{aligned} \tag{8.7}$$

where g is the usual statistical weight, A_{21}, the Einstein spontaneous emission coefficient, and T_λ, the radiation temperature of the photosphere at the wavelength of the line. If there are no other sources of excitation, we should be able to find the population of state 2 by equating the number of electrons entering the upper state to the number leaving it, obtaining the statistical equilibrium equation:

$$N_1 F_{12} = N_2 F_{21}. \tag{8.8}$$

If we neglect stimulated emissions, combining Eq. (8.7) and Eq. (8.8) gives

8.9. (left) Term diagram for the lowest states of FeXV. The Einstein A's are in parentheses, units in sec^{-1}.

8.10. (right) Term diagram for the FeXIII ground configuration.

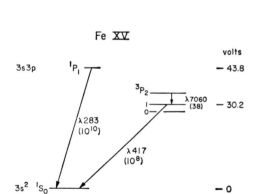

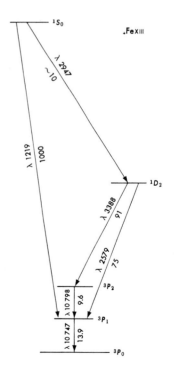

$$\frac{N_2}{N_1} = \frac{1}{2}\frac{g_2}{g_1}e^{-h\nu/kT_\lambda} = \frac{1}{2}\frac{g_2}{g_1}10^{-5040\chi(eV)/T}, \qquad (8.9)$$

which is exactly the Boltzmann distribution corresponding to the temperature of the irradiating body multiplied by a dilution factor $1/2$. This is a good example of the pseudo-Boltzmann equilibrium that we discussed at the end of Sec. 4.2. Since typically $h\nu/kT_\lambda \approx 4$, there are about 60 times fewer electrons in the upper state, if the g's were equal. The strongest forbidden line in the coronal spectrum is FeXIV $\lambda5303$, which is a $^2P_{3/2} \rightarrow {}^2P_{1/2}$ transition. Using Eq. (5.18) we find $A = 60/\text{sec}$, and Eq. (8.8) gives $F_{12} = 0.91/\text{sec}$. If we take Eq. (4.64) for the collision rate with $T = 10^6$, $g_m = 2$, and $\Omega = 0.25$ (Blaha 1969), we obtain $C_{12} = 1.1 \times 10^{-9}$; with a density $N_e = 5 \times 10^8$, the upward collision rate is $0.5/\text{sec}$. Thus for $\lambda5303$ photospheric excitation is more important than direct collisional excitation unless $N_e \geq 10^9$.

Since collision rates are proportional to the electron density, the brightness of a collisionally excited line, proportional to N_e^2, will fall with half the scale height. A line excited by photospheric radiation varies only as the apparent angular size of the solar surface, $1/R_\odot^2$. We should therefore be able to detect electron densities above 5×10^8 by high brightness gradients in $\lambda5303$. Or we can use lines of lower transition probability as indicators at lower density. But it is hard to find data on height gradients of coronal lines and only quiet Sun data may be used because of the temperature variation with height in active regions. As noted above, that gradient is along a curved flux loop, rather than a radius. So density diagnostics are most difficult in the only place where densities are high enough for these effects to be likely. The same thing takes place in flares, where the hottest and densest materials are at the loop tops.

There are a number of forbidden coronal lines which arise in upper levels or terms which cannot be excited from the ground state by photospheric radiation (Zirin 1968) and must be excited by collisions. These lines will be seen only if the density exceeds a certain value, so when we see them in a coronal condensation, we know the minimum density. One such line is the $\lambda7059$ of FeXV. The ground configuration of this ion (Fig. 8.9) is $3s^2$, which has no splitting and no forbidden coronal line. But the first excited triplet state is $3s3p\ ^3P_{2,1,0}$, and $\lambda7059$ is the $J = 2 \rightarrow J = 1$ transition. This term lies too high for excitation by photospheric radiation and therefore is excited by collisions only. The energy is only 40 eV, so it is not strongly temperature dependent except for the ionization equilibrium of FeXV. Thus anyplace that this line is observed must have a high density, typically above 5×10^8. Usually it appears only above active regions. The lower level of this line is upper level of the intersystem FeXV line $3s3p\ ^3P_1 \rightarrow 3s^2\ ^1S_0$ at 417 Å. In Fig. 7.25 we saw that this is a strong line in a postflare coronal condensation. The splitting of the ^3P term is a measure of the spin–orbit interaction which mixes the level with $3s3p\ ^1P_1$, the top level of the $\lambda283$

line. The latter is one of the strongest in the ultraviolet, and is seen in Fig. 8.12 to trace out the loop tops and other higher temperature areas. The transition probability of $\lambda417$ is determined by the mixing with $\lambda283$; it is high enough that all electrons landing in $^3P_1$ make the $\lambda417$ transition, and the intensities of $\lambda7059$ and $\lambda417$ may be related.

Our discussion has generally omitted the effects of cyclic excitation, population of the upper levels of forbidden lines after emission of permitted lines from more highly excited configurations. In the lines in the visible this is not important because of the dominance of photospheric excitation and the fairly high cross-sections for direct excitation of coronal lines. But it is important for levels like $^3P_2$, which cannot be excited directly by photospheric radiation. In the UV the photospheric excitation is weak and the permitted line cycle becomes quite important.

The p^2 configuration occurs in many astrophysically important ions. In FeXIII we observe three forbidden lines. The term structure is shown in Fig. 8.10; the ground term is $^3P_{2,1,0}$, which give rise to two lines at 10747 Å and 10798 Å in the infrared. Because only $\lambda10747$ may be excited directly from the ground state the ratio of the other two lines to it increases with density. These variations were first studied by Firor and Zirin (1962); more detailed calculations allowing for proton excitation were carried out by Chevalier and Lambert (1969, 1970) and Finn and Landman (1973). Eclipse observations by Malville and Eddy (1967) and Byard and Kissell (1971) showed good agreement with the models. Monochromatic images in the two infrared lines were obtained at eclipse by Badalyan and Shilova (1984) and used to determine the densities of different AR's.

8.11. Ionization equilibrium of Fe ions. (Jordan 1969)

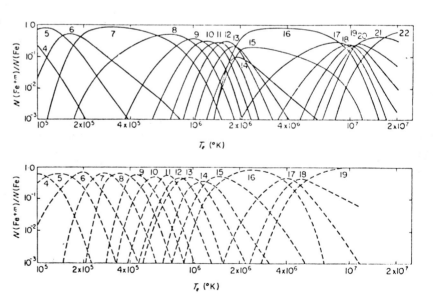

8.12. The corona and transition zone in three resonance lines formed at different temperatures. NeVII shows the details of the chromospheric network and the bases of the AR loops. It is limb-brightened as vertical structures overlap inside the limb. MgIX is more diffuse over the network and illuminates the middle of the AR loops. It brightens just above the limb when we see the corona behind. FeXV comes mainly from AR's and emerging dipoles, like that to the right of center, emitting strongly from the loop tops. The progression of AR emission in these lines reveals the temperature gradient in the loops. MgIX is most representative of the normal corona, active regions, new dipoles, and enhanced network. (NRL, N. Sheeley)

The two yellow coronal lines of CaXV, $\lambda5445$ and $\lambda5694$, bear exactly the same relation to one another as the infrared lines of FeXIII. The $J = 1 \rightarrow J = 0$ line, $\lambda5694$, is usually much stronger, because it can be excited by photospheric radiation and, at low densities, every ion making the $\lambda5445$ transitions must subsequently cascade through $\lambda5694$. At high densities upward collisions become important, the levels are populated according to their statistical weight, and the intensity of each line is proportional to gA_{ij}. In this case $\lambda5445$ should become considerably stronger than $\lambda5694$, but we have never observed a density high enough for this to occur. The calculated ratio $\lambda5694/\lambda5445$ (Zirin 1964) runs from 3.2 at $N_e = 5 \times 10^8$ to 1.5 at $N_e = 3 \times 10^{10}$. This calculation did not include proton excitation (Bahcall and Wolf 1968), which would enhance $\lambda5445$, because proton excitation is a quadrupole process and goes from $J = 0$ to $J = 2$. The observed values are in the predicted range but show no systematic trend with density.

The above techniques may be used for the many UV forbidden lines listed in Table 8.4. Feldman *et al.* (1983a) studied the density sensitivity of lines of FeXII, for which the ratios had been calculated by Flower (1977). The relative intensities of these lines, observed in the quiet corona at 1242, 1349,

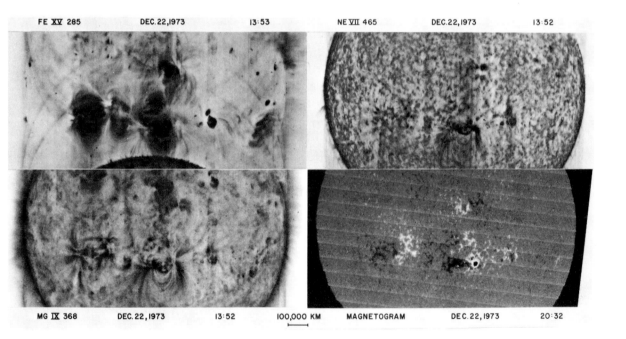

FE XV 285   DEC. 22, 1973   13:53   NE VII 465   DEC. 22, 1973   13:52

MG IX 368   DEC. 22, 1973   13:52   100,000 KM   MAGNETOGRAM   DEC. 22, 1973   20:32

2169, 2406 and 2566 Å, change at densities below 10^{11}. Feldman *et al.* found that some of the lines were distorted by blending, but obtained ratios from which reasonable densities could be determined. However, they found the lines were too bright, and an unreasonable path length was required. They point out that Dere *et al.* (1982) found higher densities from FeXII lines near 300 Å. Nothing is simple.

When an active region is observed at the limb, the coronal condensation associated with it is often dense enough for the K corona to be observed with an ordinary coronagraph. This gives an excellent measure of density, if we can assume the line-of-sight dimension to be comparable to that in the plane of the sky. Using the Thomson cross-section, we have

$$N_e = \frac{0.620 \times 10^{25}}{L} \frac{I}{I_\odot}. \qquad (8.10)$$

Using (8.10) I calculated the density in several limb flares (Zirin 1964), comparing the result to densities derived from the line/continuum ratio for $\lambda 5694$. Unfortunately the line/continuum ratios were generally a factor two higher than predicted, but the observed values did increase with the measured electron density, so the result was not too bad, given our lack of accurate Ca abundances and cross-sections. Use of Eq. (8.10) is a good technique when possible, as it is independent of unknown physics; the only error is due to density clumping, as it gives the integrated density along the line of sight.

The coronal ionization equilibrium is governed by special conditions of low density, high temperature and low radiation temperature. Fortunately we may observe a wide range of ionization stages (FeIX to FeXXVI). We must calculate a general expression for the ratio $N(i + 1)/N(i)$ of each successive pair. The theory was first worked out by Biermann (1947) and by Shklovsky (1948), then simplified and improved by Elwert (1952). But the correct application began only with the introduction of dielectronic recombination by Burgess (1964), who had taken up a suggestion by Unsöld.

Of the two common ionization mechanisms only collisional ionization is important. The typical ionization potential of coronal ions is about 300 eV, and for photoionization we have available only the photospheric radiation flux corresponding to a 6000° black body, which means the average quantum has only about 0.5 eV of energy. The corona itself is optically thin, and its emission is always below a 7000° black body.

Because of the low density in the corona, three-body recombination (the inverse of collisional ionization) can be neglected. The only remaining recombination processes are two-body and dielectronic recombination; both are proportional to N_e, but at coronal temperatures the latter is about ten times larger. Processes involving the ejection or capture of more than one electron at a time are infrequent, so we may divide the ionization problem into a series of equilibria between successive stages. This is given by:

$$\text{Ionizations} = \text{Recombinations}$$
$$N_i N_e C_i = N_{i+1} N_e R_{i+1} \tag{8.11}$$

where C_i is the ionization rate coefficient from the ground state of the ith ion and R_i is the recombination rate to that ion (counting both two-body and dielectronic recombinations, and including cascades down to the ground state). This leads to a ratio that is independent of density:

$$\frac{N_{i+1}}{N_i} = \frac{C_i}{F_i}, \tag{8.12}$$

and the ratio between distant stages of ionization is the product of the successive ratios. The most abundant ions at any temperature will be those for which $N_{i+1}/N_i \approx 1$. The ionization has been calculated in detail by Jordan (1969), allowing for dielectronic recombination (dr), excitation to autoionizing levels, *etc.* These tables have been used by most workers. They agree with House's (1964) calculations without DR and with those of Kozlovsky and Zirin (1968) for oxygen, which included DR. For those who do not have Jordan's tables handy, one can use the simple form derived by Elwert, divided by a factor ten to allow for DR:

$$\frac{N_{i+1}}{N_i} = 2.1 \times 10^4 \frac{P_i}{n} \left(\frac{\chi_H}{\chi_i}\right)^2 \frac{e^{-(\chi_i/kT)}}{\chi_i/kT}. \tag{8.13}$$

The ratios are so steeply dependent on temperature that a temperature derived from this formula will differ by only a few percent from Jordan's values. Fig. 8.11 shows Jordan's results for Fe, where the number indicates the stage of ionization. Note how the closed shell ions Fe($+7$, $+16$, $+22$ and $+24$) dominate the equilibrium, and the next lower state has a long tail. As pointed out by Kozlovsky and Zirin (1968), the high-ionization potential of these ions produces a substantial increase in dielectronic recombination to the Li-like state at high temperatures, so the latter will still be present at high temperatures. On the other hand, the Li-like ions are so easily ionized that they are never the dominant ionization stage.

The most complete observational data with which to compare the ionization theory is the coronagraph data of Firor and Zirin (1962), which included observations of FeX, XI, XIII, XIV, and XV. The ionization peaked variously at FeXI, XIII and XIV around the limb, while FeX and XV were typically about ten times fainter. Jordan's tables give FeXI, XII, XII and XIV peaking at 1.26, 1.41, 1.58 and 2 million deg respectively, with four or five ionization stages always observable. This gives much better agreement with the data than Elwert's form Eq. (8.13), which predicts a much steeper peak in ionization.

The techniques of calculating intensities of permitted coronal lines are summarized by Heroux *et al.* (1972) who calculated the intensities of the resonance and subordinate transitions from the Li-like ions OVI, NeVIII

8.13. (a) and (b) The solar UV spectrum between 160 and 290 Å from a rocket 1969 April 4. The lines are mostly coronal but transition zone and HeII lines appear at the long wavelength end. (Malinovsky and Heroux 1973) (AFCRL)

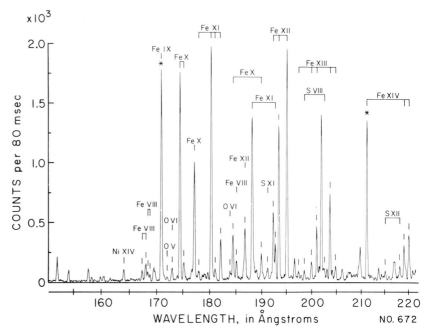

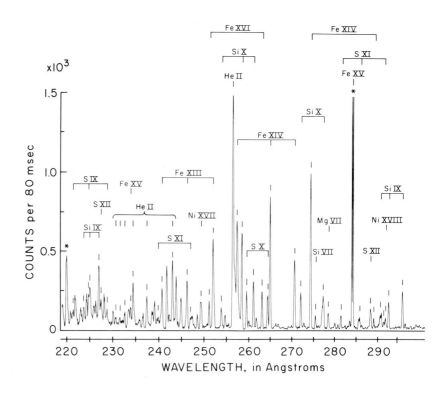

and MgX to compare with their rocket measurements. They plot the expected flux in resonance and subordinate lines as a function of temperature. One would expect to be able to use the ratios of these to measure temperatures by the methods of Sec. 7.4. But as was the case in the chromosphere, the subordinate lines weight higher temperature regions and the derived temperatures are too high. The resonance line of OVI, for example, could be explained by temperatures from 300,000° to 1,000,000°, but the subordinate lines required 1,200,000°. Heroux *et al.* found similar but smaller discrepancies for the other ions.

A solution to this problem was proposed by Withbroe (1975), who used an iterative method to define the emission measure in a line as a function of temperature. This gives a better hope of fit and, at the same time, an idea of the temperatures along the line of sight. Obviously it will work better in the corona, where the temperatures are more uniform, than in the chromosphere, where they change rapidly and we don't know what is going on anyway. He found peak intensity for the permitted lines to occur at $2\,000\,000°$. But we have seen in Fig. 8.12 that the ions of different temperature arise from different heights in the same loop. It seems hopeless to get abundances or temperatures in the corona without images in the relevant lines. One could then use Withbroe's method to define the emission measure in each temperature range, assuming each ion to radiate near its favorite temperature.

The X-ray lines from the corona present an entirely different excitation problem. The ionization is the same as before, but the high value of Z_{eff} produces large $L \cdot S$ perturbation between the terms, so intermediate coupling exists and transitions are possible between systems of different multiplicity. This has been thoroughly studied in the case of He-like ions, for which all the previously forbidden transitions are now permitted to varying degree. The $2\,^3P \to 1\,^1S$ "intersystem" transition, barely detectable in He, is quite strong, while the $2\,^3S \to 1\,^1S$ "forbidden" transition, never observed in He, is moderately strong. Neither, of course, is as intense as the permitted line $2\,^1P \to 1\,^1S$. The theory of these transitions was worked out by Gabriel and Jordan (1972). At low densities, all collisional excitations lead to a radiated photon, and the line ratios are just those of the collision rates $N_e C_{1n}$, allowing for excitation via other levels. As the density increases, a point is reached at which $N_e C_{1n} = A_{n1}$ and the population of the upper level saturates at a Boltzmann equilibrium. The lines of higher A-values reach this point at a higher density, so they keep increasing. Eventually the populations of all the lines are given by the Boltzmann formula and the intensities are proportional to $A_{n1} g_n \exp(-\chi/kT)$, but that occurs only at high densities. Unfortunately the lines of He-like ions do not begin to saturate until the density exceeds $10^{9.5}$; discussion of these is deferred to Chap. 11.

8.5. Radio emission of the quiet corona

The corona is the main source of solar radio emission at $\lambda > 10$ cm, so we can study it directly at these long wavelengths. Variation of brightness with frequency may be translated into variation of temperature with height. Radio bursts which emit at the plasma frequency are a good density diagnostic if we know their position. The scattering of waves from cosmic and man-made sources grazing the Sun, as well as direct radar observations, are additional tools to explore the corona. The radio and radar data can be obtained any day without a spacecraft, and provide height resolution.

We have already discussed the sources of opacity at radio frequencies in Chap. 7. Using Eq. (7.5), we can define an absorption frequency ν_A for the corona, at which the optical depth $\tau_{\mathrm{ff}} = 1$. For $T = 10^6$ and $H = 6 \times 10^9$, this has the surprisingly simple value:

$$\nu_A = 0.7 \frac{N_e}{10^6} \quad \text{MHz}. \tag{8.14}$$

If we set this equal to the plasma frequency from Eq. (3.33),

$$\nu_p = 9 \times 10^{-3} \sqrt{N_e} \quad \text{MHz}, \tag{8.15}$$

we find $\nu_A = \nu_p$ for $N_e = 3 \times 10^6$ cm^{-3}. For lower densities the free-free cutoff frequency is less than the plasma frequency, so the latter dominates in the upper corona. At the base of the corona ($N_e = 5 \times 10^8$) the free-free cutoff is 300 MHz, and the plasma frequency, 200 MHz.

All through the outer corona, the low frequency cutoff is determined by the plasma frequency. Since the opacity peaks sharply there, the emission will be given by the Rayleigh-Jeans formula for the temperature at the point where the plasma frequency equals the frequency we are using. The plasma frequency has the property that all frequencies above ν_p are transmitted, and all lower frequencies, reflected. In contrast with the sharp cutoff at the plasma frequency, the free-free absorption has a gradual effect. Because the brightness temperature of the opaque corona is so high, even small free-free opacities can make the major contribution to T_{br}. We see this in Table 7.1, which shows the coronal contribution dominant down to 5 cm.

Radio temperatures at lower frequencies for the quiet Sun are summarized by Lantos (1980). At meter waves, coronal holes give $T_{\mathrm{br}} = 6 \times 10^5$ deg and enhanced network areas, $T_{\mathrm{br}} = 1.1 \times 10^6$ deg. Coronal holes are definitely visible in meter waves. Thus the long wave emission maps out the coronal temperature. Because the conductivity is so high the coronal temperature does not vary much, and radio images of the corona are relatively uniform except for active regions and coronal holes.

The refractive index changes as we approach the plasma frequency, and there may be considerable bending of the ray path if we do not look exactly at the center of the Sun. This is shown in Fig. 8.15; the observer at the Earth (at the right) sees the integrated emission $S_\nu \tau_\nu$ along the refracted path to infinity. The opacity is greatest near the turning point. If we

8.14. A Map of the Sun at 21cm made with the VLA. Upper left, intensity; right, circular polarization (maximum 30%). Below left: λ10830 (KPNO); right, magnetogram (KPNO). Active regions are bright; the high density makes the corona opaque. The sign of the circular polarization is set by that of the magnetic field. (Dulk and Gary 1982)

look toward the limb, the refraction makes us see a greater height at each frequency, so the temperature : frequency : height correspondence can be mapped onto a temperature : height correspondence. Because the coronal temperature is fairly uniform, there is little limb brightening in coronal radio observations except the limb doubling.

The free-free and plasma frequencies are always present in the quiet corona. Special circumstances give rise to absorption or emission by gyroresonance, synchrotron resonance, or the Razin effect. The gyrofrequency, $\nu_g = 2.8$ MHz/gauss, is the frequency with which electrons spiral

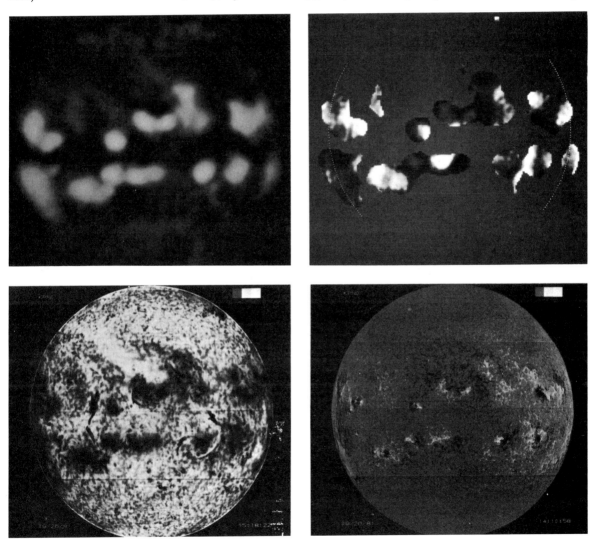

around the field lines. At low temperature absorption occurs at the gyrofrequency only, and since $B < 10$ gauss, the effect is at low frequencies. At higher temperatures gyroresonance absorption occurs at a descending series of peaks at harmonics of the gyrofrequency. The dependence of the absorption coefficient on the order of the harmonic is a complex function of temperature and field direction for which approximations are given by Zheleznyakov (1977). Typically the first four or five harmonics are important. If we are looking at the corona above an active region, the magnetic field strength varies sharply with distance from the poles, so the gyroresonance peaks are spread out to produce a continuous absorption descending with frequency, and the medium is opaque up to the third or fourth harmonic. Gyroresonance opacity in the first harmonic dominates if the field is above 120 gauss, hence gyroresonance in the higher harmonics will be important for fields above 20 gauss.

When the energy of the particles is high, the emission at higher harmonics is strong, and there are so many that a continuum appears, seen in the microwave emission from flares. For synchrotron emitting sources associated with flares, the Razin effect cuts off synchrotron radiation below the frequency:

$$\nu_R = 20\, N_e / H_\perp \quad \text{MHz.} \tag{8.16}$$

So the occurrence of a sharp cutoff in a continuum burst defines the ratio of density to magnetic field high in the corona (Ramaty 1969; Boischot and Clavelier 1967).

While the corona is normally optically thin at high frequencies, the high density above active regions increases the free-free emission and the strong magnetic fields increase the gyroresonance opacity. T_{br} can approach the coronal temperature. We can distinguish the two effects by the polarization; the gyroresonance opacity in the ordinary and extraordinary rays is different, so if there is a temperature gradient, the source will be right- or left-hand circularly polarized. Free-free absorption knows nothing of magnetic fields and gives unpolarized emission. In Fig. 8.14 we see the concentration of emission over active regions observed by Dulk and Gary (1982) at 1.4 GHz. The frequency dependence is clear in Fig. 8.15, which shows high-resolution maps of an active region obtained by Gary (1986) using the VLA during an eclipse. One uses the Moon to isolate slices of the image and eliminate sidelobes. The radio contours are superimposed on videomagnetograms so we can see the relation to magnetic structure. At 21 cm the emission is increased all over the AR and is unpolarized. The agent must be free-free absorption; Eq. (8.14) shows that a density of 2×10^9 is sufficient. Since the corona may not be optically deep, the measured T_{br} of about $1\,500\,000°$ is a lower limit for the coronal temperature above the AR. At 5 GHz the corona above the active region is transparent, and the

8.15. Maps of an active region in 21cm (top) and 6cm (bottom) superposed on a video-magnetogram. At 21cm the corona is opaque and emission is general, independent of the magnetic field. At 6 cm the corona is transparent except in the strong field regions where gyroresonance opacity occurs. (BBSO)

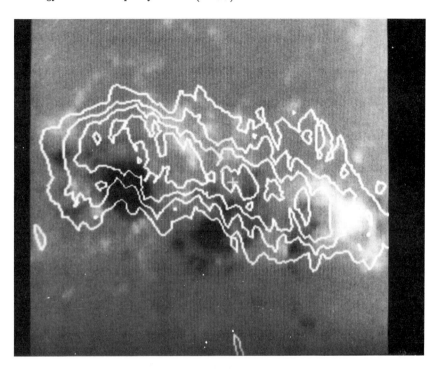

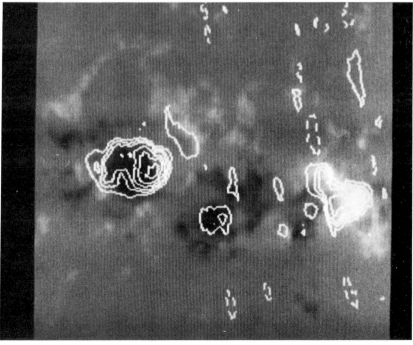

radio brightness peaks in a small strong field region above the sunspots, the only place that the gyrofrequency is high enough to produce gyroresonance opacity. The maps at 5 GHz are a perfect replica of K line images, except that the AR's are relatively much brighter, probably as a result of gyroresonance opacity.

Flares initiate disturbances which excite emission at the local plasma frequency as they pass through the corona. If we know the plasma frequency we know the density; all we need is the height of the source. So the observation of these travelling type II and III radio bursts (Sec. 13.6) makes possible determination of the run of electron density along their path. Observations with arrays such as the Culgoora radioheliograph show the source position on the sky at several frequencies. By identifying the source with a flare on the surface, we can calculate the trajectory and the source height. Since the emission in these bursts is the first or second harmonic of the plasma frequency we can fix the electron density at that height. Most analyses of type II bursts fit a density about twice that of the Newkirk model, which seems typical of AR's.

Although the ionosphere limits such observations to frequencies above 10 MHz, observations with spacecraft like the Radio Astronomy Explorer followed radio sources down to 100 kHz, the plasma frequency outside the Earth's magnetosphere, and produced models of the corona all the way to the Earth.

The fact that $\nu_A > \nu_p$ for densities above 3×10^6 (which corresponds to $\nu_p \approx 30\text{MHz}$) makes it peculiar that radio bursts at frequencies reach us at all for higher frequencies. The burst emission must be produced by plasma oscillation, but the optical depth is too high for it to escape. Either the emission is in the second harmonic or there is considerable irregularity in the corona. For the same reason, radar reflections cannot be received for

8.16. Coronal ray trajectories for 100 MHz. (Kundu, 1965).

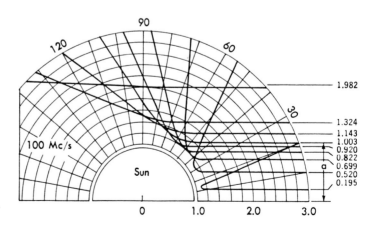

frequencies above 30 MHz; free-free absorption attenuates the signal before it reaches the reflection point.

Radar observations of the corona are an extremely interesting technique. They were carried out by Jesse James (1966, 1970) of El Campo, Texas at 38 MHz. James found the Sun's radar cross-section to be quite large, about $8\pi R_\odot^2$ near sunspot maximum, falling to $\pi R_\odot^2$ near minimum. James explains the great cross-section by wave phenomena which might enhance the reflections. Owocki *et al.* (1982) calculated various coronal models in a vain attempt to explain James's results. They found that the signals surely came from the Sun, but found little more than the variation of cross-section to be correlated with the level of activity. There was no correlation of the data with solar rotation or details of activity. In principle solar radar could reveal a great deal about the outer corona.

8.6. The solar wind

Until 1957 the link between the corona and interplanetary space was largely ignored. Biermann (1951) had noticed that many comets showed behavior – excess ionization and abrupt changes in the outflow of material in the tail – that could be explained only by the existence of a particle outflow from the Sun. He proposed this outflow to have density of $500/cm^3$ and velocity, 200 km/sec near the Earth. Since all this material would have to come from the solar surface, where the area is 214^2 times less than at 1 A.U., an outward flow of 3000 km/sec at the solar surface would be required, which of course we don't observe. Biermann had the right qualitative picture, just the wrong numbers.

The first analysis of the temperature structure of the outer corona was worked out by Chapman (1957). He assumed that above a certain height in the corona losses by emission are insignificant and only conductivity is important. The conductive flux is

Flux = area × conductivity × temperature gradient

$$F = -4\pi r^2 K \frac{dT}{dr},$$

(8.17)

where the thermal conductivity is

$$K = 7.8 \times 10^{-7} T^{5/2} = K_0 \left(\frac{T}{T_0}\right)^{5/2}$$

(8.18)

This gives a thermal flux

$$F = -4\pi R_\odot^2 K_0 T_0^{5/2} \left(\frac{r}{R_\odot}\right)^2 \left(\frac{T}{T_0}\right)^{5/2} \frac{dT}{dr} = \text{const.},$$

(8.19)

which can be made a perfect differential:

$$d(T^{7/2}) = \frac{\frac{7}{2}FT_0^{5/2}}{4\pi K_0}d\left(\frac{1}{r}\right) = C\,d\left(\frac{1}{r}\right), \tag{8.20}$$

where C is a constant. The integral is:

$$T_0^{7/2} - T^{7/2} = C\left(\frac{1}{R_0} - \frac{1}{r}\right). \tag{8.21}$$

Setting the temperature at infinity at zero, we obtain

$$C = R_0 T_0^{7/2}, \tag{8.22}$$

which fixes the total flux at

$$F = \frac{2}{7}4\pi R_0 K_0 T_0. \tag{8.23}$$

Substituting for C in Eq. (8.22), we obtain the functional dependence of temperature:

$$T = T_0\left(\frac{R_0}{r}\right)^{2/7} \propto r^{-2/7}. \tag{8.24}$$

Because of the high conductivity and absence of losses, the temperature decreases slowly. If it is $1\,000\,000°$ at the Sun it will be $219\,000°$ at the Earth.

Parker (1958) pointed out that this static model would not work because the pressure remains finite at infinity. For constant temperature the scale height increases as $1/r^2$, and if we include the temperature variation from (8.24), it goes as $r^{-12/7}$. At 100 $R_\odot$, $H = 1.6 \times 10^8$, and at 70 AU, about twice the distance of Pluto, the scale height is one light year. The mass of the solar wind would be infinite. This is analogous to the result in stellar structure that an isothermal polytrope has infinite mass.

Parker reasoned that Chapman's model of the temperature had to be correct. Since the mass in each shell is proportional to $r^2 n(r)$, he concluded that the solar wind would only have finite mass if the $n(r)$ fell off as $1/r^2$ or faster (in the Chapman model it falls off only as $r^{-2/7}$). This is possible only if the material is flowing outward, in which case the conservation of matter gives us:

$$\rho v r^2 = \rho_0 v_0 r_0^2, \tag{8.25}$$

where ρ is the density. If the flow is along a streamer of cross-section $A(r)$, we have:

$$\rho(r)v(r)A(r) = \rho_0 v_0 A_0. \tag{8.26}$$

The pressure and density in the tube are related by the adiabatic law:

$$\frac{p(r)}{p_0} = \left(\frac{\rho(r)}{\rho_0}\right)^\gamma, \tag{8.27}$$

where γ is the ratio of specific heats, 5/3 for a perfect gas. Parker argued that the gas behaved like a fluid passing through a nozzle, the internal pressure in the low corona being converted to mass motion as the gas flows out.

The total energy per gram of gas in the corona is the sum of kinetic, internal and gravitational energies:

$$\begin{aligned} E_T &= v^2 + \frac{2kT}{m_H} - \frac{GM_\odot}{r} \\ &= 0 + 1.7 \times 10^{14} - 2 \times 10^{15} \quad \text{at } R = 1R_\odot, \end{aligned} \tag{8.28}$$

where the pressure is due to electrons and protons. Although the total energy is negative, the gravitational term decreases as $1/r$ while the internal energy falls only as $r^{-2/7}$. If T drops slowly, E_T becomes positive beyond $r \approx 12R_\odot$, and as the second two terms decrease, a solution is possible in which the first term increases to balance them and outward flow ensues. Thanks to the high conductivity of the corona, a solar wind occurs. The internal energy of the corona is converted into mass motion of outward flow by the nozzle action of the gravitational field. Note that when the kinetic energy exceeds the thermal energy the flow is supersonic. The point at which this takes place is called the critical point and corresponds to the hose nozzle. Strictly speaking, the total energy of the gas bundle is negative, but the fact that the corona is permanently kept at a high temperature by the unknown heating mechanism forces the outflow.

Fig. 8.17 shows some of the solutions found by Parker (1963) for U, the ratio of kinetic energy to pressure. Note that many solutions are possible, but we are constrained to take one that has $U = 0$ at the surface and $U \neq 0$ at infinity. The asymptotic velocities depend on the assumed temperatures at the base of the corona and are around 400 km/sec for a million degrees.

The prediction of the existence of a solar wind was soon confirmed by satellite measurements, which found a density near the Earth of $N_H \approx 1$ cm^{-3} and $V \approx 300$ km/sec. This flux of 3×10^7 ions/cm^2/sec can be supplied by an outward flow of only 0.2 km/sec at the surface, far lower than Biermann's numbers. One would like to detect the outflow directly as a blue shift in the coronal lines on the disk, but the thermal broadening of 50-100 km/sec makes that impossible. Rottman *et al.* (1982) and Orrall *et al.* (1981) claimed a blue shift of $6-10$ km/sec in MgX emission in coronal holes, at least ten times greater than the amount required to supply the solar wind. Pneuman and Kopp (1978) pointed out that such measurements are relative, and the steady downflow observed in most AR's would produce a spurious relative downflow in coronal hole regions. So the source of the solar wind is not established. Deviations from hydrostatic equilibrium have not been observed, and these would be required to produce outward acceleration at low heights. Spicules cover too small a fraction of the surface to play a role.

Considerable effort has been devoted to understanding the dependence of the solar wind flow on density and temperature at the base of the corona (reviewed by Leer *et al.* 1982). The velocity increases with the temperature at the base. Similarly, it increases with decreasing base density. Curiously, the work of Durney and Hundhausen (1974) suggests that inhibition of conduction by the spiral magnetic field increases the flow velocity by forcing a higher pressure gradient.

The theory of the solar wind is not yet complete. The various studies have not explained how the thermal conduction model as outlined above can produce the high-velocity streams associated with coronal holes. Leer *et al.* state that the base density must be unrealistically low for such high speeds to occur near the Earth and suggest that either energy is added higher up in the corona or the conductivity is somewhat different from the values in (8.19). And although Parker's model gives a high dependence of solar wind speed on base temperature, the high velocity streams in fact come from the coolest corona, coronal holes.

An interesting way of observing the outward flow is the use of "Doppler dimming", proposed by Kohl *et al.* (1983). Lines such as Lyα are excited by scattered chromospheric emission; if the material is moving outward, it is shifted out of the wavelength of the exciting chromospheric line. Thus when the solar wind acceleration takes place we should see the coronal

8.17. The family of solutions of the Bernoulli equation Eq. (8.28) (Parker 1963). U is the ratio of kinetic to internal energy and ς, the distance from the surface in solar radii. Parker chose the solution where the gas is static at the origin and the kinetic energy increases asymptotically with distance at the expense of internal energy. The solutions are ordered by L, the ratio of the escape velocity to the thermal velocity at the base of the corona. The solar wind solution is forced by boundary conditions.

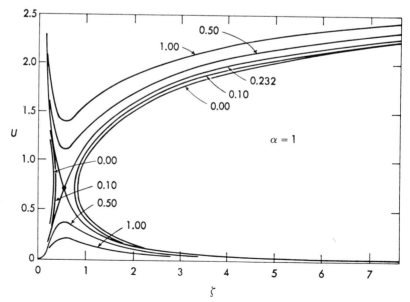

Lyα dim. The effect is especially interesting for two close strong lines, as the lines of OVI and CII at 1037.587 Å and 1037.018 Å. The OVI line moves with the corona, and when its outward velocity exceeds 120 km/sec it should be enhanced by scattered CII emission. By contrast the OVI line at 1031.8 Å will weaken, as it shifts out of the band of OVI emission from lower down.

The discovery of coronal holes demonstrated the significance of the solar wind in determining coronal structure. Like objects in Charlie Chaplin's upside-down airplane, everything not tied down by the magnetic field is carried away by the solar wind. The enhancement of coronal emission in the vicinity of an active region may be as much the result of the confining action of the closed magnetic field as the heating of the corona above sunspots.

8.7. The interplanetary magnetic field

The average speed of the solar wind is 400 km/sec, so on the first day after it leaves the Sun the gas travels 3.5×10^7 km, while the point of origin rotates 14.2°. After four days the gas reaches the Earth while the source has rotated 57°. Brandt (1970) likens the entire structure to a phonograph record: the material moves radially outward like the needle, while the groove structure rotates every 27 days with the Sun. The grooves then mark the magnetic field pattern stretched out into space. The locus of the material moving out from a point is traced by the Archimedean spiral

$$\theta - \theta_0 = \Omega t = \Omega \left(\frac{r_0 - r}{w} \right), \tag{8.29}$$

where θ is the polar angle, $\Omega = 2.7 \times 10^{-6}$ rad/sec, the angular rotation rate of the Sun; r, the distance; and w, the solar wind speed. The angle between the lines of force and the radius vector at any point is given by

$$\begin{aligned}\tan \psi &= -\frac{(r - r_0)d\phi}{dr} \\ &= \frac{(r - r_0)\Omega}{w} \quad \psi < 90°.\end{aligned} \tag{8.30}$$

As we go farther from the Sun, the angle of the spiral to the radius vector increases until the lines are almost perpendicular to it. The magnetic field lines moving out from the Sun follow the spiral, so the particles are constrained to move along the lines of force even though the plasma is moving outward carrying the field. As a result the magnetic field presents a barrier to radial conductivity as well as outward flow.

Ness and Wilcox (1964) discovered that the outward flow of the solar wind maps the large-scale solar magnetic fields on the interplanetary medium. When the Sun is active, both the surface fields and the interplanetary magnetic fields (IMF) are complex and difficult to interrelate. But near sunspot minimum the structure is simple and dominated by large-scale patterns easily detectable at the Earth. The concept of a sector structure was introduced by Wilcox and coworkers (Wilcox and Ness 1965; Wilcox 1968). For substantial periods, the azimuthal distribution of IMF is divided into two or four sectors in which the field is directed either outward or inward. The effect is shown graphically by Sheeley and Harvey (1978, 1980) who compare the sign of the IMF with the pattern of coronal holes (Fig. 8.20). The latter provide a measure of large scale solar fields at $3R_\odot$, and that pattern is reflected in the sign of the IMF. A spacecraft is not needed to measure the latter; Svalgaard and Wilcox (1975) showed it could perfectly well be measured with magnetometers at northern latitudes where the solar wind flow reaches the surface of the Earth and directly affects the terrestrial field.

As the picture of the mapping of solar fields on the IMF became clearer, the sectors were seen not to fit; the field boundaries in the photosphere may not match in the different hemispheres and do not really lie in the N-S direction. Instead Svalgaard and Wilcox (1976) proposed the idea of a warped equatorial current sheet. In the N polar region the fields point outward; in the S, inward. The field lines from pole to pole do not close, but are dragged away by the solar wind. At the solar equator there must be a sharp field reversal, which, by Eq. (3.20c), includes a current sheet. Since the fields are not symmetric, the current sheet is warped like baseball stitches, and as it rotates by the Earth, we cross back and forth into the regimes above or below it. Svalgaard and Wilcox point out that spacecraft which reached high solar latitudes never encountered boundaries because they were above the convoluted current sheet. The reader must bear in mind that this simple structure occurs only near sunspot minimum. Further, although clearly there is a corotating structure mapping out the solar fields, the kinematics of the streams is such that they can interact and modify each other in various ways, so the current sheet is not that simple.

The interaction of the solar wind and the corotating magnetic field takes different forms. Neugebauer (1983) separates the solar wind into three components: the high-velocity streams from coronal holes; the slow, dense, cool flows characteristic of sector boundaries; and transients associated with solar flares and coronal mass ejections. Their properties are summarized in Table 8.5. The high velocity streams have a higher kinetic temperature than the others, even though their ionization temperature is lower because of their origin in the cool holes. It has been suggested that the origin of the highest speeds in the coolest regions is due to the divergence of the magnetic field lines, but experts disagree.

A most remarkable aspect of the transient flows is the He enrichment; in fact Neugebauer uses the abrupt appearance of He-enrichment detection as a sign that a transient is passing the spacecraft. The low temperature of the transients, considering that many arise in flares, suggests that they are flare ejecta, particularly sprays and erupting prominences. There also may be cooling by adiabatic expansion in these events, which appear to involve closed magnetic geometry, loops of flux erupting outward from the Sun.

Abundances in the solar wind fluctuate considerably. The states of different He enrichment are discussed by Borrini *et al.* (1982, 1983). They find that the He abundance varies according to the three states listed in Table 8.5. In coronal holes it is "average" (about 6% by number) and proton velocity and temperature are high; in the sector boundaries or coronal streamers the He abundance is lower, as are the proton velocities and temperatures. In transient events (Hirshberg *et al.* 1972; Hirshberg 1972, 1974) there is a big enhancement in He associated with moderate velocity and low proton temperature in the solar wind. There is also a solar cycle dependence of He abundance, the fraction of He atoms by number dropping from about 0.05 near maximum to 0.035 near minimum. How much of this is due to the transient events is not clear.

The change in He abundance immediately reminds us of the Breneman and Stone (1985) results discussed earlier. Although obviously it is a high-FIP element, He is not included in the normal list of underabundant elements because its abundance is so variable. The high-FIP elements (Ar, Ne, O) are systematically underabundant in flare particles, which is contrary to the occasional sharp increases in He number in flares.

Once the solar wind leaves the lower corona its ionization state is frozen because of the low recombination rate. From Eq. (4.50) the recombination rate for FeXIII is about $10^{-11}N_e$; ions flowing from a density below 10^5 will reach the Earth before their ionization state changes. For this reason,

Table 8.5. Solar wind parameters in three types of flow (Neugebauer, 1983).

Property	Hole	Boundary	Transient
v, km/s	700	380	440
n, #/cm^3	4	15	10
nv, 10^8/cm^{2}s	3	5	5
nmv^2, dyne/cm^2	3	3	3
nm(mv^2/2 + mMG/r), erg/cm^{2}s	2	2	2
T_p, 10^5 deg	2.3	0.7	0.6
T_e, 10^5 deg	1.0	1.3	0.5
Ionization temperatures:(10^6 K)			
Oxygen	?	2.1	< 3.4
Iron	?	1.6	< 17
B, γ	6	3	9
Field topology	Open	?	Closed
$\beta = 8\pi nkT_p/B^2$	1	2	0.3

8.18. The life of the famous Skylab coronal hole I in 1973. Left, 6/1 (day 151), 6/28 (178), 7/25 (206); right: 8/21 (233), 9/16 (259), 10/14 (287), 11/10 (314). The effect of differential rotation can be seen to be quite weak; large scale solar magnetic fields rotate almost like on a rigid body. (AS&E)

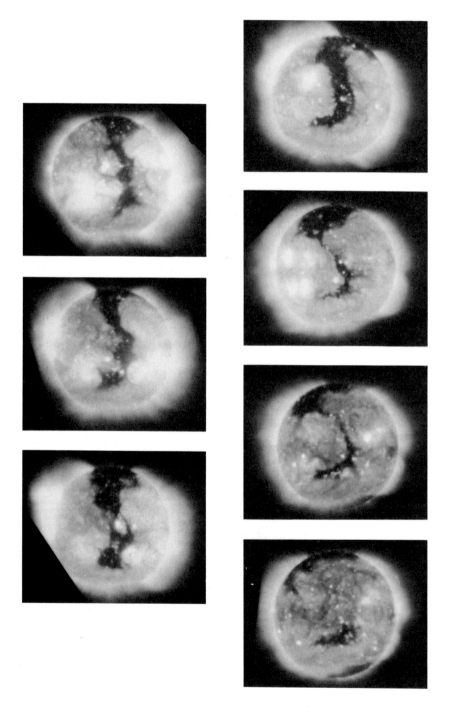

measurements of the solar wind at the Earth find the Fe ions to be in their typical coronal ionization state, with 12 or 13 electrons removed.

8.8. Coronal holes

Although the presence of coronal holes had been suggested by earlier observers, only the Skylab X-ray pictures showed their true structure. When I first heard about coronal holes, I did not understand the significance of empty regions in the corona. In fact they mark a global magnetic phenomenon of great importance. The holes are regions of open magnetic geometry which permit the free outflow of the solar wind. While we could not recognize such regions in the complexity of magnetograms, they are easily seen in coronal images. Their existence shows that dense coronal material will invariably leave the Sun if it is not held by closed magnetic geometry. They give us a chance to see if the corona has any influence on the chromosphere (it doesn't). Their long life and great latitude extent (Fig. 8.18) provide one of the best checks on the differential rotation of the Sun.

The first suggestion of the significance of coronal holes came from geomagnetism. Bartels (Chapman and Bartels 1940) postulated the existence of "M-regions" on the Sun, sources of geomagnetic disturbance with a marked 27-day recurrence. Since coincidence of transient events identified geomagnetic disturbance with sunspots and activity, the fact that the M-regions had no observable AR counterpart was a mystery. The first indication that regions of low coronal density might affect the Earth was the discovery by Bell and Glaser (1954, 1956, 1957) that peaks of geomagnetic activity were preceded a few days earlier by the central meridian passage of a minimum in FeXIV emission observed at the east limb with coronagraphs. Since many individual geomagnetic storms had been identified with flare activity, this statistical result was greeted with some skepticism. The im-

8.19. The same coronal hole recorded at (left) 160MHz by the Culgoora radioheliograph, compared with the Skylab X-ray image (right). (CSIRO)

1973 AUGUST 21

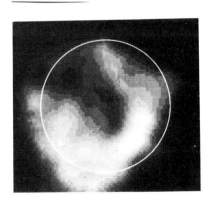 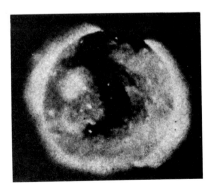

160 MHz   X-ray

8.20. A Bartels diagram (Sheeley and Harvey 1980) comparing the sign of the interplanetary field, the distribution of and magnetic field sign in coronal holes along the Earth-facing meridian, the solar wind velocity near the Earth and the geomagnetic disturbance index C9. R is the sunspot index. In the upper half the coronal holes control the pattern with a 27 day period; in the lower half a 28.5 day period becomes clear.

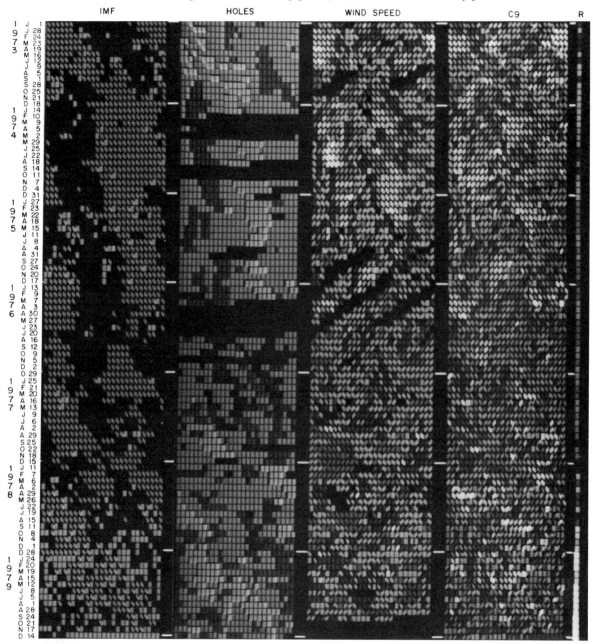

portance of open field lines was correctly noted by Billings and Roberts (1964). A study by Roelof *et al.* (1975) shows that there are two separate connections between coronal brightness and geomagnetic activity – the Bell–Glaser relation to coronal holes and the transient relation with the activity peaks associated with big AR's.

Recognition of coronal holes came with the Skylab mission and the X-ray images obtained with the Wolter telescope (Vaiana *et al.* 1973). The high contrast of the SXR images (Figs. 1.3, 8.18) made the holes quite visible and long sequences of observations were obtained. Timothy *et al.* (1975) showed that the large holes were long lived and indicated an equatorial acceleration far below that obtained from sunspot data.

The location of the holes was soon explained by calculations by Levine (see Zirker 1977, Chap. IV). He found that, if the surface magnetic field of the Sun can be approximated by a potential field in Eq. (3.20d), then a map of those field lines which intersect the surface $R = 3R_\odot$ matched

8.21. A coronal transient observed by the SMM coronagraph, showing transient running ahead of the erupting prominence. (HAO)

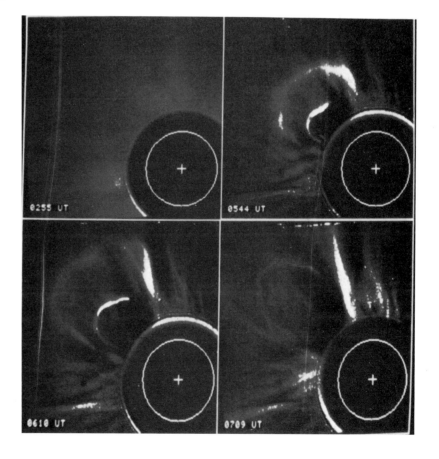

the coronal hole distribution. This meant that the coronal holes occupy regions of open magnetic field lines, places from which field lines do not form closed dipoles, but connect to distant fields. They go out so far that they are swept away by the solar wind and never do connect back. The presence of semi-permanent coronal holes at the polar caps near sunspot minimum confirms this picture; at that time of the cycle the general field is poloidal and lines of force that would go from pole to pole are instead swept into interplanetary space by the solar wind and coronal holes. One can recognize these field lines in the streamers in Fig. 8.1. Fig. 8.2 shows how closely the coronal holes match the open field lines.

Because chromospheric helium is excited by the coronal SXR emission, the spatial extent of coronal holes may be obtained from any monochromatic He picture. Coronal holes are monitored routinely in the HeI 10830 line by Harvey at Kitt Peak (Fig. 7.22). How exactly does the structure

8.22. A great eruptive prominence and coronal transient from just behind the limb, imaged in white light by the NRL Solwind coronagraph on the P78−1 spacecraft (Shot down by the US Air Force in 1985 while still taking data). The transient emerges before the prominence but is overtaken by the latter at 1812. Although the prominence is travelling at least Mach 3 (700 km/sec) it retains its shape. Transients are an important source of material for the solar wind. (NRL) (Sheeley *et al.* 1981)

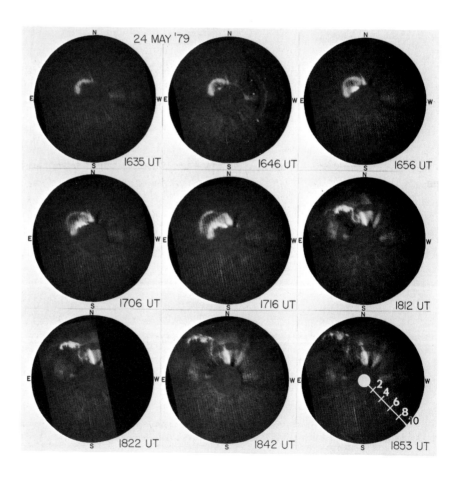

in coronal holes differ from the normal solar atmosphere? As noted in Chap. 7 the chromospheric height gradients are lower in holes. On the other hand, Doschek *et al.* (1977) could find no difference in densities determined by line ratios in and out of holes. In any event, the NeVII images show that transition line extending higher in coronal holes. As mentioned in Sec. 7.3, there is no explanation for the enhanced 8 mm emission from coronal holes. It is possible, though far-fetched, that the corona is heated by non-thermal energy passing through the chromosphere, and the absence of corona results in extra energy deposition in the transition region.

It is an accepted fact (Nolte *et al.* 1976) that coronal holes coincide with high velocity streams in the solar wind and the recurrent geomagnetic disturbances they produce. They are the long sought M-regions. The direct connection of field lines to the solar wind permits these high speeds. Fig. 8.20 presents a chart prepared by Sheeley and Harvey (1980) showing the relation between the position of coronal holes, the interplanetary field direction, the solar wind speed, and the magnitude of disturbances in the Earth's field. There is no doubt that the fields of the coronal holes extend outward into the interplanetary medium and produce the M-region storms at the Earth. The relation is not so simple at periods of high solar activity, when the field is complicated and flare shocks and CME produce irregularly spaced disturbances.

8.9. Coronal mass ejections

The possibility that material might leave the Sun was raised by observations of the great sprays associated with solar flares; but few eruptions of this type exceed the escape velocity (618 km/sec). Dunn (1968), the first to study coronal movies systematically, found little motion. After years accumulating λ5303 coronal line movies, DeMastus *et al.* (1973) classified a substantial number of coronal changes, mostly associated with apparent magnetic field rearrangements. These events could be observed only in the low corona.

The development of externally occulted coronagraphs made possible the observation of the outer corona, and to the surprise of most, a number of splendid coronal mass ejections (CME) were discovered by the HAO instruments on Skylab and SMM and the NRL Solwind coronagraphs on the P78-1 spacecraft. The data were recently reviewed by Wagner (1984). As we see in Fig. 8.21, the CME is seen as a great eruptive loop far above the surface of the Sun, majestically moving outward with speeds of 400-700 km/sec. Since they are never seen edge-on, Wagner concludes that they must be three-dimensional bubbles projected against the sky. This idea is confirmed by the Solwind observation (Howard *et al.* 1982) of a transient associated with the eruption of a large filament on the center of the disk. The material, directed at the Earth, appears as a bright halo around the Sun. Howard *et al.* reason that a bubble would produce a thin ring expanding away from the Sun, rather than a halo. The authors conclude that the eruption is constrained near its base, expanding outward near the top as in a cone. It may or may not be filled with material; many CME show a sharp edge, but others are filled.

CME are surprisingly common; Wagner quotes 0.74/day in 1973 (near minimum) and 0.9/day in 1980 (maximum). The number must be further increased by about half to account for behind-the-Sun events. Sheeley *et al.* (1982) found about two CME per day with Solwind. Because the rate does not change much during the magnetic cycle, it is clear that flares are not the major source. However these data do not include the real minimum of activity, when no filaments are present on the disk, and the tragic shooting down of P78-1 by the US Air Force precluded that observation for this cycle. The major source is erupting prominences, which have been identified with about half the observed CME. Those flares identified with CME tend to be accompanied with mass ejections in the form of surges or sprays. There is a significant group of CME which is not connected with any optical solar phenomenon, but given the weather limitations on optical observations, one cannot be sure if this is a real class. Some allowance in these estimates has been made for beyond-the-limb events.

Is the mass ejected by CME important? It is quoted by the various authors as $10^{15} - 10^{16}$ gm (10 billion metric tons) per day. By comparison the solar wind at the Earth carries 1 atom/cm^3 at 4×10^7 cm/sec through 3×10^{27} cm^2, also $\approx 10^{16}$ gm/day. So the CME represent a mass loss as important as the steady solar wind.

About half the CME are related to flares, the rest to erupting prominences. In both cases the phenomenon is associated with instabilities in prominences. As we shall see in the next chapter, prominences always erupt upward, and when flares are not involved, some sort of instability related to the so-called "melon-seed effect" must occur. Rather than being expelled upward by an outside agent, the CME are accelerated as they move outward. Numerous examples have been reported of MHD phenomena running ahead of the CME, including enhanced mass forerunners, pressure waves and rarefactions. Jackson (1981) reports brightness enhancements in the K corona some hours before the CME. These data strongly suggest that the eruption is not an explosion, but a magnetic transformation of the entire magnetic structure.

Prominences often occupy the core of a helmet streamer, separating extended unipolar regions of opposite sign, and the flux lines connecting these loop across above the boundary in the corona. Higher up the lines open out and are dragged into space by the solar wind, and the outward extension of lines of force from the two opposite polarity regions produces a boundary of sharp field reversal, which may be stabilized by a current sheet. The current sheet is subject to reconnection, with field lines snapping back in and the outer field lines reconnecting and flying outward. This is known as the Kopp-Pneuman (1976) mechanism. Material above the reconnection point is pulled outward by the fields. Unfortunately, the filament is normally below the reconnection level, so the real operation of the mechanism is unknown.

Reconnection in the current sheet would be the agent by which the fields holding the filament down become unstable. The legs of the CME remain tied to the surface during the eruption and increase in brightness as matter falls back; how the field lines behind the CME manage to reconnect and disengage from the Sun is unknown, but they manage to do it, and the CME travels off into space. The fact that the CME and its shock run ahead of the erupting filament support the idea of the breakup of the restraining field structure above the filament. Occasionally a coronal hole will form behind the CME. It is not known if this is due to the opening up of the field

8.23. A coronal transient observed by the HAO Skylab coronagraph with the NRL HeII 304 image inset. In this case the prominence did not overtake the transient. H-alpha data show a fine spray. (Poland and Munro 1976) (HAO – NRL)

AUG 21, 1973   EAST LIMB

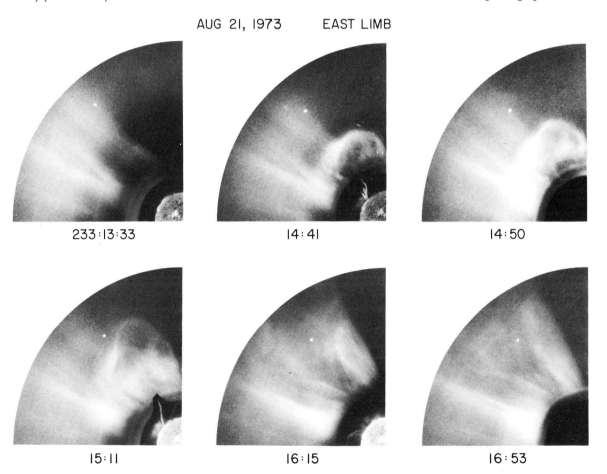

233:13:33   14:41   14:50

15:11   16:15   16:53

lines behind it, or simply the loss of the coronal material to the eruption. The new holes last for about 10 hours.

The time delay between the CME and arrival at the Earth of a shock is about two days. Because many events may occur in this interval, it was difficult to connect CME's directly with interplanetary shocks. This problem was solved by Sheeley *et al.* (1985), who compared CME's detected with Solwind with interplanetary shocks measured by the Helios spacecraft at the orbit of Mercury. Because Helios was so close to the Sun, the delay times were short and identification easier. Selecting CME's from the limb near the spacecraft, they found that 72% produced an interplanetary shock at Helios, 26% may have and 2% didn't. The solar events were generally large, low-latitude mass ejections, and the shock speeds were typically

8.24. (left) Shocks observed by Helios. Top, magnetic field; middle, velocity; bottom, coronal hole distribution position of Helios plotted.   (Burlaga 1979)

8.25. (right) A comet observed in the outer corona by Solwind. The last frame shows the coronal enhancement produced by the comet. The comet never reappeared after its encounter with the Sun. Hα pictures show no effect on the Sun. (NRL)

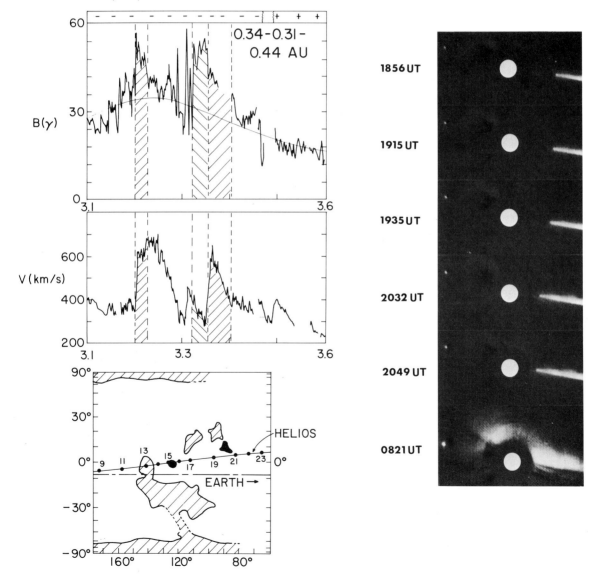

above 500 km/sec. As solar activity decreased there were fewer events, but shock speeds above 1000 km/sec became common.

The melon seed effect, or magnetic buoyancy, is thought to be responsible for the outward eruption of fields that are not tied to the surface. Early in the study of MHD, it was realized (Parker 1957; Schlüter 1957) that an isolated plasma is diamagnetic, that is, it tends to move in the direction of weaker magnetic fields. Picture a magnetic field with intensity falling off with height. If we push a cloud threaded by a separate magnetic field in the direction of higher field there will be a repulsive force which grows as we push the polar field lines apart. The "melon seed" will always want to fly away in the direction of weaker field. The force on the cloud is the magnetic gradient times the length times the area of the cloud, or may be considered as the surface integral of the external field all around the cloud. Because the pressure on the stronger field side is greater there always will be a net force on the cloud. If the magnetic field gradient is steep, there will be a buoyancy effect; as the cloud rises it expands so the internal and external pressure balance; if the external field decreases rapidly with height, the internal magnetic field will expand and the gas with it. If this expansion is greater than the adiabatic gradient, there will be a strong buoyancy effect. Magnetic buoyancy therefore occurs when the magnetic gradient exceeds the adiabatic gradient.

The operation of the melon seed effect is hindered by the fact that the magnetic fields threading everything on the Sun are attached to the surface. The buoyancy must work against the tension of the magnetic lines of force. The material cannot erupt until the field lines attached to the surface reconnect. When we watch a great eruption, we see material draining down the feet of the arches, which we seldom actually see break off or connect across the base of the arch. Some of the problems of reconnection and magnetic buoyancy of CME are addressed by Pneuman (1984).

References

Billings, D. E. 1966. *A Guide to the Solar Corona.* New York: Academic Press.

Brandt, J. C. 1970. *Introduction to the Solar Wind.* San Francisco: W. H. Freeman.

Evans, J. W. 1963. *The Solar Corona.* New York: Academic Press.

Neugebauer, M. 1983. *Solar Wind Five.* NASA Conference Publ. 2280.

Parker, E. N. 1963. *Interplanetary Dynamical Processes.* New York: Interscience.

Shklovsky, J. S. 1963. *Physics of the Solar Corona.* London: Pergamon Press.

Zirker, J. B. 1977. *Coronal Holes.* Boulder: U. of Colo. Press.

9

Prominences

9.1. Forms, motions, shapes

Any cloud of material visible above the solar surface in Hα may be called a prominence. They are the most beautiful of solar phenomena. How do they defy gravity? They may be supported by magnetic fields, they may be thrown up by a flare, or they may condense out of coronal material. Prominences also play a critical role in solar flares, which almost always involve prominences. Since they are low density objects, prominences are invisible in the continuum, but may be seen in any strong emission line. Most of the existing data is in Hα, the easiest line in which to observe them. The best films of prominences are obtained with coronagraphs, which permit observations of the faintest features against the sky. But normal Hα telescopes may be used as well.

Since prominences appear bright against a dark sky (Fig. 9.1), astronomers had some difficulty relating them to the dark filaments (Fig. 9.2) seen on the disk in Hα. But it soon was evident that the dark filaments on the disk became bright prominences at the limb, and we shall use the terms interchangeably. A typical filament is about 10% of the disk brightness in Hα centerline (or 2% of the continuum, as the central depth of Hα is 0.2); that is still much brighter than the sky. Because there is little motion, the Balmer lines are narrow in most filaments, so they rapidly disappear as we leave line center. On the disk the background brightens as we move out of the Hα absorption line, so filaments disappear even sooner. The motions in active prominences broaden the line profiles and they are easily detected by absorption more than 0.5 Å (25 km/sec) off the line. The advent of flares is often marked by the appearance of absorption in the blue wing of Hα as the filament at the flare site rises.

Why are prominences darker than the disk? They are a good example of pure scattering. The density and temperature are low enough so there is little collisional excitation; the emission we see is due to excitation of levels by the photospheric radiation field. But the photosphere fills only half the sphere visible from the prominence, and radiates into all solid angles. So the light is weakened each time it is absorbed and emitted. When we see a prominence at the limb the emission we see has been robbed from the outward-directed beam, leaving it darker than the disk. Prominences often produce a bright region (Fig. 9.2) in the underlying chromosphere (as seen in Hα) because they blanket the chromosphere and reduce its emission losses, thus increasing the brightness temperature in that line. The radiation of prominences at the limb has been taken from the outward directed beam.

Transient, energetic prominences, such as flares, surges, and loops, may appear bright against the disk. That is a sign that the collisional excitation exceeds the photospheric excitation and the intrinsic emission of the prominence exceeds that of the photosphere. It occurs at densities above 10^{12} cm^{-3} (see Sec. 11.3). Another kind of brightening occurs (Zirin 1969) when a prominence moves transverse to the line of sight. The Doppler shift relative to the photosphere causes the prominence to absorb in the wings of the photospheric Hα absorption line, where the photosphere is much brighter. Since the prominence is moving transverse to the line of sight, in our direction we see it against the darker centerline background, and the moving prominence appears in emission relative to the chromosphere.

Prominences are classified by association and morphology. The first such

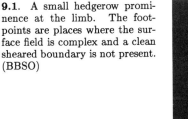

9.1. A small hedgerow prominence at the limb. The footpoints are places where the surface field is complex and a clean sheared boundary is not present. (BBSO)

scheme was that of Menzel and Evans (1953). Table 9.1 (modified from Zirin (1978c) orders the prominences by the way in which they occur.

Class 1 exhibits a low excitation spectrum and class 2, high excitation. Because the active prominences are so short lived, they make up a tiny fraction of the prominences on the Sun at any time. Quiet region prominences (QRF) take varied forms: hedgerows, arches, funnels, coronal rain, surges, sprays, loops. All can be fitted into the classes of Table 9.1 by their lifetime. Hedgerow prominences (Figs. 9.1–3) are the most common quiescent type, extensive filaments with a few points of contact with the surface. As one can deduce from Fig. 9.16, the feet of the hedgerow are places where there is some magnetic anomaly, usually a small dipole. Usually quiescent prominences end their existence by upward eruption. The eruptive prominence (Figs. 9.4, 9.5, 9.22–3), or, on the disk, *disparition brusque* (sudden disappearance), is among the most spectacular solar events, and responsible for a large fraction of coronal mass ejections (CME).

Active region prominences (ARF) differ somewhat from QRF. They are darker, smaller, and have more coherent fibril structure along their axis than QRF. A sheared magnetic field runs parallel to this axis, permitting considerable flow of material along it. ARF eruption usually produces a sizable flare (Fig. 9.6, 9.13), while QRF eruptions may produce a big CME but only weak chromospheric brightening. The ARF may erupt and reform several times (Tang 1986).

Filaments form gradually and are among the longest-lived solar features, sometimes lasting for several rotations. The first step in formation of a big filament will be the filament formed as an AR breaks up into two unipolar regions. Subsequent AR's form neutral lines which join to produce an extended filament. The average filament is smaller, lasts only one rotation

Table 9.1. Classes of prominences.

Class 1.	Quiescent (long-lived) prominences.
	(a) prominences or filaments in or near active regions (ARF)
	(b) prominences or filaments in quiet regions (QRF)
	(c) ascending prominences (they were once long-lived)
Class 2.	Active (flare-associated, transient) prominences.
	(a) limb flares.
	(b) loops and coronal rain.
	(c) surges: collimated ejected material previously not seen.
	(d) sprays, uncollimated ejecta previously visible as pre-flare elevated features.

and changes its form a little almost every day.

Because they are long and also present at high latitudes, filaments have been used to measure solar rotation. This is difficult in practice because the variable vertical extent of the filament changes the apparent position measured near the limb. Returning filaments seldom have the same form in the successive rotations. There is a popular picture that filaments form in a N–S channel and are stretched out in the E–W direction by the equatorial acceleration, and indeed some behave that way. But many are not tilted that way, since the position of the filament is really the result of the successive eruption of active regions and diffusion of the unipolar regions left behind.

Filaments usually disappear by eruption (Figs. 9.4–7). Remarkably, they always erupt *upward*; despite all the work people have done trying to understand what supports them, that is no problem for the Sun; it has trouble holding them *down*. Any reasonable model of filaments must not only support them but also render them stable against upward perturbation. The initial phase of eruption may take days, the material slowly rising to greater height and the absorption increasing as Hα broadens. Virtually every prominence that rises above 50 000 km will erupt in 48 hours. ARF

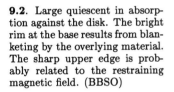

9.2. Large quiescent in absorption against the disk. The bright rim at the base results from blanketing by the overlying material. The sharp upper edge is probably related to the restraining magnetic field. (BBSO)

eruptions are fast (10–20 min) and a large flare occurs; QRF erupt slowly (hours) and only a little brightening occurs. In some cases the filament simply appears to diffuse away as it rises. Often the filament forms again. In active regions, there is enough material on hand for re-formation to be fairly rapid. But the new filament is seldom as large as the first; the eruption, after all, involves a change, usually for the worse, in the magnetic stabilizing system.

All solar flares are associated with filaments. These are not simple eruptions; a sudden heating, brightening, and expulsion of the prominence occurs inside the active area (Fig. 9.6). The exact role of the prominence is not known, but the erupting filaments usually consist of material accumulating in sheared magnetic fields. Often the erupting filament overlies a sunspot. Sunspots must have vertical fields and filaments, horizontal ones. So there must be a sharp turn in the field, indicating a curl and a current. This is the source of the flare energy. How did it get that way? Probably the sunspot pushed under the plage as it grew.

Many filaments have a sharp upper bound, perhaps reflecting the fact that they are held down, with arches reaching raggedly to the surface; but the connection of the arches to surface features has not been estab-

9.3. Another hedgerow, showing the horizontal chromospheric structure underlying. The bending of spicules near the filament suggests that the magnetic field is axial. (BBSO)

lished. Usually filaments show little or no motion; one must guard against a tendency to remember the most interesting films with the most motion. Liszka (1970) studied limb spectra of 100 prominences and found only a few per cent with velocities above 7 km/sec. Engvold (1972) found from spectra that the trunks of hedgerows showed little motion, but the irregular parts showed shifts from 1–4 km/sec. Big Bear data confirm this. An unpublished study by Margaret Liggett of our extensive films showed no large-scale motion in half the cases, with perhaps one fourth gradually draining down and one fourth flowing upwards. Most people are confused by the fact that post-flare loops all pour down as material condenses from the corona, and some of the best early films showed downward motion.

Remarkably, some prominences rotate. Although this was known to be common in erupting events, it was first recognized in spectra of quiescents by Öhman *et al.* (1968). Sometimes the whole prominence slowly rotates, one half rising and the other falling (Fig. 9.8). Liggett and Zirin (1984) have described a number of interesting cases, including some where Doppler shifts give the three-dimensional rotation. Sometimes (Fig. 9.9) one part of a hedgerow is in violent rotating motion. Since the lines of force are attached to the surface and the prominence must be supported by the

9.4. A quiescent about to erupt. Once an arch breaks loose or rises beyond 30,000 km, it will soon erupt. Spicules appear in the chromosphere below. (BBSO)

field, it is hard to understand how gas threaded by field lines can rotate continuously without twisting them until they resist further motion. In fact, each rotating element only goes around about once.

The prominences of Class 2 are spectacular. The flare itself may be a bright blob (Fig. 9.10) elevated above the limb. There is no official level at which flare intensity is reached but the threshold is about twice the chromospheric Hα centerline brightness. Sometimes a small prominence in which the flare starts will rise and expand (Fig. 9.6) majestically in a great eruptive prominence, usually a loop, while intense emission appears on the surface. Or the flare may blow away overlying material, producing a spray, bits of material flying out in a wide angle. This occurs in the most violent flares only. If the material ejected is well collimated, it is called a surge (Figs. 9.11–13). Surges often return to their point of origin and occasionally produce a splash when they come back down. They are due to small flares in strong, uniform fields, and typically occur near a dominant sunspot, which produces the required configuration. The velocities are 50–100 km/sec and the flares involved are usually small. Often the surge will start out bright and become dark as temperature and density are reduced. There is a basic difference between surges and loops as compared

9.5. A big eruptive photographed with the Climax coronagraph. Events of this sort begin as big filaments and are usually thermal, with a soft X-ray spectrum. (HAO)

9.6. A filament overlying an active region rises for 12 minutes, then erupts outward in a graceful loop. Much of the material is low excitation and still absorbing.

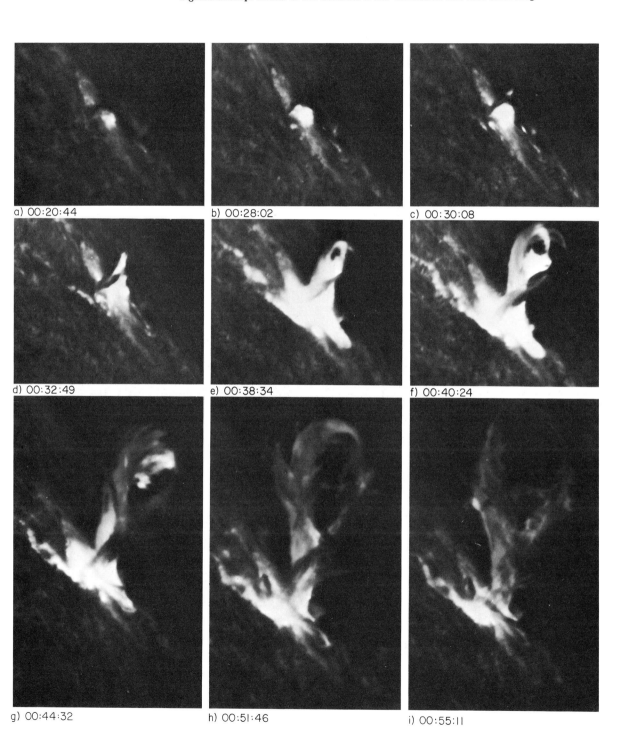

a) 00:20:44

b) 00:28:02

c) 00:30:08

d) 00:32:49

e) 00:38:34

f) 00:40:24

g) 00:44:32

h) 00:51:46

i) 00:55:11

to other prominences. Surges and loops are simply material moving along force lines, driven by either gravity or pressure. By contrast, erupting filaments, sprays and other ejecta are driven by magnetic forces in the fields which thread them.

The rapid motion that we see in prominence films is only visible because the playback of stop-motion films is usually accelerated several hundred times. If the smallest resolution element is 1 arc sec (730 km), we might expect to see the motion in real time if it covered one element per second,

9.7. (a) A flare near the limb ejects a loop, which is better seen in the (b) overexposed frame. (BBSO)

or 700 km/sec in real time. Even such speeds would be hard to detect and are quite rare; I known of only two in all my years of observing. Typical surge velocities are 50–200 km/sec; flare sprays are usually 200 km/sec, but can reach 2000 km/sec. By contrast, loop prominences rain down at about 10 km/sec, and quiescents show motions <0.5 km/sec.

The physics of material ejection has not been explored. Although the flare is a magnetic phenomenon, and surges certainly are channeled by the fields, it is not established whether the gas is accelerated by magnetic or gas pressure effects. The highest temperatures observed in major flares are 30 to 50 million deg, which corresponds to a mass velocity of 600 km/sec. If this energy were channeled it would account for almost all ejecta, and it is a reasonable explanation for surges. But most of the eruptive events behave like magnetic instabilities rather than pressure-driven explosions. The ejecta typically show constant velocity, indicating that the expelling force, gas or magnetic, is present outside the originating region. Surges, however, often decelerate and fall back.

Loop prominences appear (Figs. 9.14–15) after great flares. The hot material brought into the corona by the flare cools after the energy input ceases and rains down to the surface. The dense, hot post-flare loops have sufficient excitation to appear in emission against the disk.

Prominences exist only where mechanical forces can support cool, heavy material above the surface and restrain it from leaving the Sun. This can occur in three ways, given in Table 9.2. By hydrostatic support we mean that the material is lifted into the corona by flare heating and condenses and falls as permitted by viscosity. By ballistic support we mean that the material is supported by the inertia of its initial velocity (although magnetic effects may occur). By magnetic support we mean that the material is held up by magnetic fields. The last is the most common class of prominence and the only stable one. The magnetic support takes place only along horizontal field lines and may be unstable against upward displacements.

Table 9.2. Prominences classified by type of support.

Support Mechanism	Type of Prominence
Hydrostatic support	Loop prominences Coronal rain
Ballistic support	Sprays Surges
Magnetic support	All filaments

9.2. Prominence magnetic fields

When the Babcocks made their first magnetograms, they immediately saw that filaments marked lines of magnetic field reversal. Subsequent observations showed that this was almost invariably the case, which made sense because the field can hold up the material only if it is horizontal. All the great unipolar regions are separated by filaments from their neighbors of opposite polarity, and every time we find a polarity inversion we find a filament (Figs. 7.6, 7.13). In most cases the field lines are parallel to the inversion line. Why don't they connect directly across the polarity inversion line? In some cases they do; some examples of "field transition arches" (FTA) are indicated in Fig. 7.13. Post-flare loops also connect opposite fields directly across the boundary. But in all other cases the connecting lines of force run along the boundary, and all substantial and stable prominences reside in such a sheared boundary. Bright X-ray arches parallel to the filament (Fig. 8.4) indicate that coronal fields just above it also run along the inversion line, but on a large scale Fig. 8.2 shows that the loops go across the boundary. Thus the shear boundary is a relatively shallow phenomenon which somehow stabilizes the boundary.

How do we know that the field is sheared? Hα and UV observations show a broad channel about 50,000 km wide in which filaments lie. This channel is free of chromospheric network or any sign of chromospheric or coronal heating. The Hα fibril structure in the filaments and the underlying chromosphere runs parallel to the filament axis. Foukal (1971) pointed out that spicules on either side of the channel turn in opposite directions as they approach it, depending on what polarity they arise in.

In quiet region filaments the Hα effects are not so well marked; often there is no channel in the chromosphere. But HeII λ304 shows a channel underlying almost all filaments. Serio *et al.* (1978) show X-ray arches parallel to an X-ray cavity along the filament axis. After the filament erupts, a fine set of X-ray and Hα loops appears arching across the cavity about 100,000 km above the surface, the peak emission at the loop tops.

The flares produced by filament eruption (Fig. 9.13) give a three-dimensional picture of the field structure. The filament channel shows the usual magnetic field running along it, and the Hα structure shows no postflare change; but postflare loops in Hα and X-ray are perpendicular to the channel and terminate in two bright strands running along either side of it.

How does the shear boundary form? The process takes place over a fairly long time, since the unipolar regions require several rotations to grow. Martin (1973; see also Smith 1968) has shown how they appear as channels of growing horizontal field. Fig. 10.1 has an example of the formation of such channels. Far from the emerging dipole the returning field lines in the photosphere are parallel to the dipole axis. When the dipole encounters a unipolar region of opposite local sign, the field lines will join across the boundary, but preserve their horizontal form. As the region decays, the closely bound field lines connect to distant fields across the growing boundary. The filaments are somewhat unstable as the region grows – the one in Fig. 10.1 erupted several times – but the sheared channel remains for a long time. The configuration is evidently a stable one, the arched fields over the top restraining the buoyancy of the filament.

Even inside an active region, where the sunspots are originally linked by FTA, a shear boundary forms. This may be due to a more diffusive process, the fields stretching along the inversion line as they spread. Because of their great length and non-potential form, the flux lines form a trap for cool material and a prominence forms.

The turn that field lines take as they go into the filament channel requires a current to stabilize it; the sheared state must be a force-free field. A critical question is whether the filament is simply material that accumulates in this trap, like petroleum along a fault, or whether the presence of

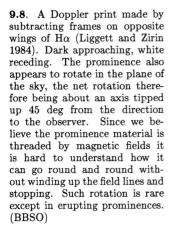

9.8. A Doppler print made by subtracting frames on opposite wings of Hα (Liggett and Zirin 1984). Dark approaching, white receding. The prominence also appears to rotate in the plane of the sky, the net rotation therefore being about an axis tipped up 45 deg from the direction to the observer. Since we believe the prominence material is threaded by magnetic fields it is hard to understand how it can go round and round without winding up the field lines and stopping. Such rotation is rare except in erupting prominences. (BBSO)

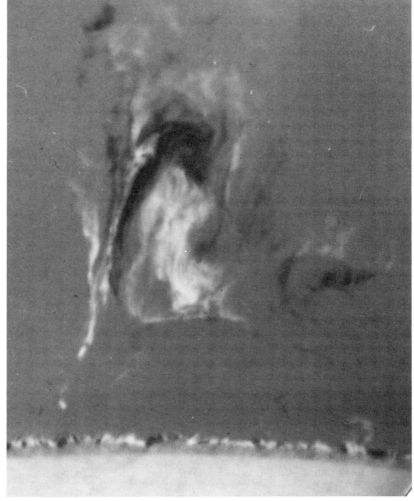

the filament is necessary to stabilize the boundary. Possibly the mass of cool material in the filament discourages magnetic buoyancy; in any event magnetic boundaries are rarely found without a filament.

Beside the deduction of field patterns from morphology, prominence magnetic fields may be measured by conventional Babcock magnetographs or by the change in polarization of scattered photospheric light. The first measurements were made in Hβ by Zirin and Severny (1961; also Zirin 1961) with a conventional magnetograph; later, special self-calibrating instruments were built for the purpose (Lee, Rust, and Zirin 1965) and accurate measurements were made by Harvey, Rust, Malville and Kotov (reviewed by Leroy, 1979). The measurements are difficult because prominences are faint and the fields relatively weak. But the various determinations agree on fields of 5 or 10 gauss in quiescent prominences and 100 gauss or more in active ones. The prominence shows a characteristic reversal of Zeeman signal relative to the disk, indicating that we are measuring an emission line. The magnetic field pressure far exceeds that of the gas, but is anchored in the photosphere.

In recent years attention has turned to measuring the field by the Hanle effect. Hanle (1924) observed rotation of the plane of polarization of lines from a mercury source in a weak magnetic field. The theory was worked out by House (1970a,b; 1971) and applied to prominences by Bommier

9.9. A hedgerow prominence with one rotating part R (Liggett and Zirin 1985). The vortex went around several times, each time breaking up after a single rotation. (BBSO)

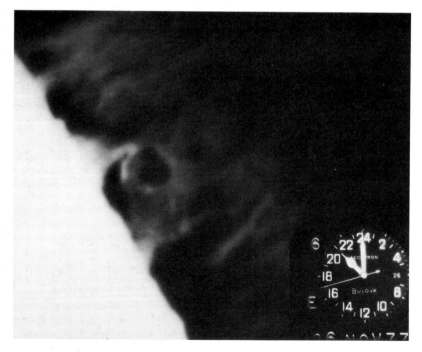

and Sahal-Brechot (1978). Normally the prominence radiates by resonance scattering of photospheric light in a particular transition, say $2\,^3P_0 \rightarrow 3\,^3D_1$ for HeI D3, and re-emitting the same. Polarization parallel to the surface results. But for fields above 15–20 gauss, the Zeeman splitting exceeds the spin–orbit splitting, and the J-levels are mixed; the light is re-emitted in a transition involving different J-levels and forgets its original polarization. For fields below 20 gauss the polarization is simply reduced and rotated, the rotation occurring because the different m_J levels are mixed. Since the polarization of the $\Delta m = \pm 1$ components is $\pm \lambda/4$, the mixing of the sublevels produces varying rotation of the plane of polarization for different field values. In D3 this occurs above 9 gauss. By measuring the angle between the plane of polarization and the tangent to the limb along with the reduction of the polarization relative to the theoretical maximum, we can calculate the field strength. Leroy (1979) summarizes results of Hanle effect measurements, which also give some idea of field direction. The magnitudes are similar to the data obtained with direct magnetograph observation.

There has been considerable attention paid to the problem of magnetic support of prominences, although, as I mentioned above, the Sun mostly has difficulty in holding them down. The first serious filament model, still endorsed by many, is the Kippenhahn–Schlüter (1957) model, in which the filament is supported by sagging arches perpendicular to the filament axis at the base of a helmet streamer. Even if such a configuration existed, it

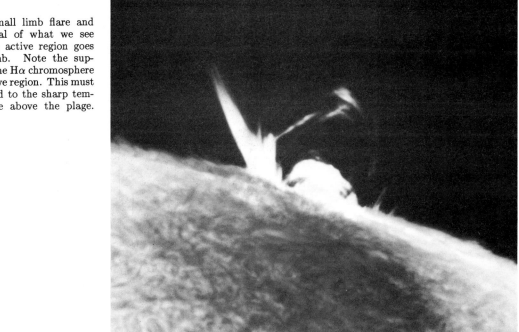

9.10. A small limb flare and surge, typical of what we see as a modest active region goes over the limb. Note the suppression of the Hα chromosphere over the active region. This must be attributed to the sharp temperature rise above the plage. (BBSO)

would be unstable because plasma cannot be contained by field lines convex to it. A lower energy state exists where the filament slips through the lines of force which shorten behind it to a potential configuration. In the real Sun the Kippenhahn–Schlüter model does not occur; the observed fields run *parallel* to the filament and filaments never fall down. Furthermore, all the field lines are rooted in the surface and may be expected to be convex upward; the problem is to understand what keeps the material from running down the convex field lines. Anzer (1979) and Priest (1982) discuss all the models in detail.

Kuperus and Raadu (1974) proposed a model in which the prominence forms in a neutral sheet suspended above the surface. In this model, convex loops crossing the neutral line underlie the current sheet. In the real filament, as outlined earlier, the loops cross the neutral line above the filament. Van Hoven and Mok (1984) analyzed the condensation of material in a horizontally sheared field, showing how thermal equilibrium could occur, but pointing out that the gravitational stability problem always exists. Other attacks on this difficult problem have been made; unfortunately few treat the actually observed sheared configuration. The observers don't know how to calculate and the calculators don't know what is observed.

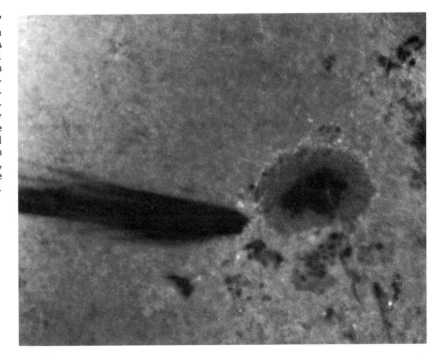

9.11. A comet hitting the Sun? No, a dark surge ejected from an emerging flux region near a sunspot. The photo is at Hα − 1 Å, so the line of sight velocity is at least 50 km/sec. The previous photos show ejecta and filaments as complex, twisted objects which must be threaded by field lines. But most surges are linear, gas moving along field lines. Because the *p* spot field is dominant and regular and open, flares in front of them produce well-collimated surges. (BBSO).

**9.3. Prominence spec-
tra**

What do we expect to be the state of a cloud of gas suspended above the solar surface? We know most of the energy inputs and outputs and should be able to predict what physical conditions will occur.

First we must distinguish between transient and stable prominences, as in Table 9.1. While a transient prominence cannot change much from its formation temperature during its brief lifetime, a stable one will reflect the equilibrium produced by its environment. The possible energy sources for a quiescent prominence are: photospheric radiation, conduction from the corona, gravitational energy release and magnetohydrodynamic heating. Of these only the first two are known to be important, and the first is the dominant factor.

Order-of-magnitude calculations quickly show the dominance of photospheric excitation in the scheme of things. From Eq. (8.7) the upward photo-excitation rate is $1/2$ times the Einstein A value times the Boltzmann factor, which is 0.0223 for 5770° with dilution factor 0.5 and absorption line depth 0.2. But in this case the A value is 0.6×10^8 instead of 100. For Hα, for example, this is about 2×10^5 absorptions/sec. If we apply Eq. (4.64) with $\chi = 1.89$, $f = 1$ and $P \approx 0.1$, we find the collisional excitation rate $N_e C_{12} \approx 10^{-7} N_e$ at 10 000° and ten times higher at 40 000°. For normal densities $N_e \approx 10^{10}$ collisional excitation is several orders of magnitude below radiative. At temperatures above 20 000° and $N > 10^{12}$ collisions will dominate and the prominence will appear bright.

Little of the photon energy heats the gas, because it is re-radiated; thus the photospheric irradiance mainly determines the excitation. The photospheric photons affect the kinetic temperature through photo-excitation; if that were the dominant factor it would produce a Maxwellian distribution at the effective temperature of the chromosphere in the various continua, and there should be a balance between this heating and the energy losses of the electrons by collisional excitation of lines.

Without going into a big calculation, we can estimate what the temperature of a prominence in pure radiative equilibrium with the photosphere will be. It absorbs in 2π steradians (except for minor limb darkening corrections) and emits into 4π. So the temperature is obtained by equating

$$\frac{\sigma T_{\text{eff}}^4}{2} = \sigma T_{\text{prom}}^4 \tag{9.1}$$

with $T_{\text{eff}} = 5770°$ the effective temperature of the photosphere (Eq. 6.32). The prominence temperature is obviously $5770/\sqrt[4]{2} = 4850°$. We know this is a simplistic model, because the prominence is transparent in the continuum and primarily absorbs and emits radiation in the hydrogen and helium lines and the Lyman continuum. But the brightness temperature of the photosphere in those radiations is even less than 5770°, so 4850° is an upper limit, not far from the truth. Only quiescent prominences are cool; active prominences, reflecting their origin, can be quite hot.

Since the prominence is immersed in the million-degree corona, why doesn't it heat up? First, the prominence is so dense compared to the corona that it would radiate away the energy as fast as it came in. In fact, the absence of corona inside helmet streamers is attributed to such a "refrigerator effect". Second, the prominence is supported by magnetic field lines that go to the surface, so hot coronal ions would have to cross the field lines to heat it. Lastly, most prominences are situated in the middle of a region of horizontal field where there is little chromospheric or coronal heating. However, all prominences emit weak ultraviolet lines; the intensity was estimated by Orrall and Schmahl (1976) as 1/4 that of the intranetwork transition region. This emission must be produced by limited conduction and UV irradiation from the corona.

The temperature in the various continua differs somewhat. The effective temperature of the chromosphere in the Balmer continuum is about 5500°; Noyes and Kalkofen (1970) found it to be 6400° in the Lyman continuum, but the slope of that continuum corresponded to 9000°. This implies that Lyα comes mainly from hot spots, while we know prominences are in cool channels. So the main continua heating the prominence could both reasonably be 5500°. We can derive the photo-ionization rate from the

9.12. A fine limb spray; these are associated with larger flares and less dominant fields than surges. (BBSO)

9.13. (a) A two-ribbon flare resulting from the eruption of the main filament of the active region of Figs. 8.4 and 10.9–10.12. Skylab X-ray images showed intense emission from the loop tops which drain down to these ribbons. Note that the fibrils marking the lines of force still run parallel to the filament axis, while the post-flare loops cross it. (b) Four hours later, the ribbons have moved apart and the filament has reformed. The seeing is better because it is later in the day, the first frame is near dawn. (BBSO)

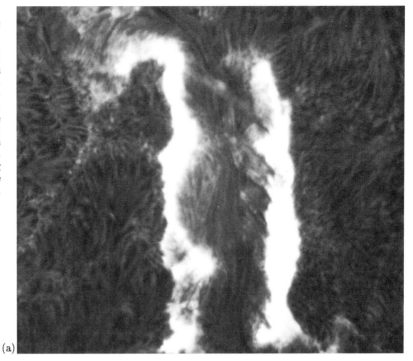

(a)

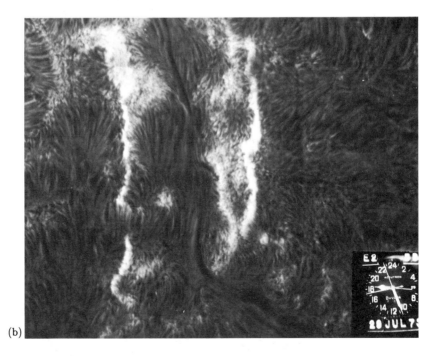

(b)

recombination rate by Eq. (4.45). From Eq. (4.50) the recombination is $R = 7 \times 10^{-13}$. The photo-ionization is thus

$$F_{1k} = \frac{1}{2} \left(\frac{N_2 N_e}{N_1} \right) R_{1k} \qquad (9.2)$$

where a dilution factor $1/2$ has been introduced. The Saha equation (Eq. 4.33) for this temperature gives $N_2 N_e / N_1 = 4 \times 10^8$, so the photo-ionization rate is .0015/H atom/sec, or once every 10 min. Thus heating by photoionization is unimportant. The effect of conduction from the corona is not clear, but the data of Orrall and Schmahl suggest it is not great.

Quiescent prominences are mechanically and thermally stable, and it is likely that they are in a temperature plateau. If the emission is a steep function of temperature, a small increase in T will produce a great increase in emission and cool the system; a decrease in the temperature will re-duce the energy losses so sharply that the normal heat input mechanisms (whatever they are) will raise the temperature back to its previous value. Although the main energy source, photospheric radiation, is constant, that energy is rapidly re-radiated, while the cooling hydrogen emission increases rapidly with temperature.

The existence of temperature plateaus in the solar atmosphere was first proposed by Athay and Thomas (1956), who pointed out that large changes in the emissivity of a plasma occur when hydrogen or helium are partially ionized. Similarly, there are unstable regions where the increasing ioniza-

9.14. A very large loop display; the loops go higher and higher in the thermal cooling of the flare. The use of a coronagraph en-hances the brightness. (SPO)

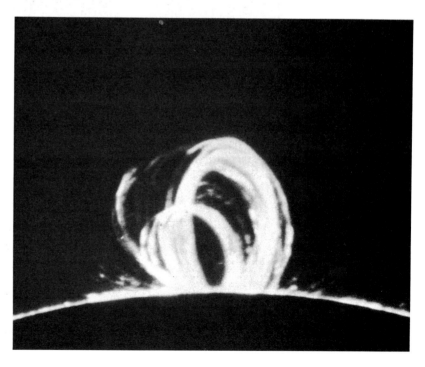

tion robs the electrons of hydrogen atoms to excite, so that the hydrogen emission per atom begins to fall with increasing temperature. The gas is then thermally unstable, and the temperature rises until another dominant means of energy loss turns up. There is some evidence for this behavior, as we shall see from the spectral observations.

Spectroscopy is a powerful tool to investigate the physical conditions in prominences; and because we may see the whole structure of the prominence suspended above the surface, the picture is simpler than the chromosphere, where we cannot disentangle the chromospheric radiation from that of the photosphere and spicules. The prominences are faint, and it was only when Waldmeier observed their spectra with a coronagraph–spectrograph

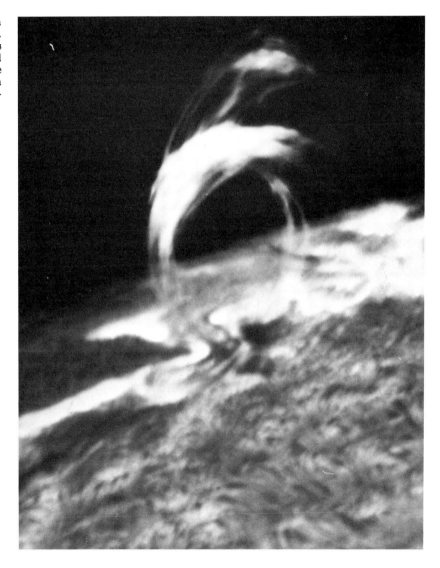

9.15. Post flare loops ending in the flare ribbons inside the limb. The soft X-ray emission comes from the loop tops. A second arcade is forming above. These loops are as bright as those on the previous frame, but the exposure is for the disk. (BBSO)

9.16. An (a) Hα – (b) videomagnetogram pair (6/6/81, 19:00 UT) showing how a filament separates fields in enhanced network. Beside the large filament there is a small ring shaped filament (F) surrounding a bit of black polarity of unknown provenance. Some magnetic elements are numbered to aid matching. Note that knot 4 is on the wrong side of the filament; this is probably a projection effect, the filament lying higher on a tilted neutral plane. The field lines run along the filament channel; spicules from black areas (such as 3) turn west (left) into the filament channel, those from white polarity turn east. (BBSO)

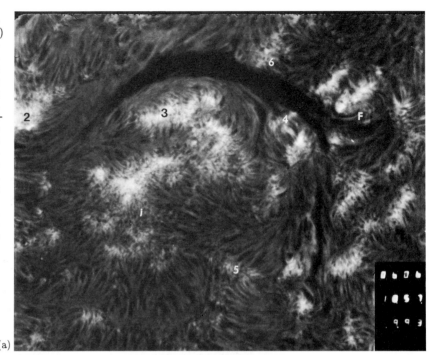

(a)

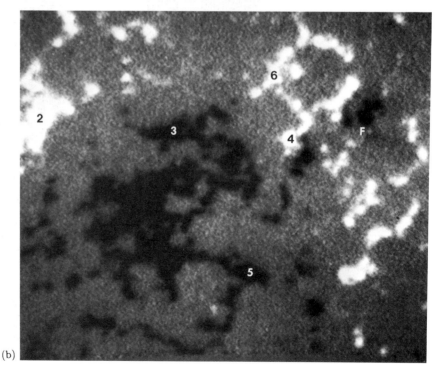

(b)

9.17. Spectra of a limb flare, showing strong H and He line emission. Weak, Doppler-shifted metallic line emission from flare ejecta are also seen. A weak electron-scattering continuum passes through the core of the flare, which displays considerable Doppler broadening. (Acton and Zirin 1967) (HAO).

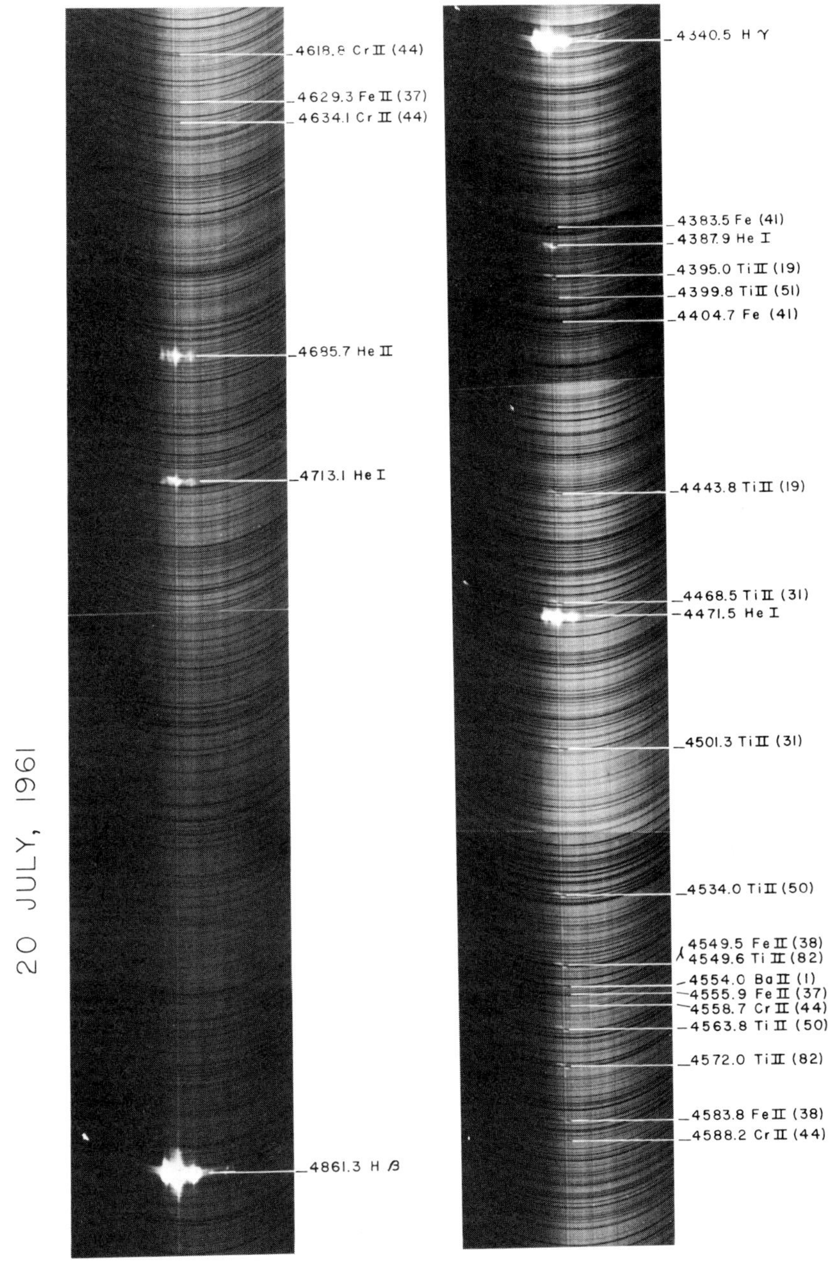
20 JULY, 1961
4618.8 Cr II (44)
4629.3 Fe II (37)
4634.1 Cr II (44)
4685.7 He II
4713.1 He I
4861.3 H β
4340.5 H γ
4383.5 Fe (41)
4387.9 He I
4395.0 Ti II (19)
4399.8 Ti II (51)
4404.7 Fe (41)
4443.8 Ti II (19)
4468.5 Ti II (31)
4471.5 He I
4501.3 Ti II (31)
4534.0 Ti II (50)
4549.5 Fe II (38)
4549.6 Ti II (82)
4554.0 Ba II (1)
4555.9 Fe II (37)
4558.7 Cr II (44)
4563.8 Ti II (50)
4572.0 Ti II (82)
4583.8 Fe II (38)
4588.2 Cr II (44)

that the weaker lines could be studied. The conventional solar telescope scatters so much light that only the stronger lines may be photographed. Even with the coronagraph we can obtain complete spectra of the brighter prominences only, since we are limited by the sky brightness.

Waldmeier (1951) classified prominence spectra by the intensity of the FeII line at 5169 Å (b_3) relative to the lines of the neutral magnesium triplet at λ 5184, 5173, and 5167 (designated b_1, b_2, b_4 by Fraunhofer). The theoretical ratio of the Mg lines is 5:3:1, and this ratio is usually found. The FeII line usually has an intensity 1.5 on this scale, but can vary. Waldmeier (Table 9.3) used the three MgI components as a scale for the increasing intensity of the FeII line.

Waldmeier found that almost all of the prominences he studied were in Class III. He found no clear-cut correlation with the nature of the prominence. The fact that the overwhelming majority of prominences are QRF

9.18. Evolution of an active region filament in two rotations: (a) June 7, 82: the filament curls in to the main sunspot of a highly active region. (b) One rotation later the activity has died down and the filament, fed by material from the active region, has grown considerably. One would think that with time the poles of the active region would draw together and sink back below the surface, but unknown forces prevent this. The filament may play a role in this limitation. The region, which was an interloper from the other hemisphere or else completely inverted polarity, returned once more in similar form, then both polarities expanded away from the neutral line. Other developments in this region appear in Figs. 11.11–13. (BBSO)

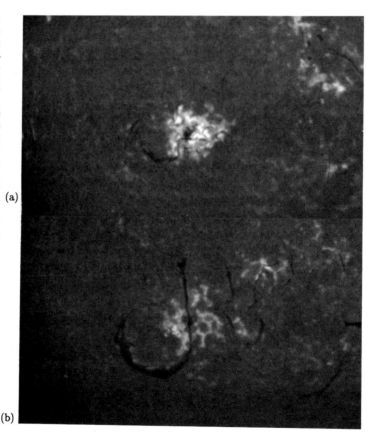

(a)

(b)

means that we usually get the typical Class III ratio. A direct comparison with the classification by Table 9.1 would be informative but does not exist.

Tandberg-Hanssen (1963) found that FeII $\lambda 4924$, which belongs to the same multiplet as b_3, is enhanced relative to TiII λ 4572 in some flares and loop prominences. Since the lines of a multiplet must change in concert, we may expect that the FeII line in the Mg triplet also changed, therefore producing a change in the Waldmeier criterion. So although the relative intensities of the metallic lines in prominences do not change substantially, the FeII lines may be enhanced in hot prominences. This is found in higher excitation stellar sources such as quasars. Detailed photometric measures of the H and He line strengths were made by Landman and Illing (1976). Landman (1984, 1985) and coworkers have made accurate measurements of many prominence lines, along with analyses; unfortunately no comparable data are available for active prominences.

Zirin and Tandberg-Hanssen (1959) found a sharp distinction between the spectra of active and quiescent prominences.Quiescents radiate a cool spectrum with weak or absent HeII and relatively strong ionized metallic lines. Active prominences radiate a hot spectrum marked by relatively strong HeII lines and weak or absent metallic lines. The H and HeI lines are greatly enhanced; CaII H and K remain bright, but do not increase as much as H and He. The same is true in stellar flares. The spectrum lines in hot prominences are always greatly broadened by thermal and macroscopic motions; those in cool prominences are narrow unless they are erupting. The difference is clear in Fig. 9.19. The quiescent shows a number of lines due to singly ionized metals, which are absent in the surge. Hot prominences (Fig. 9.20) show strong lines of HeII: $4{\rightarrow}3$ (4686 Å), $7{\rightarrow}4$ (5411 Å) and $9{\rightarrow}4$ (4942 Å). A temperature of at least $30\,000°$ is required to excite these lines, which have an excitation potential of 50 eV. Because the HeI line $\lambda 4713$ is close to $\lambda 4686$ and about the same intensity in the hot prominence, their ratio is an excellent discriminator of temperature. Another good pair to watch are CaII H *vs.* Hϵ; the latter is always enhanced in flares. The characteristics of the two classes are summarized in Table 9.4 along with several line ratios that may be used as criteria.

Table 9.3. Waldmeier's prominence spectrum classification.

Class	Mgb line ratios
I	$b_3 > b_1$
II	$b_1 > b_3 > b_2$
III	$b_2 > b_3 > b_4$
IV	$b_3 = b_4$
V	$b_4 > b_3$

The characteristics mentioned apply in all regions of the spectrum. Lines of ionized or neutral metals as intense as the neutral helium lines in quiescents disappear in the high temperatures of active prominences. Within the two types of spectra there is little variation. The "cool" spectrum reproduces itself monotonously in prominence after prominence, with only minor variations in the relative intensities of the lines. The line ratios also vary little with height or position in a given prominence. We can use the term cool because the metallic lines that disappear have low excitation and must be emitted by a relatively cool gas.

Although the metallic lines are comparatively weak in hot prominences, they may still be seen in the brightest flares and loops. The relative intensities of these weak metallic lines among themselves is essentially preserved, as though there is a certain cool state that a small amount of the material has reached. This implies that there is a range of temperature in the prominence, which makes sense if the material is cooling. Of course the effect may just be line-of sight confusion. Whether there are definite temperature plateaus which emit the hot and cool spectra is not certain. For example, the quiescent prominences radiate some UV lines of high ionization, presumably from the prominence-corona interface.

Clearly the prominence spectra are related to their dynamics. The loop prominences condense directly out of the million-degree corona, so that they must represent the first radiation in the visible spectrum from the cooling material. Conversely, the material in a surge or spray is in the process of being converted to coronal material as it is expelled outward. So it should show the radiation of material at subcoronal temperatures. All of these are transient phenomena, but they show the same spectrum because they radiate principally at the temperatures at which hydrogen is being ionized. There is a little variation in the hot spectrum, depending on the phenomenon; ionized He lines are not quite so strong in surges.

Widing *et al.* (1986) recently studied the EUV spectrum of a great erupting prominence observed with the overlappograph. The spectrum is iden-

9.19. Spectrum of a flare surge (S) with an ARF (A) right and a quiescent prominence (Q) left photographed with the SPO coronagraph in the spectral range 3540 Å to 4515 Å. A curved slit was used so the absorption lines, which are the Fraunhofer spectrum scattered by the sky, are curved; the prominence emission lines appear on this background. The Balmer series lines (H), are broad in the surge, somewhat less in the ARF and narrow in the quiescent; they converge to the Balmer limit at top right. Further to the blue the Balmer continuum radiation starts to the red of the Balmer limit because the H lines are merged by Doppler and Stark broadening. The Balmer series in A can be followed to n=26, giving log N_e <12. The narrow metallic lines (*e.g.* SrII λ4077) are absent in the surge because it is too hot. In the middle band we have CaII H and K in the center, Hδ at bottom and H7 at top. Hϵ, next to CaII H, is relatively enhanced in the surge. At the bottom Hγ appears as well as HeI λ4471. All of the fainter lines in the surge are helium lines: λ4026 in the center spectrum and λ4471 in the left, near the λ4101 marking. The lines at 3864 Å, 4121 Å and 4144 Å also appear. There is a faint continuum associated with S, probably electron scattering of photospheric light. (SPO)

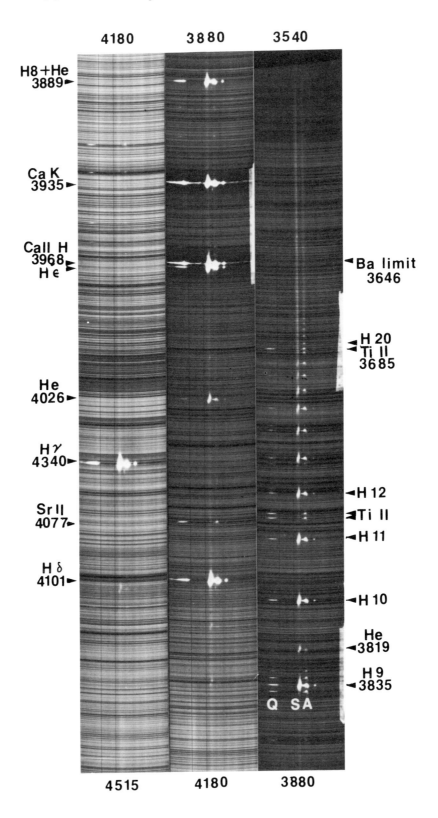

tical with that of the transition zone, except that only lower ionization coronal lines appear as the erupting material is heated. The spectrum shows the entire range of lines from Hα through HeII 304 to MgIX. Widing *et al.* find relatively little change in the emission measures from one ionization stage to the next. Since they observe five ionization stages of Ne and four of Mg, they can get reliable abundances; they find little weakening of high-FIP lines. So the returns are not yet in on where the high-FIP lines are filtered out.

Various other line ratios can be used to deduce the state of excitation or the density; these are summarized by Hirayama (1979). The singlet : triplet ratio in helium can be used as a diagnostic. The lines $\lambda5678$ (D3) and $\lambda6678$ come from corresponding transitions in HeI. In LTE (high collision rate) or in pure recombination, they should occur in the ratio 3 : 1, and in normal prominences D3 should be much stronger because it is populated from the metastable $2\,^3S$ state. Zirin (1956) calculated ratios from 50 : 1 to 20 : 1, proposing that the ratio could be used as a temperature diagnostic. In quiescent prominences Landman and Illing (1976) find a ratio around 11 : 1, without much variation (so much for my ability to calculate). But in flares the ratio may reach or pass the LTE value; Lites *et al.* (1986) measured a ratio of only 2 : 1 in a disk flare. I am dubious of such a small ratio, but it would be possible if the lines approach the Planck maximum. The ratio in active prominences at the limb has not been measured.

Another ratio of interest is that of the helium line $\lambda3889$ ($3\,^3P \rightarrow 2\,^3S$), which is the next series member above $\lambda10830$, to the neighboring Hς, at 3888 Å. Since these two lines are close they give a good opportunity to compare H and He line strengths. But different studies have given contradictory results.

Table 9.4. *Characteristics of "cool" and "hot" prominence spectra.*

"Cool"	"Hot"
Narrow lines	Broad lines
Strong hydrogen	Strong hydrogen
Strong HeI	Strong HeI
Weak HeII	Moderate HeII
Strong metals	Weak metals
Criteria	
HeI $\lambda4026 \sim$ SrII $\lambda4077$	HeI $\lambda4026 \gg$ SrII $\lambda4077$
HeI $\lambda4713 \sim$ TiII $\lambda4572$	HeI $\lambda4713 \gg$ TiII $\lambda4572$
HeI $\lambda4713 \gg$ HeII $\lambda4686$	HeI $\lambda4713 \sim$ HeII $\lambda4686$

Ultraviolet observations give us critical data on the high-temperature part of prominences, which, for quiescents, is the prominence–corona (P–C) interface. Orrall and Schmahl (1976, 1980) analyzed a rich store of data on UV lines and the Lyman continuum (Lyc) obtained with the Harvard spectrometer on Skylab. They found a brightness temperature for Lyc of $6316 \pm 75°$ and a slope corresponding to $7524 \pm 739°$. While the errors are no doubt greater than they suggest, these numbers are probably a good estimate of the temperature in hotter parts of the prominences studied. Since the slope is due to the velocity distribution of recombining electrons, it measures the true kinetic temperature of the region emitting the Lyc; but this technique preferentially picks out the hotter parts of the prominence that are ionized and that emit the Lyc. Unfortunately no UV data for hot prominences have been published.

9.20. Coronagraph spectra of loop prominences following a limb flare, with slit jaw frames at right. The intense scattered light continuum from the loop is coincident with the CaXV lines and the HeII line. Note the particularly strong continuum in a narrow core in the λ5694 frame. The FeXIV and FeX lines are only slightly enhanced; it is too hot. The last frame shows continuum from the two branches of the loop. (SPO)

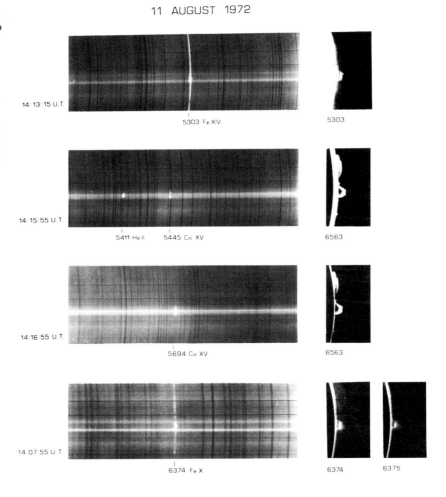

11 AUGUST 1972

Another clue to the kinetic temperature in prominences is the intensity of HeII 304 Å. This line appears to come from throughout the prominence. Its intensity relative to the chromosphere can be judged from Skylab images such as Fig. 9.23. In λ304 prominences at the limb are about as bright as the intranetwork regions; thus on the disk there is little contrast, because the brightness matches the background. Only the ARF are visible in absorption in λ304. λ304 emission has a steep temperature dependence, so active prominences are quite bright. Every filament, even in quiet regions, is marked by an extensive dark channel in λ304, due to the absence of coronal excitation. Heasley and Milkey (1976, 1978) found considerable dependence of various line intensities, particularly He, on the exact UV irradiation. This requires careful determination of the UV flux from the dark filament channels.

The enhanced UV in the active region results in greater λ304 optical depth. For this to produce an increase in filament darkness, the excitation temperature in the prominence must be less than the photosphere, even though the ionization temperature is higher. Even if the prominence temperature exceeds that of the photosphere, the low density produces a brightness temperature much less than the LTE value.

Simple application of Eq. (5.31) to measured line widths yields firm upper limits on quiescent prominence temperatures. As smaller and smaller elements are studied and mass motions and self absorption eliminated, narrower lines are found; temperatures from 4500 to 8500°, averaging 6000°, are cited by Hirayama (1979) in his review of prominence spectra. For a long-lived filament the dominant heating effect is the enormous flux of photospheric radiation and possibly conduction. Electron temperatures are dominated by Balmer and triplet He photo-ionization continua, which are generally below 6000° in color temperature because they come from the temperature minimum. So a temperature of 5000–6000° is most likely.

The integrated electron density along the line of sight may be obtained by applying a variant of Eq. (8.3) to the observed intensity of the electron scattering continuum (Zirin 1964); the scattered intensity per electron, allowing for limb-darkening, is

$$I_{scatt} = 1.61 \times 10^{-25} N_e L I_{\odot}. \tag{9.3}$$

Solving for N_e, we have:

$$N_e = \frac{0.62 \times 10^{25}}{L} \frac{I}{I_{\odot}}. \tag{9.4}$$

For a large quiescent this is typically 2×10^{-5}, giving $N_e L \approx 6 \times 10^{19}$. While we don't really know the thickness of prominences strands, L must certainly be less than 10^9, hence the density must be above 6×10^{10}. If the ionization is as low as that recently obtained by Landman (1984), namely

HII/HI = 0.07, the total density reaches 10^{12}, which is too high. From the same data Landman concludes $N_H = 10^{11}$ by comparing Balmer-line intensities with metallic-line intensities from Yeh (1961) and Yakovkin and Zel'dina (1964).

Hirayama (1985) discusses the determination of electron density by the Stark effect, which he regards as the most reliable method. He finds $N_e = 10^{11}$ for five hedgerow prominences and $N_e < 10^{10.5}$ for two old curtain-like prominences. It is difficult to measure lower values with this technique because one has to detect ever higher Balmer lines. He points out that there are significant discrepancies between these values and those from excitation calculations. It is hard to see how the Stark effect results could be wrong, while excitation and ionization are subject to error.

The Lyα emission cannot exceed the Planck function. Prominences are observed in Lyα (Orrall and Schmahl 1976) to be as bright as the disk. The disk intensity is quoted by Vidal-Madjar (1977) as about 3×10^{12} phot/cm^2/ster near line center, corresponding to a brightness temperature about $5000°$. This gives a rough lower limit for the temperature (of course much is scattered chromospheric emission) and is again in the right range. It is significant that prominences are no brighter than the disk in any line,

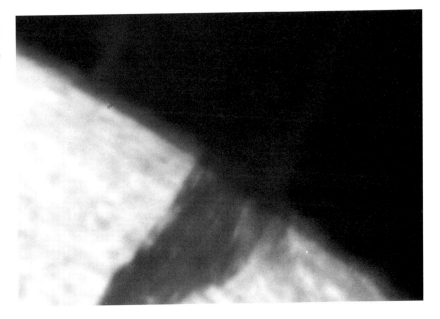

9.21. A filament seen against the limb obscures the chromosphere in Hα. Because the chromospheric spectrum is free of continuum, such observations can be used to place an upper limit on the extent of filamentation in prominences. (BBSO)

which suggests that their temperature is below 6000°.

The ratio of Lyα to Hα in prominences is about unity. The flux from the whole Sun at the earth in Lyα is seven ergs/cm²/sec. We can evaluate the intensity of the whole chromosphere in the center of Hα by multiplying the photospheric flux (167 erg/cm²/Å/sec) by a line width 0.5 Å and a centerline depth 0.1, giving the chromospheric Hα flux at the earth as eight ergs/cm²/sec. Thus the Lyα and Hα intensities from prominences are about equal. Since the energy of Hα photons is five times less than that of Lyα photons, there must be five Hα photons for every Lyα photon. For unknown reasons, this seems to be the case for most astrophysical sources (Zirin 1978b). One would expect more photons in Lyα. For example, each recombination to $n = 3$ gives Hα and then Lyα; but those to $n = 2$, only Lyα. Collisional excitation always produces more Lyα. The only possibility is that in the long optical path many Lyα photons are destroyed; but why should the result just equal the Hα flux?

In analyzing prominence spectra we must account for the obvious fine structure. We measure integral quantities which surely are coming from individual structures denser or hotter than the mean values for the prominence. Because we all like to show our best data, the familiar images of prominences usually show fine threads, and it has been generally supposed that there is a great deal of fine structure. But I have many prominence images of excellent quality that are amorphous and homogeneous. An objective measurement of the degree of fine structure is to measure the degree of extinction of chromospheric emission by a prominence in front of the chromospheric limb band. This cannot be done on the disk, because continuum emission leaks through. But because the chromosphere at the limb behaves like a monochromatic source in Hα, the contrast behind and away from the prominence is a good measure of the transmission. As we see in Fig. 9.21, the chromosphere appears to be completely blocked by the main body of the prominence footpoints. The transmission is so low that it is difficult to measure, but measurements of the jump in brightness in the prominence at the edge of the chromosphere edge indicate that less than a few percent of the chromospheric Hα is transmitted; hence whatever structure there is inside the prominence must cover at least 98% of the line of sight. This limits the density contrast of inhomogeneities, since they overlap at random to cover the line of sight. Monte Carlo calculations show that if the ratio of fibril thickness to prominence thickness is

$$a = r/R, \tag{9.5}$$

then about $4/a$ fibrils along the line of sight are required for the observed upper limit of transmission. These effects were studied in the Lyman continuum by Orrall and Schmahl (1980). They compare the attenuation by Lyc absorption in the prominence of emission lines from areas behind it with the limb emission. Despite a thorough analysis of the radiative transfer, they were unable really to determine the limits of fine structure, although they suggest a ratio between 4 : 1 and 10 : 1. The problem might be solved if one could observe the same prominence in enough different ways. Liggett and Zirin (unpublished) measured a number of filaments in Hα and found the transmission to be $< 1\%$, and probably zero. As in the case of the Lyc, one cannot fix the filling factor, but strong inhomogeneities, cases where the density in the fibrils is ten times the mean, can be ruled out.

In conclusion, the density and temperature in prominences can be estimated as follows:

1. The temperature of quiescent prominences is between 4800° and 6000°, probably closer to the lower.
2. Densities are of the order $10^{11} \mathrm{cm}^{-3}$.
3. Prominences of class 1 have temperatures above 30 000°.
4. Quiescent prominences are fainter than the chromosphere in almost all spectral lines, including the UV, but their spectra are similar.

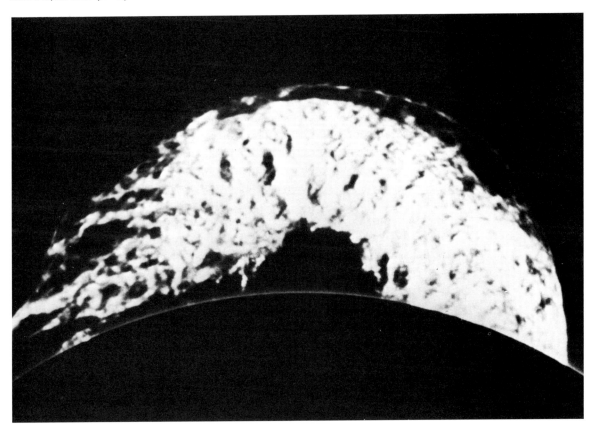

9.22. "Grandpa", the greatest eruptive ever seen, photographed in June 1964. Before erupting, it was a huge filament covering the polar region and rising more than 100,000 km. (HAO)

9.4. Active prominences   We have already mentioned that almost all flares result from instability of ARF located at a highly sheared inversion of the longitudinal field. Sprays, surges and loops result from the flare. The behavior of these different prominences is an important field of study, both for the physical processes they reveal and the light they shed on the flare phenomenon.

Postflare loops are an elegant feature of most large flare. It is not known if they occur in small flares because of visibility problems. But they are absent from some explosive flares, and may require a particular geometry. Their appearance coincides with the end of impulsive brightening and the peak of hard X-rays (HXR); as soon as the great flare energy input ends, the loops appear, and they coincide with the SXR maximum. This does not mean that energy input ceases; as we shall see in Chap. 11, the thermal emission in the loop phase requires continued energy input. In Hα material appears to condense at the top and rain down the sides of the loop; spectra show that these are real motions. Observations of coronal lines, continuum and SXR images confirm that the density peaks at the loop tops, in defiance of hydrostatic equilibrium. That is why the Hα-emitting material rains down. The first postflare loops form quite low, while succeeding loops form higher and higher in the atmosphere. The base of the loops is in the two bright ribbons (Fig. 9.17) of the flare; if one of these strands is near a sunspot, it will be considerably smaller, since the force lines converge. If the shear is great, the loops will first form at about 45° to the neutral line, and the successive loops will be more and more orthogonal to the inversion line. They trace out the overlying, less sheared field.

It is known that the loops form in an extremely hot (40 million deg) and dense corona cloud produced by the flare. When the heat input falls, this cloud can no longer be sustained, the highly ionized material cools and condenses. However some feel that ongoing magnetic reconnection is taking place at the loop tops, feeding in new energy. So as we see the loops form higher and higher, we are seeing not only a progression of recombination, but one of reconnection. Compression of the material at the loop tops, either by thermal instability or by reconnection, will leave the loop top density out of hydrostatic equilibrium and that is why the material comes down. But since even the hottest material shows peak density at the tops, there must either be mechanical compression, or an even hotter, unseen cloud, out of which the new material appears. It is not surprising that the loop tops are hottest; they are farthest from the cool photosphere and come out hottest in any conduction analysis.

There have been a number of calculations of the rate of condensation of loops from the corona. The early model of Lüst and Zirin (1960) depended on suppression of conduction; it showed that high densities and temperatures were required. Moore *et al.* (1980) show that the enormous soft X-ray flux in the post flare regime can produce the desired cooling rate.

We can estimate these values by the following: the free-free emission is

$$I_{ff} = 1.435 \times 10^{-27} Z^2 T^{1/2} g N_e^2$$
$$= 1.1 \times 10^{-23} N_e^2 \quad \text{erg/cm}^3/\text{sec}$$

(9.6)

for $T = 3 \times 10^7$ deg, $Z_{\text{eff}}^2 = 1.4$, and $g = 1$. This is similar to the complete energy output found by Tucker and Koren (1971). The cooling time is the thermal energy divided by the emission:

9.23. A great eruptive prominence photographed by Skylab in the HeII 304 line. The twisted magnetic fields are easily apparent. The emission is probably chromospheric 304 scattered by HeII in its ground state.

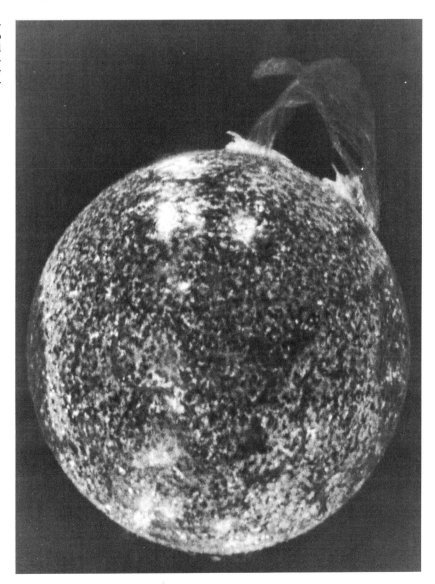

$$t = \frac{N_e kT}{I_{\text{ff}}}$$
$$= \frac{3.4 \times 10^{14}}{N_e} \quad \text{sec.}$$

(9.7)

At the very high densities which we now believe occur in post-flare coronal condensations, cooling in times of the order 100 sec is possible. Note that the morphology of the loop favors radiative cooling, a process proportional to N_e^2. Conduction would cool the feet of the loop first, which is not observed. When enough ions recombine, line emission becomes important and radiative cooling becomes a thermally unstable process, while conduction slows as the loop cools.

The downward velocity of loop elements is of the order of 25 km/sec, and it is not free-fall. Viscosity is produced by other material in the tube, which must be pushed out of the way. Besides the flow of material, there must be considerable downward conduction in the loop, because the footpoints are always bright.

The term "coronal rain" is used for loop-like material without an obvious condensation at the top. This tends to occur in the late stages of the flare. X-ray images show a coronal cloud at the top, invisible in Hα. In the late stages of flares the two ribbons of emission in the chromosphere appear thinner and more separated as the loops become higher and weaker. Hours after a flare we can still recognize that a great event has occurred.

Sprays are the ejecta of flares. They normally are filament material lying in the inversion channel or just above it, ejected by the flare process. Often a small filament ten or twenty thousand km long, erupts into a spray covering a significant fraction of the Sun. This is a manifestation of hydrostatic equilibrium; one scale height has as much material as all the overlying layers, and can have a big effect in the corona. Spray velocities can be huge, up to 2000 km/sec, but most sprays have velocities of 200–300 km/sec. Usually the filament is somewhat elevated at first, then it explodes outward in the spray. The acceleration is so great that no plausible explanation has been given. The appearance on films is as if a mighty force pushes up on the filament until it suddenly breaks the overlying fields and blows out. The material is so broken up that we cannot estimate the total mass or energy. A shock wave normally accompanies the spray, and the evidence supports a close connection with coronal mass ejections.

A spray can disrupt the overlying magnetic fields; a surge follows them. Surges are collimated eruptions, normally produced by small flares at their base. Helen Dodson-Prince always felt the surge had to retrace its steps in order to qualify as a surge but nowadays we are not so strict. When the surge does fall back we usually see a splash in the chromosphere. But often the material just disappears as the density drops. The surge usually occurs in a special set of circumstances; a small satellite spot of opposite

9.24. A great surge (Zirin 1976) erupts out of an active region, forming a filament near points a, b, and c, complete with bright rim. After about 20 minutes the whole thing changes its mind and is pulled back by unseen magnetic lines to return to the active region, producing some brightening as it falls in. (BBSO)

BIG BEAR SOLAR OBSERVATORY   **SEPT. 17, 1971**

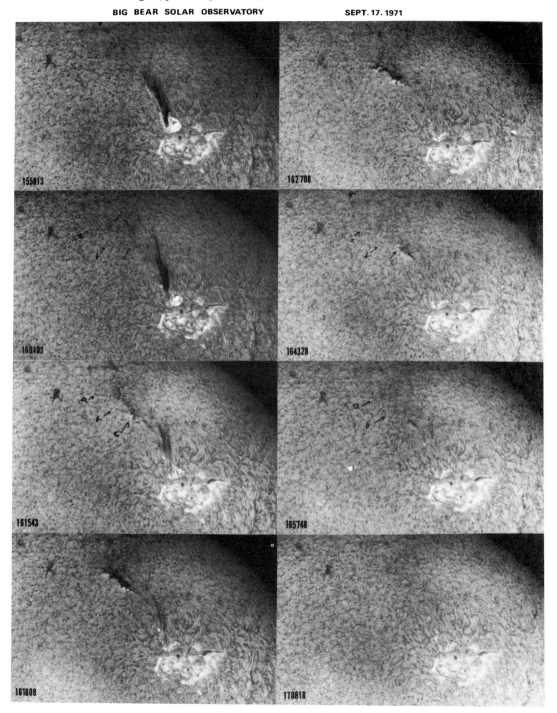

polarity comes up in the penumbra of a big, regular sunspot (Fig. 9.11). A small flare near the satellite spot results is an outward ejection of material collimated by the dominant spot field. Almost all surges start out bright – either bright base or entirely bright – and turn to absorption as the density drops. Surge velocities are lower than sprays, typically 100 km/sec, often less. While a spray is a single ejection, the surge shows continuous flow along its axis for minutes. There is a tendency for the surges to carry off some of the flare energy; the SXR flux from a flare with a surge often drops sharply when the surge starts. Other field anomalies – local field swept up and sheared by spot motion, or new flux emergence – may also spawn surges; the one requirement is a dominant field collimating the outward flow of material.

Can we explain the surge acceleration? More specifically, is the flare at the base the source, or is there a further acceleration as it moves outward? Because the material is tightly confined, as in a gun barrel, it is possible for

9.25. A filament in Hα (left) on two successive days, with the magnetic map of the same region. The filament divides the two magnetic polarities. Note that spicules coming out of the magnetic elements below turn right, while those above turn left. On the second day the emergence of a dipole at left disrupts the support of part of the filament. (BBSO-Sara Martin)

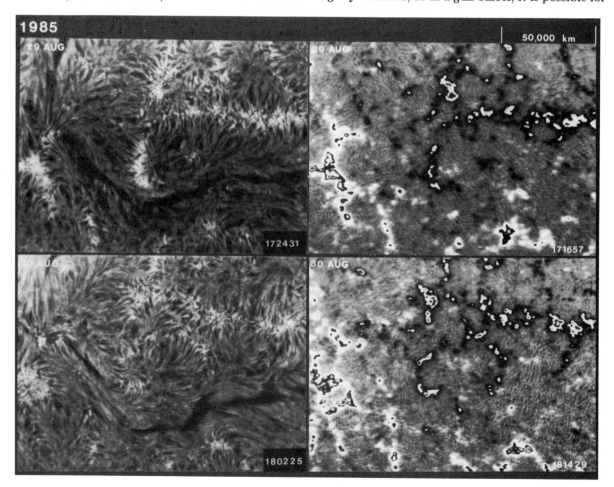

1985

the internal flare energy to be converted to linear motion. Surge velocities appear to be constant, but they are high enough that deceleration should not be important, at least at lower heights. Often the surge material does not appear to just flow along field lines, but it may have a more complex connection. In many cases backflow occurs along the path simultaneous with outflow. In some cases a surge goes up and across an arched loop, then returns in the curved oath. The most extreme case is in Fig. 9.24 (Zirin 1976) where we see how a surge flew halfway across the Sun, was trapped in a magnetic pocket to form a prominence lasting half an hour, and then rose again to arch back across the Sun to its source, as though it had been held by elastic bands. Brightening occurred as it came down. These ties show a continuing magnetic connection of the surge and its source. This implies that the surge did not simply flow along the field lines, but magnetic field moved with it.

The modest brightening which occurs when surge material falls to the surface gives an experimental test of the amount of surface heating produced by falling material. The brightening is about that of a weak plage. This evidence concludes a bizarre phase in flare theory. On the basis of the connection of flares with filaments and the observation of red shifts in some flares, Hyder (1967a,b) suggested that flares might be caused by falling prominences. This model gained great favor in spite of the fact that it disagreed with most of the data. Prominences have never been seen to fall; downward velocities of hundreds of km/sec would be required to produce the flare temperatures, and the rise times of flares are an order of magnitude more rapid than the times of prominence change. But the best test, even of a foolish model, is a solar experiment: drop the prominence on the chromosphere and see what happens. And that we now have seen.

We have already discussed the *disparition brusque* flares and need add only a few minor points. One can often foretell the eruption of a filament by its height above the surface. The filaments will slowly rise over a period of days. Once the filament rises above 50 000 km, it will not remain long. The eruption of QRF are somewhat slower than ARF, taking a few hours. Very little brightening is seen in the quiet region eruptions – just a few points will brighten. Why? Probably because fields are weak in these filaments, so there is not much energy.

From an extensive study of filament eruptions Tang (1986) recently concluded that there are two types. In the "classic" filament eruption the entire filament leaves the Sun. This is normal for all quiet region filaments and most eruptions of small filaments in active regions (as Fig. 9.6). But the big filaments marking shear boundaries in active regions, even old spotless ones, normally do not completely erupt; only an upper layer peels off, and a low-lying remnant remains behind in the chromosphere. This may occur no matter how violent the flare disruption.

The manner of prominence disruption is full of evidence on prominence structure. If an emerging flux region appears under a quiescent prominence far from other activity, it will erupt within a day, sometimes by halves, sometimes all at once. This indicates that the filament is a huge single magnetic structure from one end to another. Similarly, polar QRF will erupt as a whole when new magnetic connections intrude on their neutral line. Yet large polar QRF will, in about half the cases, disappear bit by bit as the magnetic boundary in which they lie is broken up by magnetic changes. The footpoints of hedgerow filaments are always magnetic anomalies, intrusions of mixed polarity that break up the neutral line and permit access to the surface.

We have seen that prominences play major roles in solar magnetic activity. They form the boundary between opposite magnetic fields and may well be the locus of extensive submergence of flux of opposite polarity. And almost all solar flares, down to the very smallest, involve filaments. We now turn to the activity itself.

References

Jensen, E. *et al.* 1979. *I.A.U.Coll. No. 44* Oslo: Inst. Theo. Astr..
Hirayama, T. 1986. *Solar Phys.* **100**, 415.
Öhman, Y., ed. 1968. *Mass Motion in Solar Prominences.* Nobel Symp. vol. 9.
Tandberg-Hanssen, E. T. 1974. *Solar Prominences.* Dordrecht: Reidel.

10

Solar Activity

The Sun's magnetic field has the remarkable property that it is not distributed uniformly, but concentrated in flux ropes which appear on the surface as sunspots, plages and network. All the magnetic flux important for the general field structure emerges in the form of sunspots and plages, which are not only clumped, but restricted to latitudes $< \pm40°$.

The sunspot cycle is remarkable in its elegant complexity and regularity. The three major aspects of the cycle are the 11-year period of sunspot number, the Hale–Nicholson law of sunspot polarity, and the reversal of the general field. The mean annual sunspot number rises sharply and falls gradually. When Schwabe (1849, 1851) announced his discovery of the sunspot cycle, Wolf (1858) searched the fragmentary early data and produced the sequence in Fig. 10.1 (with modern improvements added by Eddy back to the first observations in 1610). Carrington (1858) reported that the mean latitude of the sunspots shifted with time, a conclusion later confirmed by Wolf and by Sporer. The first spots of a cycle appear near 30° N and S latitudes, and the last ones near the equator. The variation in latitude is often called Sporer's law, but its regularity is best demonstrated by Maunder's (1922) "butterfly diagram," Fig. 10.2. It is the *locus* of the successive sunspots that changes, while the individual spot groups typically last only one 27-day rotation or less and move very little relative to the photosphere. There are few spots above 30° latitude.

The Zurich sunspot number plotted in Fig. 10.1 is not really the number of spots, but a number introduced by Wolf, defined as

$$R = K(10g + f) \qquad (10.1)$$

where g is the number of spot groups, f is the number of individual spots, and K is a correction factor applied to the observations from each observatory to allow for the size of the telescope, atmospheric conditions and relative enthusiasm of the observer making the counts. If there is one sunspot on the Sun, both g and $f = 1$, and if $K = 1$, then the sunspot number $R = 11$. This system gives extra weight to isolated spots, which is fair because they tend to be large.

The Zurich number underestimates the number of spots because it is not measured with the most powerful instruments; high-resolution photos of active regions reveal hundreds of short-lived tiny spots, which are called pores. But we are more interested in the variation of R, so as long as the number is always measured the same way there is no problem. A bigger problem is the fact that number is a less significant measure of the level of activity than area.

Since the numbers before 1840 were reconstructed by Wolf, they should be used with caution. In addition to the obvious 11-year cycle, there may be a longer-term fluctuation with a period of about 80 years. The maximum in 1957–58 was the greatest on record, with a peak monthly mean of 254 in October, 1957. It showed the rapid rise to maximum typical of high cycles. By contrast, the lower cycles rise slowly. The preceding minimum, in 1954, was one of the lowest on record, with eight monthly means below three. How does one get a mean of three when one sunspot gives $R = 11$? Because there were many days with no sunspots at all.

10.1. The sunspot number since 1610 as compiled by J. A. Eddy. The sunspot areas usually peak a year or two after the sunspot numbers. (HAO)

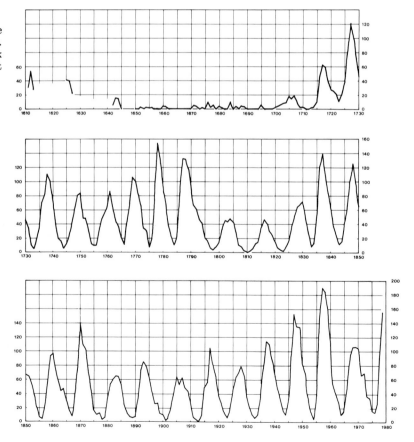

The yearly relative sunspot number is a running mean. Sunspots do not erupt at random; they behave like Bose–Einstein particles and have a strong tendency to break out all over the Sun at once. The probability that new spots will emerge near old ones is ten times as great as the chance they will come up in an equal area far from AR's (Liggett and Zirin (1985). Because of this clumping, there is a strong 27-day fluctuation in the weekly numbers through the year. Often the activity is limited to one side of the Sun; in 1985–86 almost all activity was in one 60° longitude band. Even if we eliminate this longitude dependence by trying to guess the number of spots on the far side of the Sun, we still find a sizable short-term fluctuation in the "whole Sun" relative sunspot number. In 1984, for instance, the monthly number varied between 13 and 84; and in December, 1984, the daily number varied between 0 and 63. There are long-term patterns in the activity in the two hemispheres. In the 1958 and 1969 cycles, the northern hemisphere strongly dominated, while currently the two are comparable.

If we measure solar activity by the frequency and size of solar flares, along with their X-ray and radio flux, sunspot areas are a better measure of activity than the Zurich number, peaking about a year after the maximum of R. The term "sunspot maximum" usually refers to the number maximum, a time at which the Sun is covered with many medium-sized groups of limited interest. The greatest sunspots usually occur several years before or after the number maximum. The greatest flares of the 1958 maximum occurred in 1960; peak activity of the 1969 maximum was in 1972–74, and of the 1980 cycle, in 1982–84. If observations could resolve all the little spots in these big groups, the Zurich number might well peak then also. For some

10.2. Butterfly diagram (compiled by J. A. Eddy) showing the drift toward the equator of sunspot position. The length of each cycle is about 12 years, causing the old and new cycles to overlap. (HAO)

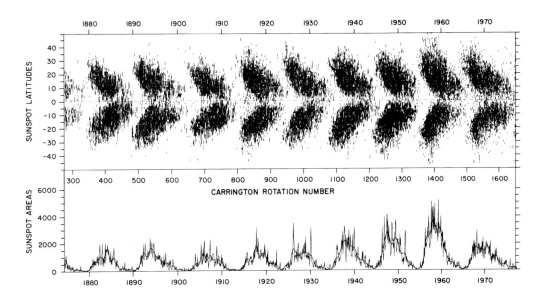

reason, international campaigns and spacecraft to study the sunspot maximum are usually keyed to the Zurich number maximum; the campaigns are on time and thus miss the greatest activity, the spacecraft are usually late and catch it. The SMM spacecraft was on time, and did not see much great activity, but its repaired version was turned on the day before the greatest flare of the cycle in 1984.

The sunspot area and the Zurich number are roughly related by

$$F \text{(in millionths of the disk)} \approx 16.7R, \qquad (10.2)$$

but obviously it is not close or the two would peak together. The largest spot areas ever measured were about 1% of the disk (including penumbra).

Other indices of activity frequently used are the CaII K-line plage areas and the 2800 MHz flux. Comprehensive data on solar activity are published monthly by the U. S. National Oceanographic and Atmospheric Administration (NOAA) in *Solar Geophysical Data* and by the Academy of Sciences of the USSR in *Solnechnie Dannie* (Solar Data). The Tokyo Astronomical Observatory publishes the *Quarterly Bulletin of Solar Activity*, which contains data on sunspot areas also. Sunspots are numbered by Mt Wilson, Ca plages by Big Bear and AR's by NOAA.

10.3. A full-disk magnetogram of the daily series obtained for many years by J. W. Harvey at KPNO, obtained near the peak of the spot cycle. The bipolar active regions in the northern hemisphere show white polarity leading, and the opposite is true of the southern. The poles have already assumed the trailing polarity of each hemisphere. The unipolar regions drift polewards and backwards from the centers of activity, but the prominent ones in the south show only limited differential rotation; in the several rotations they take to form, they should be 6 or more days behind at their polar extremity. N top, W right (KPNO)

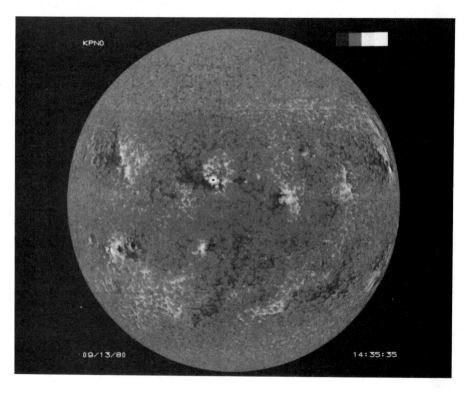

Hale (1912) discovered the magnetic fields of sunspots by the Zeeman effect. Hale *et al.* (1919), in a paper all solar physicists should read, showed that in virtually all the sunspot groups in a given hemisphere, one magnetic polarity leads and the opposite follows, while in the opposite hemisphere the situation is reversed (by lead or precede we mean the west spot, which leads in the solar rotation, and by following or trailing we mean the east spot; we will henceforth use p and f for the preceding and following polarity for that hemisphere and that 11-year cycle). They also announced that the polarities were reversed in the new cycle spots that appeared in 1912; this is called the Hale–Nicholson law. At first Hale thought it was a latitude effect, since the new spots were at high latitudes.

10.4. A synoptic map of magnetic fields for one rotation made by Harvey. The upper chart is actual flux, the lower is a normalized measurement which indicates the polarity. The growth of unipolar regions from active centers is evident. (KPNO)

In the same paper the Mt Wilson magnetic classification, still in use today, was set forth. The results of a study by Joy of 2633 spot groups drawn by Carrington and by Sporer were also given; the magnetic axis showed an average tilt of 5.6° to the E-W line, with the p spot closer to the equator. We call this "Joy's law". The tilt varies from 3° at the

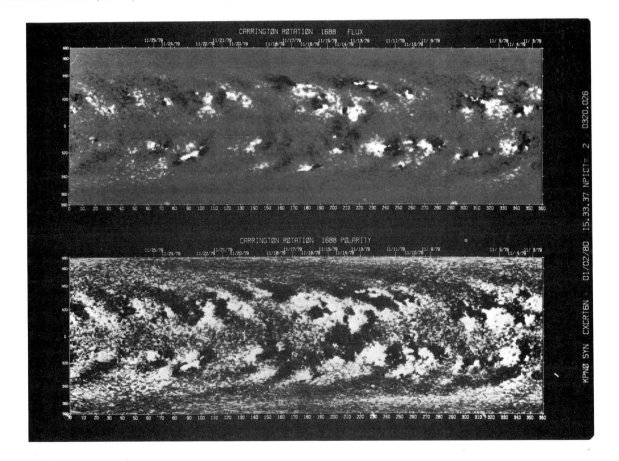

equator to 11° at 30° N or S. Hale and Nicholson (1925, 1938) confirmed this result and further polarity reversal in the subsequent cycles. Thus the Hale–Nicholson law of sunspot polarities follows a 22-year magnetic cycle. Spectroheliograms showed that if the p or f sunspot of a bipolar group was missing it was replaced by a plage. The polarity distribution is easily seen in Fig. 10.3, where the northern hemisphere groups all show white polarity leading, and in the southern, dark precedes. So strong is the Hale–Nicholson law that occasional new spot groups with the wrong polarities will die out rapidly or else undergo great activity as the p spot struggles to get from the back to the front. If a dipole emerges at a substantial tilt to the E-W line it will often rotate until parallel to the equator.

Hale searched for the general solar field for many years and claimed to have detected a 50 gauss poloidal field. Babcock and Babcock (1955), however, used the magnetograph to confirm that there was a polar field, but about ten times weaker than that found by Hale, and in 1959 H. D. Babcock detected its reversal in the new cycle. Babcock (1961) developed a model in which the fragments of f polarity migrated poleward to reverse the existing fields. Bumba and Howard (1965) showed how the spot fields broke up and spread into great unipolar regions which drifted toward the pole. The accumulation of f polarity reversed the polar field a few years after the onset of the new cycle. The unipolar regions and their sources are particularly evident in the synoptic chart in Fig. 10.4. The distribution of filaments on the disk reflects the progress of this reversal. At minimum the filament distribution peaks at 40–50°. As the new cycle develops, the filaments associated with the growing activity move equatorward, while another group moves poleward. The new belt of polar-cap filaments separates the old polar field from the new equatorial fields. The polar field, as can be seen in Fig. 10.3, is not so much a big field at the poles as a clustering of fields of one sign at one pole and the opposite sign at the other pole.

Since all magnetic flux erupts on the Sun in dipoles, it is remarkable that the field can be organized into such large unipolar entities. Only the active regions are large enough to separate fields on such a scale. In Fig. 10.4 we see how the flux drifts poleward from the active regions into unipolar regions which are spread out in longitude by the differential rotation, but are not much wider than the active region dipoles. At higher latitudes, the following polarities dominate, eventually forming the polar cap. The unipolar regions do not arise from a single spot group, but from complexes of activity in which a series of active regions recurs over 5–10 rotations. The coronal holes and the large scale fields in general do not exhibit as much differential rotation as we find in sunspots. This may be because of the "stiffness" of the fields, which produce a viscosity between adjoining regions that inhibits differential rotation.

A detailed kinematic model of the 22-year cycle was developed by Leighton (1964, 1969), based on the random walk of fields diffusing from the spot groups. If individual elements of flux move a distance L before changing direction in a time τ, then their random walk will carry them a distance

$$r = L \left(\frac{t}{\tau}\right)^{1/2} \tag{10.3}$$

in a time t. The area filled by these elements will be

$$A \approx \pi L^2 t / \tau. \tag{10.4}$$

The spreading of the density n of magnetic elements is given by a diffusion equation

$$\frac{\partial n}{\partial t} = D \, \nabla^2 \, n, \tag{10.5}$$

where the diffusion constant D is given by:

$$D = \frac{1}{2} \frac{L^2}{\tau}. \tag{10.6}$$

Leighton chose size L and lifetime τ of a supergranule cell as the diffusion parameters. With $L = 15\,000$ km and $\tau = 7 \times 10^4$ sec (22 hours), he obtained a diffusion constant of 1600 km^2/sec. He found the poleward drift rate of field could be matched by D = 1000 km^2/sec. This behavior alone would not reproduce the polar field separation. Leighton had to introduce Joy's law of spot tilt to explain the dominance of the f field; the slight poleward displacement of the f spots results in dominance of that polarity at the pole. Observations do not confirm Leighton's diffusion rate. Mosher (1977) and Zirin (1985) found $D = 300$ km^2/sec, one-third Leighton's value, from the actual motion of network elements. The enhanced network moves even more slowly than the general network. Marsh (1978), however, suggested that the reconnection of ephemeral regions with the network would produce a net diffusion rate of 800 km^2/sec.

Devore *et al.* (1985) produced excellent models of the evolution of magnetic regions by starting with an observed magnetogram and reproducing the field distribution observed a month later. They used the diffusion with $D = 300$ km^2/sec, Snodgrass's (1983) measured rotation, and a meridional flow rate varying significantly from case to case. Obviously it helps to have many free parameters, but the diffusion model is close to predicting the evolution of fields, leaving only the flux emergence pattern, Joy's law and the meridional flow to be explained. It may be that small-scale phenomena are the key, but it may also be that the change happens differently. In 1987, less than a year after minimum, the new fields are already approaching the pole. After the polar field reverses, the overall solar field takes on a poloidal aspect, emphasized by the reappearance of the polar coronal holes. Far from the Sun, the poloidal field appears in the solar wind (Chap. 8).

10.5. Formation of a sunspot by pore merging, July 12, 1983. Each spot formation has a dark arch. Two rows of pores merge to form a larger, S-shaped spot which finally forms a nice round one. The blue wing frames (a, b, c) show the tops of the AFS, while the red wing (d) shows the downflowing material. The centerline (e,f) show the AFS and the bright new plage. (BBSO)

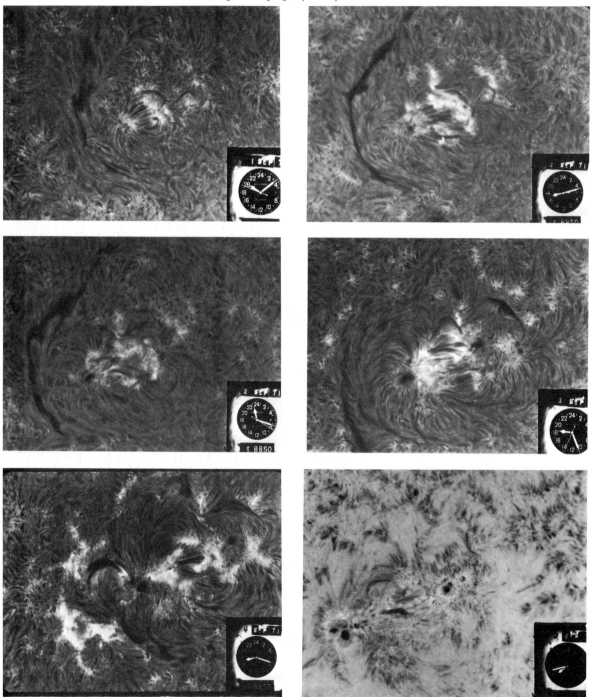

10.6. Continuation of 10.5: in the next few hours the spots merge into a single leader, and the next day (f) we see a mature spot (A) with penumbra.

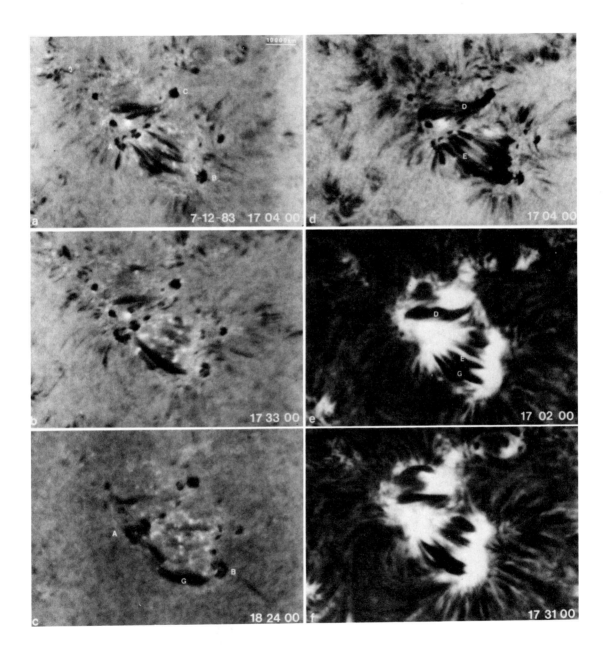

The field diffusion can take remarkable forms. In addition to the unipolar regions coming from the active regions (AR) and the mixed polarity regions elsewhere, great zones of one polarity will stretch almost from pole to pole, covering 50 or more degrees of longitude. These do not always have a clear counterpart in opposite polarity. Bumba (1976) has emphasized the importance of such "supergiant cells" in the development of the cycle. Scherrer *et al.* (1977) have shown that the visible hemisphere can have a net monopolar field. The great scale of coronal holes and the tendency for activity to erupt all over the Sun at once suggest large scale patterns driven from inside the Sun.

Howard and LaBonte (1980), analyzing many years of velocity data from the Mt Wilson magnetograph, found evidence for a torsional oscillation of the Sun in synchronism with the 11-year cycle. The amplitude is about 3 m/sec (0.167% of the rotational velocity) relative to the differentially rotating surface. Zones of faster rotation originate at the poles and drift to the equator in 22 years. The effect of these bands on the evolution of large scale fields has not yet been explored.

Harvey and Martin (1973) pointed out the existence of large numbers of short-lived dipoles called ephemeral active regions (ER), small bipolar flux elements which never have sunspots and live for about a day (Fig. 6.13). The presence of such regions had been noticed previously, but this was the first recognition of their global importance. In contrast to sunspots, they appear all over the Sun, even at the poles, and they do not follow the Hale–Nicholson law. However there are considerably fewer ER at the pole than the equator. They show the typical separation rate of growing active regions – 1 km/sec. This pattern suggests that they are the result of local, rather than global, dynamo action. There may be some dependence of their number on the sunspot cycle – this might be due to the greater amount of flux spread around the Sun which can be intensified by local dynamos. Because they are small dipoles, magnetically neutral, they can contribute only to the total field but not the net field. Harvey (1985) concluded that the X-ray bright points (Fig. 1.3) are sites where ER's are reconnecting with existing fields of opposite polarity, but I am not sure. We found over 90% of the bright points to correspond to bright Hα regions,

10.7. Growth of an EFR in Hα over 5 days Sept 1–5 1971: (a) Sept 1: The EFR emerges; weak arch filaments (AFS) connect the small spots; a filament on the left (W) separates the *p* spot from surrounding *f* polarity. (b) Sept 2: the *p* spot has moved forward and its spreading field has pushed the filament westward. (c) 6 hrs. later, the poles have spread further and the spots have grown. (d) Sept 3; the filament has partly erupted as the *p* spot pushes into it; another *f* spot has appeared. (e) Sept 4: The filament has erupted, leaving flare emission (left); the *f* spots have merged, and AFS at the center of the group show renewed emergence. (f) Sept 5: In Hα − 0.7 Å the continued growth is visible, the *p* spots are moving forward. Both this group and the EFR in Fig. 7.11 came up near each other on an otherwise fairly blank sun, showing the clustering of EFR emergence. (BBSO)

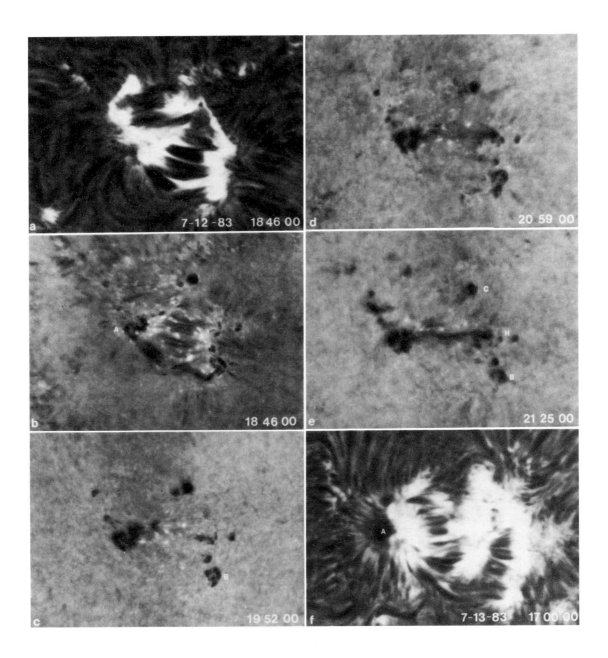

but the exact nature is not clear. Our best radio maps show little or no microwave emission from the ER's, whether reconnecting or growing. But they may produce microflares and tiny radio bursts; the smallest we have seen is 9×10^{-4} sfu.

The ER's are readily identified as small bipolar structures in quiet Sun magnetograms; as Fig. 6.13 shows, the poles separate at about 0.6 km/sec. Harvey *et al.* (1975) estimate 360 on the whole Sun in 1970, near sunspot maximum; and 178 in 1973, near minimum. The average flux is stated to be 10^{20} maxwells (average AR, 4×10^{22} mx). The regions live for a day and typically give one tiny flare. Unlike growing active regions (Sec. 10.2), they do not show arch filaments (although there may be a dark fibril or two) and they never develop pores.

Because the ER's last such a short time, new ones have to erupt every day, and in fact there are at least a thousand times more ER's than active regions of all sizes. If these numbers are right, the total flux carried to the surface exceeds that in the sunspots by a factor of five. So the contribution to the total $\mathbf{B}^2$ is big, but to the net $\mathbf{B}$, small.

Are the ephemeral regions just tiny active regions? We know that they do not obey the law of sunspot polarity or Sporer's law. Tang *et al.* (1984) studied 15 years of Mt Wilson active region data and found an exponential dependence of the integral number of active regions down a size A of about three square degrees:

$$N(A) = 4788 \exp{-(A/175)}. \qquad (10.7)$$

This equation gives an asymptotic value of 4788 for the total number of

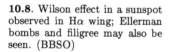

10.8. Wilson effect in a sunspot observed in Hα wing; Ellerman bombs and filigree may also be seen. (BBSO)

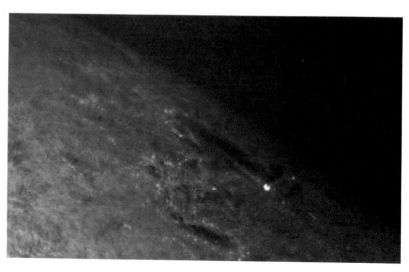

10.9. This beautiful spot group occurred during the Skylab mission in July 1973 and gave some of the finest images ever. It is a simple αp and the X-ray image appears in Fig. 8.4. The extensive moat around the p spot is filled with K-line emission and is also bright in the UV lines but dark in Hα; the weaker plage is bright in K but in absorption in Hα. (BBSO)

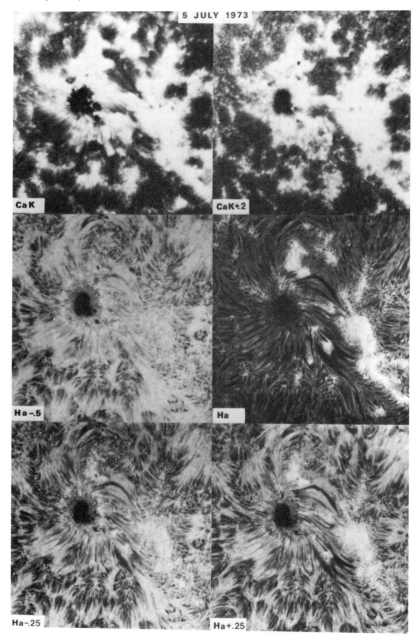

active regions of all sizes in the 15-year period. By contrast the number of ER during this period, using the Harvey *et al.* (1975) values, was 1000 times greater. Since (10.7) gives an almost perfect fit to observations, a 1000-fold jump just below the Tang *et al.* threshold would be unreasonable. Combining this with the fact that they occur all over the Sun, we conclude that ER's are a locally generated phenomenon. This idea is confirmed by the fact that they are radio quiet.

Recently Eddy (1976, 1980) has drawn renewed attention to the discovery by Sporer and Maunder of evidence for a prolonged absence of sunspots between 1645 and 1715 (also known to Lalande and others), now called the Maunder minimum. Because regular observations started in 1826, the evidence for this minimum is fragmentary, yet convincing. Eddy has introduced other evidence, primarily auroral and C^{14} data, to support its reality. This period coincided with a "little ice age" in Europe, when the Thames froze over and Norwegian farmers applied to the Danish king for indemnity for land overrun by the advancing glaciers. During the Maunder minimum the 11-year cycle continued, but with very low amplitude. There were competent, sharp-eyed observers who reported that they had seen no spots during this period, yet the 11-year cycle was not discovered for a hundred years afterwards, so sunspot observations cannot have been too thorough. The coincidence of the absence of sunspots with a cold spell is a single event, hardly enough to establish a rule; the climate in the orient, for example, appears to have been normal. But it emphasizes the possibility of long-term changes and there is some evidence for other such episodes in the past.

Another remarkable effect is the discovery by Williams (1985) of periodic variations of the thickness of Precambrian lake deposits from glacial runoff in Australia. He found a steady cyclicity over 1580 12-varve cycles. Associating these with annual runoff cycles, Williams concluded that these 680-million-year-old deposits are due to sunspot cycle modulation of the climate. If the thickness variation was indeed due to sunspots, it tells us that there was a 12-year cycle controlling the climate in Precambrian times.

10.2. Sunspot groups and active regions

Hale *et al.* (1919) introduced the Mt Wilson magnetic classification:

α: A single dominant spot, usually connected to a plage of opposite magnetic polarity.

β: A pair of dominant spots of opposite polarity.

γ: Complex groups with irregular distribution of polarities.

$\beta\gamma$:   Bipolar groups with no marked north-south inversion line.

δ: Umbrae of opposite polarity in a single penumbra.

In addition, the suffixes f and p are used when the preceding or following spot, respectively, is dominant. In the majority of groups the p spot is dominant; Hale and Nicholson found 57% of the groups to be βp and αp, and only 13% with f dominant. Even if the polarity is inverted the p spot is a big round one, moving westward. In the standard ordinary group the EFR gives small spots of opposite polarity, the f spot soon fades, and we are left (Fig. 10.6) with a p spot and f plage. This is a fundamental asymmetry in the sunspot process. The majority of the large spot groups are βp, $\beta \gamma$ or δ. This is because large single dipoles are fairly rare; normally several must erupt to form a big AR. If the polarity is inverted the p spot will still usually be the larger. In almost all cases other small spots of both polarities are present, each the result of a separate dipole emergence. Only occasionally will a stable lone ("naked") spot without plage be seen, the result of reconnection of the flux of the other sign (Liggett and Zirin 1983) with background fields. The effect of the background fields on AR evolution is not known, but probably it is small.

The Zurich classification (Waldmeier 1947; Kiepenheuer 1953) depends on size and complexity, ranging from A for the smallest single spot to F for the largest and most complex groups, with additional class G for bipolar groups with the following spot larger, and H and J for groups dominated by a single spot. In general, this gives the history of most large spot groups, which start out as A spots, grow to maximum size as E or F spots, and end in class J.

The emergence (Figs. 7.11, 10.5–7) and subsequent growth of sunspot groups gives strong evidence that a magnetic flux tube is emerging from below. Pores appear, clumped in bipolar form, and intense X-ray or Hα emission is seen. Since pores are never seen by themselves outside of AR's, the emerging flux region (Zirin 1972) or EFR is easily recognized in white light images as two little clusters of small pores. The p spot runs forward at about 1 km/sec, while the f spot is stationary or moves slowly eastward. The bright Hα or CaII K plage is crossed by groups of dark fibrils connecting the two ends. The fibrils, called an arch filament system (AFS) by Bruzek (1967), always show upward motion at the top of the arch and downward flow at the ends. Thus an H$\alpha + 0.7$Å picture shows just the ends of the loops and H$\alpha - 0.7$Å shows the tops. The arches grow particularly dark when a spot is emerging, and erupt when the spot appears. As more pores appear at the two poles, they merge into larger single spots. Some of the flux emerges as plage and never becomes dark.

The EFR is heralded by a small, bright Hα region with little surges.

10.10. (following page, top) The best K-line picture we ever made, enlarged from 10.9. The spicules show bright, following the field lines. The emission over the umbra oscillates strongly in the movies with a 3 min. period. The two main regions of f plage are marked.

10.11. (bottom) An enlarged frame in the wing of Hα on the next day. The spicules lie flat near the spot because of the dominance of the flux loops from the spot. They point in the direction of the coronal loops of Fig. 8.4. Two small transient pores are marked p. (BBSO)

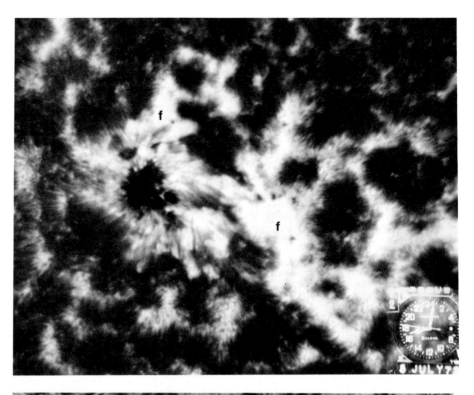

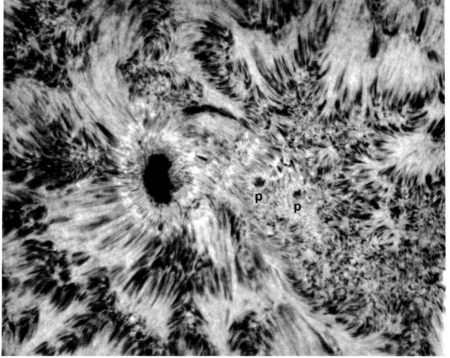

10.12. On July 8 (a: Hα; b: Hα + 1Å) a new dipole came up in the middle of the group and rapidly expanded to the accompaniment of numerous flares. By the morning of July 9 (c,d) it had slid past the main spot and formed a twin spot. The flares occurred as the new *p* spot pushed into *f* polarity above the main spot. When the group returned July 29, only plage and a big filament remained; the filament erupted to cause the flare of Fig. 9.17. (BBSO)

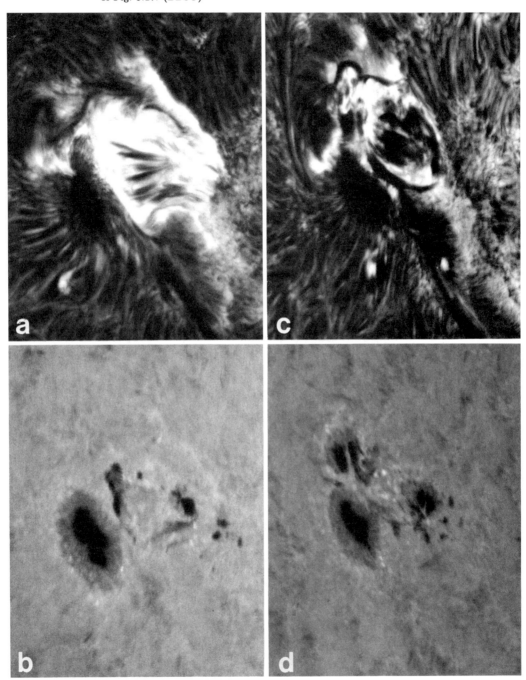

Then the pores, the AFS superposed on very bright plage, and the X-ray enhancement all appear within a few hours. Growth is rapid: we see the spots in Figs. 10.5–6 form in a few hours. When the first spots form penumbras, the arch filaments connected to them disappear but the p spot continues to move forward, leaving a limited AFS in the center so long as new flux continues to erupt. In Fig. 10.5, larger p and f spots with penumbras appear on the second day; the AFS connecting them disappear as soon as the penumbras form, but an AFS remains in the middle of the group, where flux continues to emerge. By the third day the p spot has slowed its motion; few single EFR dipoles ever exceed 50 000 km in length. Extended groups can result from the eruption of several dipoles end to end.

The behavior of the arch filaments has recently been summarized by Chou and Zirin (1987). As we see in Fig. 10.5–7, the tops of the arches are blue-shifted, indicating ascending material with velocity 10–20 km/sec. The feet are marked by red shifts due to falling material with $v = 20 - 40$ km/sec. Measurements at the limb show the arches to rise at about 15 km/sec for about 1000 sec, erupting at heights above 15 000 km as the material drains out. The loss of material produces an instability, leaving the loop buoyant and erupting upward. The downfall velocity exceeds the upward motion of the arch because of gravitational acceleration. So the whole loop is rising, but on the disk we see the red-shifted material pouring out.

Vrabec (1974) studied a growing active region with satellite p spots flowing into and merging with the leading p spot. He suggests a rising "tree trunk" model; the satellite spots represent the intersection with the surface of branches of a submerged tree which gradually merge as it rises. This process is more applicable to separate EFR's. In a single EFR (Figs. 10.5-7) several pores come up and the space between them fills in. I have searched for, but never found, small flux elements spreading to the two ends of the dipole; subsequent loops are independent entities. It may be that the sunspot formation draws in nearby field, the larger spot being cooler and more stable. On the Sun, like fields often attract. The converse of this is that spots of opposite polarity never merge, but form a δ spot (Sec. 10.4).

If the flux tube is to rise, the height gradient of magnetic field must be steeper than the adiabatic. Because EFR's obey the Hale–Nicholson law, they must come from the global fields involved in the sunspot cycle; thus the field may increase to great values far below the surface, at least down to $R = 0.8R_\odot$. Although the surface fields are in individual bundles, we may assume that their intensity increases at a rate greater than the density; this implies a field greater than 10 000 gauss at $R = 0.8R_\odot$.

During the peak of the cycle one or two EFR's per day emerge over the visible Sun outside active regions; but nearly as many emerge in the tiny fraction of the surface covered by AR's. The tendency of new dipoles to pop up in the middle of active regions leads to one form of the complex γ

and δ spots, the major sources of solar flares. But the most important δ spots (Sec. 10.4) emerge as extremely complex EFR's which do not depend on accidental collisions.

The introduction of new magnetic flux by an EFR distorts the ambient field. If that field has the same polarity as the EFR, they may merge; if they are opposite, reconnection follows. Field lines that previously connected the sibling spots of the EFR now connect to the surroundings. Such connections can appear in hours. The combination of magnetic buoyancy and the Hale–Nicholson force, pushes the new fields into the poles directly to the west, setting up substantial magnetic shear and leading to flares. Most solar flares result from the stresses produced by flux emergence (Fig. 10.12). While it is obvious that remarkable effects occur when EFR's come up in strong flux, the role of weaker indigenous flux has not been studied. Do EFR's evolve differently if they come up in a unipolar region of the same or opposite polarity?

The EFR is a messenger from below the surface, which can tell us much about the invisible controlling magnetic structure. The greatest mystery is the dominance of the p spots. What force pulls them forward and organizes them in the big stable spot? The surviving inverted polarity groups invariably produce flare activity. When the p spot is near the back (E) of a region it invariably pushes forward; the sub-surface fields pull powerfully on the p spots, and always E-W. Weart found that the majority of EFR's come up with the correct polarity; the small fraction with axes strongly tilted to the equator straighten out or die out early. Even if the axis is tilted, the p spot drifts straight west, often producing a hooked shape in

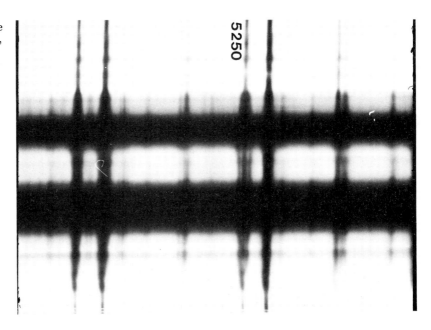

10.13. Zeeman splitting in a large spot. FeI $\lambda5250$, which is marked, has a g-factor of 2.5 and is the line most commonly used for Zeeman measurements; the splitting of 0.12Å corresponds to 3600 gauss. Dark horizontal streaks are due to the sunspot continua, and the bright area in between is penumbra. The field is almost as strong in the penumbra as the umbra. The central π component appears in the umbra; this is not because there is transverse field there, but scattered photospheric spectrum superposed on the dark umbra. (BBSO)

the growing region.

The form of the EFR gives us some insight into the sub-surface fields. The emerging regions are always elongated, so we have a nested, flat bundle of magnetic loops lying in an E-W plane up from below. It is remarkable that the convective zone does not twist these loops – while there is glib talk about twisted flux loops, at the highest resolution the AFS are always straight, untwisted, and reasonably parallel to one another. A succeeding EFR may come up at an angle, but its arches, too, are regular and parallel. The complexity in most regions is due to successive emergence of new dipoles.

Once flux emergence ceases, the p spot stops moving and a penumbra forms. Mature spots hardly ever move; the spot motion that leads to magnetic shear and flares is always in an EFR. The spots of the dipole will usually interact with local fields and other spots, behaving as though they are no longer attached to one another. It is curious that the dipole does not pull back together: the sub-surface fields still pull them apart. We do not know what stops the flux emergence. In a few cases (Zirin 1985) we have seen the EFR turn around and go back down. Most of these regions are small, but in one case (Big Bear region 19 427, Aug. 1984) the region reached moderate size. The opposite magnetic polarities come together and sink beneath the waves.

A remarkable feature of EFR's is the frequent Ellerman bombs – bright points with very broad Hα wings (HW $\approx$ 5A) that are low in the atmosphere so they are not visible in Hα centerline. Severny (1957) called them *usi* or moustaches, because their spectra look like wide moustaches with a gap in the middle. It is hard to see how such broad wings are produced in such a tiny feature, and why they are tied to emerging flux. Shklovsky (1958) discussed the problems of confining such high velocities in tiny elements. They typically occur at the center of the EFR or on the edges of spots – places where the field is breaking the surface.

The normal EFR and the development discussed above occurs 90 percent of the time, producing a peaceful spot group which may have a few minor flares and decay after a few weeks to a diffuse pair of unipolar areas. Fig. 10.12, by contrast, shows what happens when an EFR erupts in a mature αp region. The new flux must reach an accommodation with the old, forming a new magnetic structure. Since the lines of force are sticky and slow to change, the new flux pushes up or in until the field is sufficiently stressed to change to a more stable state. If this change, which usually involves reconnection, happens discontinuously, we have a flare; if the adjustment is slow, we have simple heating. Each large p spot in a mature region is the result of a separate EFR. The majority of solar flares are directly due to emerging flux disrupting and shearing fields.

10.3. Sunspot structure

Large stable sunspots always display the characteristic round umbra-penumbra form. The umbral field is vertical, typically around 3000 gauss (4000 in the largest spots). In the penumbra, the field turns horizontal and spreads radially. Sunspots like to be round and are most stable in that form. But as the various illustrations show, they are often distorted from that shape, primarily by the proximity of other strong flux. The penumbra will disappear if there is strong shear along one side. Sometimes (Fig. 10.25) we get long, thin spots in these shear regions, or the penumbral structure will be twisted. The shear results from the motion of nearby spots of opposite magnetic polarity. Shear only occurs between spots of opposite polarity, where field lines connecting them may be stretched horizontally by the relative motion.

In Table 6.4 we found that a field of 1400 gauss would produce magnetic pressure equal to the gas pressure, so horizontal fields of that order must produce the non-radial penumbral structure in Fig. 10.14. As the Zeeman splitting in Fig. 10.13 shows, the field strength in the penumbra may be close to that of the umbra. This can happen only if there is no field divergence. Often we find a light bridge, a piece of photosphere crossing the very middle of the spot. The cause of light bridges is not known; in some cases it appears connected with the close approach of an opposite polarity.

In 1769 Wilson observed a spot near the limb and saw that it appeared to have a dish-shaped depression, the near penumbra appearing much narrower than that on the far side (Fig. 7.29, 10.8). Bray and Loughhead (1964) attribute the Wilson effect to increased transparency of the umbra, and in fact the H$^-$ absorption decreases with falling temperature. It is interesting that the level we see in the Wilson effect is that at which the spot is in lateral equilibrium, and above this height the umbral field pushes the spot walls out to form a dish.

To explain the sharp distinction between umbra and penumbra, Danielson (1961a,b) proposed that the penumbral fibrils are convection rolls. A long strand of gas rotates about the horizontal magnetic field, bringing heat up from below. Thus the convective regime in the penumbra differs qualitatively from the umbra and the photosphere. Moore (1981), on the other hand, argued, using Fig. 6.10 as his example, that the penumbral fibrils are slightly elevated above the photosphere, since they appear to cover granules. The fibrils need not lie above the photosphere; because of the Wilson depression they may lie above the "surface" and still not appear as elevated structures above the limb.

We can understand the low temperature of the spot by considering the pressure balance between the spot and the surrounding photosphere:

$$\frac{B^2}{8\pi} + n_{sp}kT_{sp} = n_{ph}kT_{ph}. \tag{10.8}$$

The densities inside and out of the spot are roughly equal because hydrostatic equilibrium must be maintained. In every case $T_{\mathrm{sp}} < T_{\mathrm{ph}}$ because the magnetic pressure is non-zero. Indeed, if we insert 3000 gauss for the field, the typical value in spots, it can only be balanced by the photosphere at densities above 5×10^{17}. Since $T_{\mathrm{sp}} \approx 3500°$, equilibrium can only be reached for $n = 1.3 \times 10^{18}$, about 400 km below $\tau = 1$ in the photosphere.

In 1909 Evershed found that the spectrum of a sunspot 50° from central meridian passage showed a violet displacement on the side nearer us and a displacement to the red on the limb side. This outward flow disappears at the edge of the penumbra. It is possible that the effect is associated with the RP waves (see below). Subsequent measurements of the Evershed effect (St. John 1913; Kinman 1952, 1953) showed that at photospheric levels there is a radial outflow from the spot which probably increases with depth, *i.e.* the effect is greater in weaker absorption lines. The magnitude of the flow is around 1 km/sec. By contrast, the strong absorption lines show an inflow of material (Fig. 10.17). So material flows in at higher levels and out at lower, but the density gradient is so steep that the net flow is outward. Studies of the Evershed effect have been contradictory, probably because of the difficulty in resolving spot structure (Bray and Loughhead

10.14. A magnificent sunspot picture by Müller from the Pic du Midi. The formation of an EFR at lower right has stretched the penumbral fibrils along the sheared field lines. Elongated granules are also seen. (PdM)

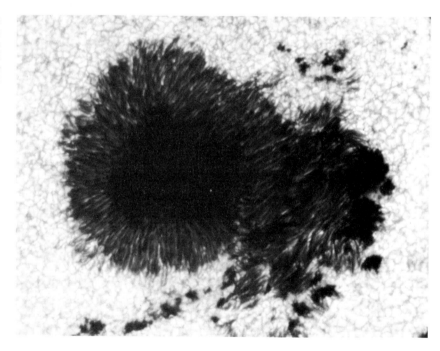

1964). Bumba (1965) proposed that the Evershed effect is really a twinning of Fraunhofer lines, with satellites (which he terms "flags") appearing near the edge of the penumbra, Doppler shifted by the outflow along the lines of force. To confuse the issue, high-resolution continuum observations show an *inflow* of bright structures at 1 km/sec along the penumbral fibrils, directly opposite the Evershed flow. Maltby (1975) subtracted filtergrams of the spot in Fig. 10.10 to measure the Hα flow and found an inflow of about 0.7 km/sec along the Hα arches surrounding the spot. This is similar to what we see in Fig. 10.17, where the velocity is much higher. It is possible that both the Hα and continuum inflows are limited to penumbral fibrils above the surface, as argued by Moore.

Outside the penumbra symmetric spots are surrounded by a "moat" of roughly twice the penumbral width in which the field is usually radial and marked by horizontal Hα fibrils. Inside the moat there is steady symmetric outward streaming of bright dots with velocity 0.4–0.8 km/sec, first seen in the low chromosphere in the CN band by Sheeley (1972). Subsequent magnetograph observations (Vrabec 1974; Roy and Michalitsanos 1974) showed that these dots correspond to moving magnetic features of mixed polarity which carry considerable flux away from the spot. Our most recent data show steady outflow in almost every spot, even smaller ones. At any one time all the outflowing magnetic elements have the same polarity, but that may be the same as the spot, or opposite, or small dipoles. If they have the opposite polarity to the umbra, which is most frequent, the total spot field is increased by the flow; and this may be the dominant mode of spot growth. We are just beginning to gather comprehensive data on this important phenomenon.

Sunspots are the locus of considerable oscillation. In the K line, Beckers and Tallant (1969) discovered a 150 second oscillation that they called "umbral flashes". The K line in umbrae is a single narrow K3 peak; this is seen in spectrum movies to oscillate from red to blue with amplitude of 0.3Å. In K-line movies, small points brighten every 150 sec, the peak coming in the blue wing. In Hα a similar vertical oscillation is observed, but in absorption. In the penumbra, "running penumbral waves" (RP) (Giovanelli 1973; Zirin and Stein 1972) are observed in Hα running concentrically outward across the penumbra at 300 sec intervals. The RP waves are the only wave motion of this sort seen on the Sun. Their shape and period are regular, their velocity is constant, and they can only be seen in regular and large penumbras. The waves are not really sinusoidal; the separation between peaks appears greater than the width of each peak. They are only seen in Hα. The RP waves are thought to be Alfvén waves but the matter is not settled. All the oscillations observed so far are in the chromosphere above the sunspot.

The high level of oscillation in sunspots is surprising in view of the fact that the 5-minute oscillation virtually disappears in active regions, at least in chromospheric movies, because the material is frozen in by the strong fields. The symmetry of the sunspot and the restoring force inherent in the magnetic field must be responsible.

The sunspot is dark in all wavelengths of the UV including SXR. However, in the UV lines the plage brightening crowds in on the spot as it does in the K line, with the moat appearing as bright as the rest of the plage. Thus all levels of the atmosphere are heated above the AR. In lines of CI and CII In the 1200–1900Å range Cheng *et al.* (1976) found a relatively low turbulent velocity of 13 km/sec in lines produced at 100 000° above a sunspot. The line widths were somewhat narrower than in quiet areas, reflecting the role of the strong field in suppressing the 5-min oscillation. They found the spectrum of the region above sunspots similar to other regions of the Sun, with absolute intensities similar to the network boundaries. The result is somewhat weakened by the fact that the slit used in the NRL spectrograph was 60 arc sec long and the spot only 30 arc sec, so the data refers to sunspot plus surrounding plage, and is useful if we remember what it represents. Judging by the dark spots on UV images, intensities over the spot must be low.

Athay *et al.* (1985a,b; 1986) studied the structure of active regions in

10.15. A stable sunspot in the center of Hα, showing curved penumbral fibrils. The *f* polarity elements are marked; all are connected to the *p* umbra by elevated fibrils. The fibril–FTA connections from the small *f*-plage at left are focussed in a small area where they distort the umbra into a heart shape. Presumably the connection to a strong nearby pole pulls the penumbral fibrils in that direction instead of the normal spread of Fig. 6.10. (BBSO)

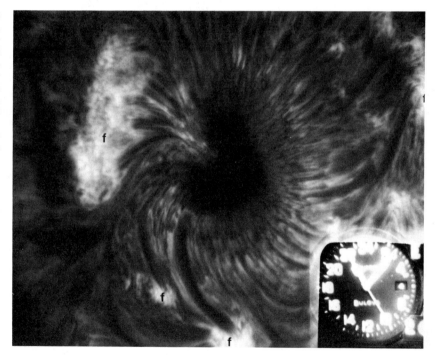

the CIV line, which is generally bright over the active region, except for the spots. The region was divided by a long and complex filament, from which a continuous downflow in either direction took place during the entire period of observation. For a long time it has been known that there is a steady downflow of coronal material above active regions. On the limb we see coronal rain, as well as the Evershed inflow. In Doppler measurements of the photosphere, the same takes place. Even the elements of the network show downflows of about 100 m/sec.

What makes sunspots dark? The most popular answer still seems to be Biermann's (1941) idea. Biermann pointed out that the strong sunspot magnetic field would strongly impede the convective currents carrying energy in the convective zone, because they could move only along the field lines. Even if the field were purely radial and the gas could move freely up and down, it would be difficult for convective heat transport to take

10.16. An EFR (top) distorts the structure near a spot, producing a shear boundary which becomes a filament (F) at right. We see the contrast between this structure and the FTA boundary (A) which connects opposite polarity above. The bright area above F is a plage with vertical field; the area below F is clearly horizontal field. There is a 90° turn in the field across F and a current must flow. (BBSO)

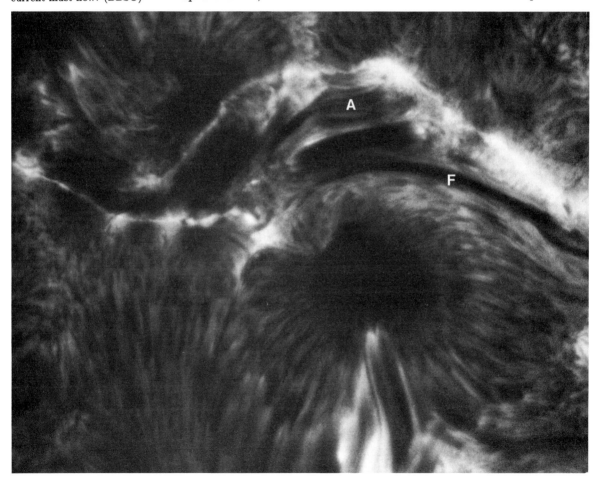

place, because there always must be some horizontal circulation associated with vertical convection. When an element of gas becomes convectively unstable, it can only rise if other gas flows in horizontally to take its place. Thus even a vertical magnetic field would inhibit convection. In a sunspot, therefore, the flow of energy from the interior must be interrupted, and the sunspot region is cool and dark compared to the rest of the surface.

Biermann's theory pointed the way for the explanation of the coolness and darkness of sunspots but, like all theories, it runs into difficulties with the complexity of natural phenomena. While the lively oscillatory phenomena in spots are so far only observed in chromospheric lines, Howard (1958) found increased turbulence in the umbral spectrum. We must also explain the difference between spots and plages. Over most of the Sun's surface, the magnetic network goes with increased temperature, so we must explain how the temperature increases for medium fields and decreases for stronger ones.

But the most serious problem for any sunspot model is posed by the discovery (Fig. 6.24) that the solar constant is reduced by the presence of sunspots in proportion to their area. Obviously if there are spots on the sun the visible spectrum should be decreased. But it was always felt that

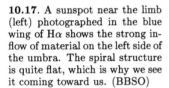

10.17. A sunspot near the limb (left) photographed in the blue wing of Hα shows the strong inflow of material on the left side of the umbra. The spiral structure is quite flat, which is why we see it coming toward us. (BBSO)

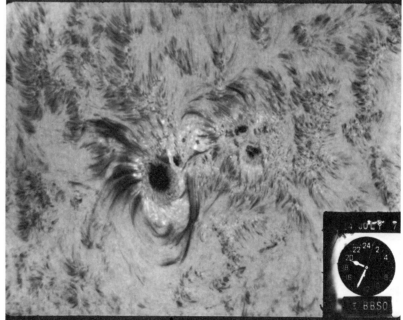

this energy would escape somehow, in other modes of radiation. After all, the energy currently leaving the Sun was generated at its center millions of years ago, and there is no obvious way to store it if a sunspot restricts its escape. Further, the solar constant is measured with a cavity radiometer that measures *all* radiant energy coming from the Sun. Foukal and others have suggested that the excess radiation from plages makes up over the long term for the energy taken out of the solar constant by spots. If that is so the missing energy somehow must be stored in the magnetic field.

There are other sunspot models. Parker (1979) shows that sunspots are overstable and must emit MHD waves which cool them. Although Parker's model oscillates below the surface, the idea is supported by the high level of chromospheric oscillation in spots. But these models do not yet explain the great stability of these strong clumps of field, nor why the p spots are larger and more stable than f spots. Nor have we found where the energy

10.18. A remarkable EFR (July 8, 1985) photographed in $H\alpha - 1$ Å. The complex chain of p and f umbrae are the ends of EFR loops that began emerging four days earlier directly above the old p spot to the lower right. The elongation (shear) of some umbrae are a characteristic of these EFR's, unrelated to collision with the old naked p spot at lower right. Note that the penumbra here, as in all such eruptions, comprises the horizontal flux region of the EFR. (BBSO)

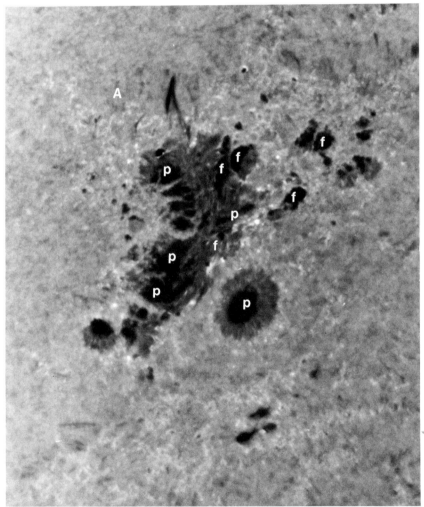

goes. The energy associated with activity is small compared to the missing radiant spot energy, and concentrated in the plage, often far from the spot.

Chou (1986) studied the relative brightness of new and old sunspots of the same field strength. He found that younger sunspots are not so dark as their older brethren; they require some time to cool down. Chou found that 15 stable spots closely tracked a line of darkness *vs.* field (Fig. 10.19), and the growing spots seemed to cool in 0.5 to 9 hrs. But both the theories mentioned above give cooling times around three minutes, essentially the time for the convection or Alfvén wave to cross the spot. This means the models are too optimistic; that is not so bad; even if they only work with 10% efficiency they will cool the spot. But the factor ten difference tells us something else is going on.

What is the threshold at which a sunspot forms? There is abundant evidence for 1000–1500 gauss fields in plages, light bridges and sheared penumbra. Conversely, measurements of the Zeeman splitting in sunspots, while they normally give values above 1000 gauss, often record fields of as little as 400 gauss. In Fig. 10.11 we see two small pores which had just developed in the f plage. Such pores often occur without EFR – just a thermal instability in the field – and seldom last more than a day. The transient pores produce no change in magnetograph signal (except for brightness), and may be a temporary instability. The easy transition between plage and pore suggests a close similarity, and the outflow of field across the moat suggests that the difference is one of field concentration. Spots and pores are fairly uniform, while plages have a smaller filling factor.

Round sunspots are among the most stable of solar phenomena, some-

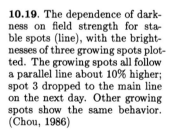

10.19. The dependence of darkness on field strength for stable spots (line), with the brightnesses of three growing spots plotted. The growing spots all follow a parallel line about 10% higher; spot 3 dropped to the main line on the next day. Other growing spots show the same behavior. (Chou, 1986)

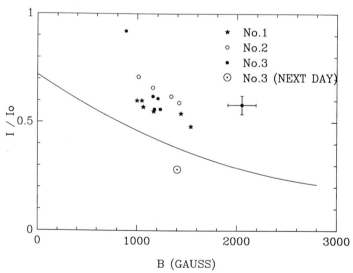

times lasting several rotations. On the other hand, small pores will come and go in hours. Most of the models have been addressed to the former, which are easy and interesting to study. The magnetic field in stable spots is vertical and fairly uniform, spreading outward to nearly horizontal form in the penumbra. The spectrum of spots is full of molecular lines which, since they do not occur in the photosphere, are not affected by scattered light; numerous measurements give rotational temperatures around 3300°. The umbra is full of structure, called umbral dots; the penumbra consists of radial fibrils, which show structure at high resolution.

Zwaan (1966) discusses the various models and problems. He finds the curve of growth for atomic lines relative to the photosphere to be fitted by a shift $\Delta\theta \approx 0.48(\theta = 5040/T)$, or $T \approx 3800°$. His model (Table 10.1) has a surface temperature 3000° and is based on empirical measurements from 80–151 km below the surface, extrapolated to 7750° at -240 km by radiative equilibrium. Note that only the four points between 80–151 km are actually measured (but Table 10.2 below has no measured points at all). The rest of Table 10.1 is extrapolated.

Chitre and Shaviv (1967) pointed out that the gas pressure increases exponentially with the scale height, while the magnetic field increases only with a scale similar to that of the sunspot. Thus the gas pressure increases

Table 10.1. Model of a sunspot umbra (Zwaan 1966).

τ_{5000}	z (km)	T	$\log P_g$	$\log P_e$	$\log P_H$
0.001	-82	3000	3.91	-1.60	3.75
0.004	-32	3060	4.27	-1.25	4.08
0.01	0	3135	4.51	-1.00	4.30
0.02	24	3205	4.68	-0.81	4.48
0.04	48	3300	4.85	-0.62	4.66
0.07	67	3390	4.99	-0.45	4.80
0.10	80	3455	5.07	-0.34	4.90
0.20	104	3609	5.23	-0.13	5.07
0.40	130	3770	5.38	+0.10	5.24
0.70	151	3945	5.49	0.32	5.38
(0.98)	163	4220	5.55	0.57	5.47
(1.37)	174	4590	5.60	0.87	5.53
(1.93)	186	5080	5.65	1.15	5.58
(2.73)	199	5650	5.70	1.50	5.63
(4.36)	214	6310	5.75	2.00	5.69
(8.61)	230	7110	5.80	2.65	5.73
(15.4)	240	7750	5.825	3.23	5.76

rapidly relative to the magnetic pressure, the temperature perturbation by the field becomes small, and sunspot cooling is a shallow phenomenon. Chitre and Shaviv produce a self-consistent model by fitting the efficiency of convection so that it disappears at 2335 km below the surface, where the field strength is 11 000 gauss. They also require a Wilson depression of 400 km, much more than Zwaan. Their model is given in Table 10.2, and one can see the substantial difference between the two models. The Chitre-Shaviv model has no empirical points, but tries to fit the possible physics; one might improve things by fitting it to Zwaan's observations.

The fact that magnetic buoyancy occurs suggests that magnetic fields may increase more rapidly below the surface, which would change the whole picture. Spruit (1977) has shown that an obstruction (by magnetic fields) to heat flow at a depth of 2000 km produces reasonable spot models and spreads the missing flux over very large areas. Table 10.2 shows the field 2000 km down to increase by a factor three; the spot must be reduced by $\sqrt{3}$ in radius. A sunspot 10 000 km in diameter would shrink to 5 700 km, and the sub-surface field lines under the Wilson dish would be almost horizontal.

The contradictory arguments of the previous paragraph leave us without a clear picture of the sub-surface field. If it does not increase as fast as the gas pressure, we can expect extremely thin flux loops below the surface, which may be peeling off from a great *ur*-field somewhere. Yet the fields must be strong enough not to be twisted up in the hydrogen convective zone, for most EFR's are simple in structure and obey the Hale–Nicholson law. If the fields increase in strength faster than the adiabatic gradient, the picture might be far different. Once they break the surface and stabilize,

Table 10.2. Theoretical umbral model (Chitre and Shaviv 1967).

h(km)	P_g(dyne/cm^2)	T	H(gauss)	$F/F_\odot$	β
400	2.09×10^4	4500	3550	0.330	0.0099
444	3.22×10^4	4534	3696	0.343	0.014
570	1.05×10^5	4955	4077	0.378	0.038
638	1.86×10^5	5577	4194	0.389	0.063
732	3.60×10^5	8950	4272	0.396	0.134
826	5.36×10^5	12060	4433	0.411	0.325
1077	1.18×10^6	13179	4934	0.457	0.568
1260	1.97×10^6	13981	5030	0.466	0.792
1765	6.75×10^6	16208	6783	0.627	0.906
2216	1.68×10^7	18242	10111	0.934	0.985
2335	2.00×10^7	18790	11027	1.000	1.000

the fields behave (with the exception of the running p spots) as though they have lost all sub-surface connections. Indeed, Leighton's random walk model implies that the surface fields behave like wood chips on the sea. Yet the connections must be maintained or the fields would all leave the Sun by magnetic buoyancy. But if the connections are maintained, the process of changing the polar field would leave everything all tangled up. Other questions, such as the possibility of sub-surface flares or reconnection, await future progress in helioseismology.

10.4. Magnetic structure of active regions

The vast majority of active regions are small and simple dipoles resulting from the emergence of one or two EFR's. Larger groups may develop from single large EFR's or the merger of multiple flux emergence. As the region grows, the main spot will form magnetic connections to distant fields and a moat will form. If no more EFR's come up, there will be a fairly regular boundary of field transition arches (FTA) separating the polarities. Around the perimeter horizontal field lines connecting the p and f polarities appear. Plage is rarely seen ahead of the p spots unless satellite opposite polarity is emerging. Once the region reaches maturity, it depends on successive flux emergence to keep it active, or else it will fade away or become a naked spot. If small EFR's emerge offset from the FTA line, a peninsula of plage forms extending to the new pole until it reconnects. If the EFR comes up wholly in a plage, the opposite polarity spot will be surrounded by FTA fibrils connecting to the plage.

The July 1973 region (Figs. 10.9–12, 8.4, 9.17) was quite a simple αp group. The main spot was surrounded by a moat and separated from the f polarity by extended FTA as well as a filament, and there was some f polarity directly above the spot. The spicules all pointed away from the group, following flux loops connected to distant regions; the loops are visible in the X-ray pictures. The K-line frame (Fig. 10.10) shows some of the emitting material in the umbra which oscillated with 150-sec period. Running penumbral waves were easily seen in the regular penumbra. Although the plage was uniform in the more intense parts, at the edges it was already broken up into network-like areas. On July 8 new flux erupted (Fig. 10.12) and completely changed the character of the region. A large EFR of the same sign erupted directly in the center of the region and pushed past the main spot. There it ran into some f polarity and produced several large flares as these reconnected. The new p spot took its place alongside the old and the region rotated off as a double spot; this is how regions with double p spots form. The development of EFR and filament produced a sheared state that remained when the region returned, giving a fine spotless flare on July 29, 1973 (Fig. 9.17). The sheared configuration remained even after the flare, the chromospheric fibrils running parallel to the inversion line. Only in the ensuing days did the region decay. The magnetic lines of force can change only slowly, and fields remain connected to old siblings in peculiar ways. It is interesting that this was the only active region on the Sun and the only EFR in a week came up right in the middle of it. The odds are 5000 to 1 that it was not a random encounter; the EFR must have been part of a deep-seated complex or guided to the same place by the sub-surface field.

Most of the active region area is occupied by plage. In the αp spot group the field lines clumped in the p spot return to the surface in a spread out region of fairly strong fields that does not form a sunspot. Thus we can surmise that the density of flux in the plage is lower than the spot by the area ratio. But considerable atmospheric heating takes place in the plage; it is bright in everything from $H\alpha$ through K and all the UV lines up to the coronal ions. This heating no doubt accounts for the absence of spicules. While the spicules are absent over the plage, they are prominent around its edges. In high-resolution plage images we see a granular structure, possibly due to individual flux tubes. In $H\alpha$ and broad-band K-line images, almost all the plage is granular; in K3, the more active areas are bright and diffuse. It may be that the overlying material is not arranged in discrete flux

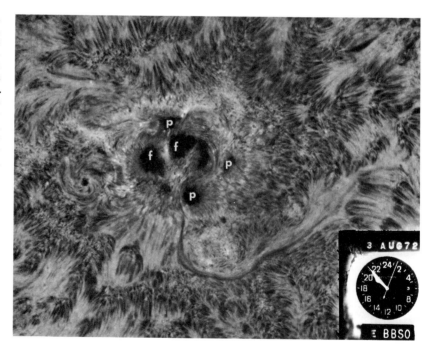

10.20. The Aug. 1972 group, grand champion active region, epitome of the "island δ". It emerged on July 12 with inverted polarity; when it returned July 30 it had grown big spots which were completely backward. The direction of solar rotation is to the left, with the f spots leading. As the p spot below pushed its way through the white polarity, a series of great flares occurred along the inversion line on Aug. 2, 4, 7 and 9. (BBSO)

elements. One also sees bright threads coming out of the edges; whether the dark or bright features are spicules, we do not know. It is not really certain if the fields in the p spot and the f plage balance. The flattened fields around the p spot seem to limit heating, and bright plage is usually seen near the p spot only if a satellite EFR appears.

Around the umbra we find bright plage in the area commonly called the moat, where there is little vertical field and arching overlying fibrils from the active region. This bright area is matched in all the UV lines (see Athay *et al.* 1986 for good comparisons). Umbrae that are covered by K-line or Hα emission will also be bright in the UV lines. In the upper right frame of Fig. 10.9, we see how the plage is graduated in intensity, especially at the edges where spicules appear.

The flux emergence in Fig. 10.12 is of the same polarity as the existing spot, so there is no great anomaly. The flares were due to the new p spot pushing its way through the f-polarity areas near the leader. If the new dipole comes up directly behind the first dipole, the p spot literally crashes into the f spot, linking up to it in what is called a δ configuration.

The δ configuration, the most active of active regions, was first defined by Kunzel (1960) as two or more umbrae of opposite magnetic polarity in a single penumbra. Kunzel showed that the δ groups were far more active than the other magnetic classifications, a conclusion soon confirmed by others. Rust (1968) pointed out that δ configurations formed by satellite spots of opposite polarity near a large umbra were a common source of flares. Tanaka (1975, 1980) showed that 90% of δ groups with inverted polarity were associated with great activity.

We have learned to recognize δ spots as the portents of great solar activity associated with their complex structure. δ spots are almost always large; conversely, large spot groups are more likely than smaller ones to show the δ property. δ groups are responsible for almost all great flares. Zirin and Liggett (1987) summarize their characteristics as follows:

1. δ spots form by joining opposite polarity spots from different dipoles.
2. δ spots rarely last more than one rotation and are shorter-lived than other spots of the same size, but new δ spots may emerge in the same complex (April-May 1984, MtW 24030, 24057).
3. The polarity is generally inverted as compared to the Hale-Nicholson Law.
4. δ spots are never observed to separate (although rarely umbrae are ejected from the group), but die out locked together.
5. Components of δ spots are not connected by direct lines of force, but by sheared magnetic field lines running parallel to the inversion line.

When active, the δ regions are marked by continued flux emergence and bright Hα emission over the umbrae. When we study their origin (Zirin and Liggett 1987) we find they occur in three ways:

10.21.   Daily magnetograms of the Aug 1972 region.   Near the limbs the polarity is distorted by projection effects.   Black is preceding, white following.   The black field pushes through the white.   This great spot lasted bu one rotation.(BBSO)

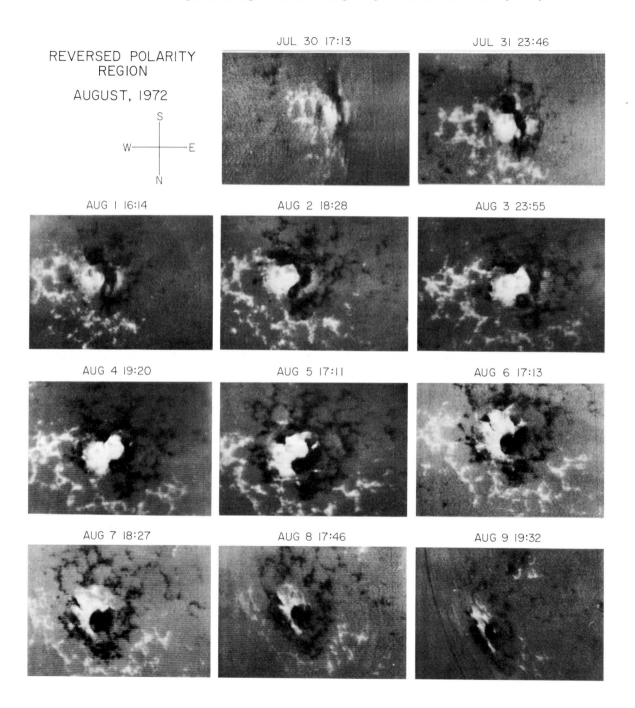

1. A single complex emerges at once with dipoles intertwined and polarities reversed from the Hale–Nicholson rules (Fig. 10.20). We will call this the "island δ".
2. Large satellite dipoles emerge close to existing spots so that the expansion of the EFR pushes a p spot into an f spot or *vice versa.*
3. A growing bipolar spot group collides with another dipole so that opposite polarities are pushed together. Examples of these are given by Kunzel *et al.* (1961) and Tang (1983).

The third type is the most frequent, occurring whenever an EFR comes up in the wrong (or right) place. It only forms from emerging umbrae, not plage. If the dipole only expands into plage, modest flares occur, without δ spot formation, as in Fig. 10.12. If it collides with an umbra of opposite polarity, the delta spot forms and bigger flares occur. The biggest flares, however, occur in the first two classes of D formation. If it collides with an umbra of the same polarity, the two do not merge, but coexist peacefully.

The epitome of the island δ was the August 1972 group (Fig. 10.20–1), which produced some of the greatest flares in history (Zirin and Tanaka 1973). It emerged with a large p spot behind a ragged f plage; this evolved on the far side of the Sun into a completely inverted δ configuration, with a large new p spot pushing through its center. There was no place for the p spot to go, and great shear built up as it pushed forward, producing several great flares (Fig. 11.2). A group of the second type was responsible for the flare in Fig. 11.4, one of a series of large homologous flares. A cluster of dipoles emerged at once in such a way as to leave a growing p spot pushing through an f spot. Was it an accident?

The sheared inversion line of the δ spot is the scene of most of the energy release, although smaller flares often occur in the surrounding tangle of spots. It is marked in Hα by elongated fibrils parallel to a neutral line, leading out to a filament. Often the Hα emission covers the umbrae, so we know that the field turns 90° between the vertical umbral state and the horizontal fibrils. In the continuum (Fig. 10.18) we see the penumbra has been sheared off by the δ stress, and even elongated umbrae often appear. Spot groups of great physical extent usually have lots of room, low field gradients and few flares.

Normal pairs of opposite polarity umbrae are some distance apart, so that field lines form a graceful arch across the neutral line (Fig. 10.9). δ spots (Figs. 10.20–10.23) squeeze together so closely that the magnetic field must make a sharp 180° turn to connect the two. The problem is resolved by formation of a shear boundary. The steep field gradients must be stabilized by large currents, so considerable energy is stored in the magnetic field. Even though the two spots emerged as part a single group, they were originally part of different dipoles. Their previous mates are left to "go home from the dance with another partner". Reconnection of the poles permits a much lower energy state, and the energy difference is apparently

10.22. A complex active region, July 13, 1982 in (top) videomagnetogram (bottom) $H\alpha - 1$ Å. White is preceding polarity. In centerline $H\alpha$ the fibrils cover most of the spots along the strongly sheared neutral line. This kind of highly active region grows by the development and interaction of numerous EFR's. Because there was no direct spot collision it produced many flares, but no great ones. (BBSO)

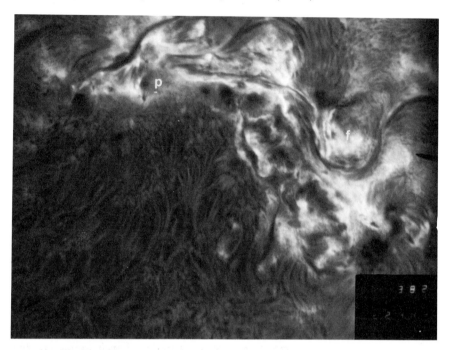

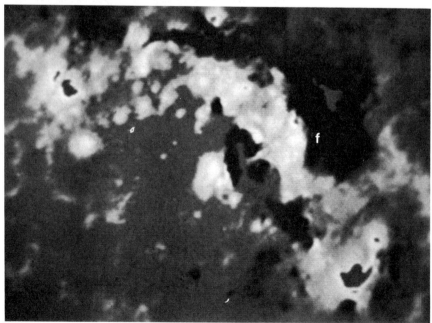

released in flares. The compression of the magnetic flux of big umbrae in the narrow channel of the inversion line means the sheared field may be very strong, perhaps stronger than in the umbrae. Note that this field is only directed one way and does not display the oppositely directed nature of popular flare models. It is, of course, turned 90° from the umbral fields.

When we watch δ spots form in the third mode we see (Zirin 1982) the opposite poles connect and then shear the connections as the motion continues. Repeated reconnections across the shear produce flares. The driving energy of the buoyant EFR may accomplish all this, but it is hard to understand how the initial connections form. It may be that reconnection occurs deep inside the Sun and the effects propagate to the surface.

In another common case (Vrabec 1974 – Figs. 2 and 7) an EFR comes up behind an older region; as the p spot plows through the f polarity, the flux lines formerly going to its sibling f spot reconnect to the indigenous f polarity. A sheared filament separates the p spot and the old f field, and a series of flares occur. The situation (which I call a "back-climber") is readily recognized by the presence of the stable, round p spot directly behind a more ragged f spot. If the EFR erupts in front of the p spot, a similar situation occurs.

True δ spots have never been seen to break apart, they just fade away in each other's arms. So from that point of view they are "solar black holes", surrounded by energy release. The δ spots lose magnetic field rapidly, and even the largest are rarely observed for a second rotation. If the spots collide at a grazing angle a δ spot forms, but the magnetic gradient is not so steep and fewer flares occur. Fig. 10.22 shows a spot of the second class, very complex and active, but with no direct collisions of spots. It had many flares but no great ones.

The appearance of the island δ raises the question: why didn't it simplify below the surface? It is possible that all active regions are formed complex, and most are simplified before we see them. Or, conversely, they were formed simple and these regions got twisted up in the convective zone. The fact that they occur away from the number maximum and that they are typically quite large suggests they were produced in the complex form. No matter, the Hale–Nicholson force pulls the p spots forward. The "black hole" picture is confirmed by our study (Zirin and Liggett 1987) of 21 δ groups, none of which ever separated, although two ejected a sunspot.

Smaller spot groups show the effects of flux emergence. Small spots appearing near the inversion line will distort it, creating peninsulas of one polarity or the other. Flares occur and the irregularities disappear. The end product is two old plages, connected by potential-like FTA. An X-ray picture would show a little cloud of corona at the top of the FTA. Great spotless flares occur only when the shear remains (as evidenced by a filament). In every active region the filaments mark the shear regions and the FTA, the stable ones. Often the filament goes right to the edge of a sunspot; the sunspot has one polarity, but the filament must (and does) separate magnetic polarities, so the opposite polarity must creep right up to the edge of the spot. The difference between filament and FTA is fairly easy to tell: FTA are thin and not very dark; filaments are much thicker and quite dark. In the wing of the line, the chromosphere underlying the AR filament is blank or threaded by dark structures parallel to the boundary; the FTA usually have plage or granular structure underneath.

Various attempts have been made to model the three-dimensional magnetic structure of the active region by taking the photospheric magnetic field as measured and fitting a three-dimensional potential or force-free solutions. Satisfactory fits to the Hα or X-ray patterns are obtained for simple configurations. Sakurai and Uchida (1977) showed that the magnetic field of sunspots was well fitted by the potential field of two sub-surface solenoids coaxial with the spots. Since the potential field has no energy available for release, flare models require force-free models, obeying the equation

$$\nabla \times \mathbf{B} = \alpha \mathbf{B}. \tag{10.9}$$

Nakagawa *et al.* (1973) showed that if the z component of the field on the surface is known, then if α is constant, the field can be uniquely determined.

10.23. A filament eruption in the same region two days earlier when the region was near the limb. The three-dimensional nature of the boundary is clear, as is the long double row of spots along the neutral line. The main body of the filament is to the right, off the frame of 10.21. (BBSO)

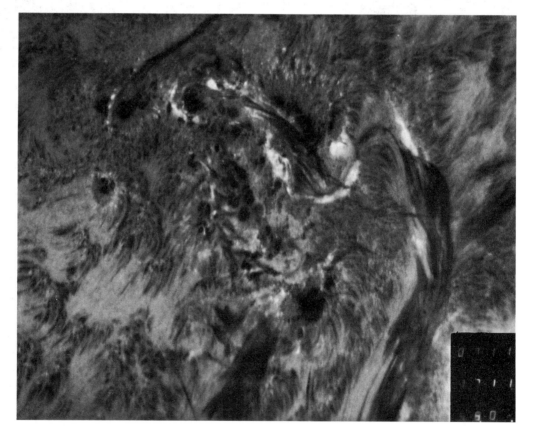

If $\alpha = 0$, we have a potential field. Tanaka and Nakagawa (1973) and Rust and Bar (1973) were able to fit the fibril structure reasonably well. The total magnetic energy will depend on α and, of course, on the degree of shear.

The value of α must be constant along a field line but can vary from one field line to the next. Sakurai (1981) developed a technique for calculating variable-α configurations, and shows some examples of the twists that can occur. But the variable-α calculation is tedious, and when he has to compute models, Sakurai (1985) still uses the potential approximation. The development of transverse magnetographs has increased interest in these force-free models.

Even in the absence of energetic phenomena, AR's radiate thermal radio emission, which can be quite intense if gyroresonance is important. But when high activity is present, ARs produce intense radio noise storms (Benz and Zlobec 1982) at meter wavelengths. These storms are all associated with frequently flaring large sunspot groups. The emission, which may fluctuate wildly, is at the plasma frequency, and is circularly polarized with the sense of the p spot. The mechanism of emission must be non-thermal, because the apparent temperatures are 100 million degrees or greater. The

10.24. An active region near the limb, photographed in Hα by Müller at the Pic du Midi. In this splendid picture the Hα fibrils outline clearly the flat angle at which lines of force from the penumbra return to the surface. But there are invisible vertical field lines from the umbra which connect to more distant places. The normal spicule structure is replaced by the bright plage material or the tilted fibrils. (PdM)

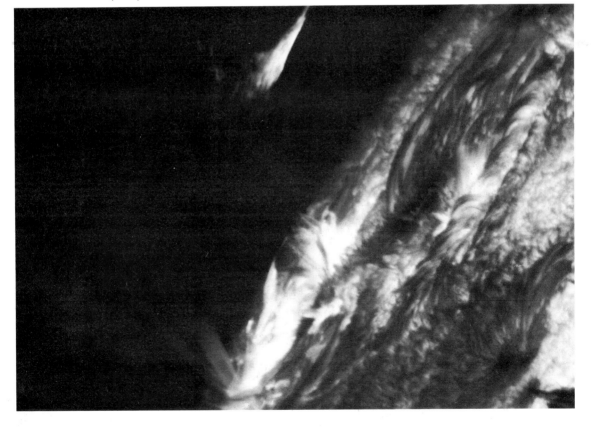

long wavelength means the source is high in the atmosphere.

It is commonly thought that the agent of radio noise storms is high-energy particles trapped in high loops. In fact, no one has a clue what is happening. The storms are often accompanied by clusters of fast-drift type III bursts (Sec. 11.4) and are a good sign of an extremely active spot group. There may be a connection with CME.

We discussed in Chap. 8 how the normally transparent corona over an active region becomes opaque to microwaves through gyroresonance absorption. The opacity in the harmonics varies with the temperature and density, so with model fitting it is possible to get a rough idea of the fields overlying the AR. Hurford *et al.* (1985) compared the observed microwave spectrum of a spot and its polarization with an integrated sunspot model and found a good fit with a field of 1300 gauss above the spot. This powerful technique will be more useful as the microwave arrays improve.

What causes the heating of the atmosphere above active regions? Examination of coronal pictures and radio maps, which pick out the hot corona, shows that in almost every case the heating is tied to flux emergence and subsequent magnetic reconnection. So long as flux is emerging, the region is bright in Hα and a coronal condensation exists. When the sunspot ages and loses its plage, the coronal condensation disappears.

As we have indicated a number of times, the most exciting feature of an active region is the solar flare, to which we devote the next chapter.

References

Bray, R.J., and R.E. Loughhead 1964. *Sunspots*. London: Chapman and Hall.

Benz, A. O. and P. Zlobec ed. 1982. *Solar Radio Storms*. Trieste: Osservatorio Astronomico.

Bumba, V., and J. Kleczek, ed. 1976. *IAU Symposium No. 71, Basic Mechanisms of Solar Activity*. Dordrecht: Reidel.

Cram, L. E. and Thomas, J. H. 1981. *The Physics of Sunspots*. Sunspot: Sacramento Peak Observatory.

Eddy, J. A. 1980. *The Ancient Sun*. New York: Pergamon.

Elgarøy, E. O. 1977. *Solar Noise Storms*. Oxford: Pergamon.

Howard, R., ed. 1971. *IAU Symposium No. 43, Solar Magnetic Fields*. Dordrecht: Reidel.

Kiepenheuer, K. O. 1953. *The Sun*, ed. Kuiper. Chicago: Univ. of Chi. Pr., p.322.

Orrall, F. Q. 1981. *Solar Active Regions (Skylab Workshop III)* Boulder: Colorado U. Press.

Tandberg-Hanssen, E. 1967. *Solar Activity*. Dordrecht: Reidel.

11

Solar Flares

11.1. Introduction

Solar flares are intense, abrupt releases of energy which occur in areas where the magnetic field is changing because of flux emergence or sunspot motion deriving from flux emergence. Since the lines of force can rearrange only slowly in response to these changes, stresses build up which are released in solar flares. This occurs most frequently at neutral lines (Severny 1958) where a filament is supported by horizontal sheared field lines. Along this boundary lines of force can shorten and release magnetic energy; afterwards loop prominences mark field lines perpendicular to the boundary. It may be that the filament lowers the conductivity and shortens the diffusion time (Eq. 3.27) so that magnetic reconnection can take place more rapidly; in any event this process can only take place along a magnetic inversion line. An enormous flux of energetic particles is produced, with electrons up to 10 MeV and nucleons to hundreds of MeV. The particles rapidly thermalize to a very hot (about 40 million degrees) plasma. Photospheric fields change little during the flare, but the Hα fibril structure connecting them is substantially altered, indicating that the connections between the moving photospheric fields have assumed a new, lower-energy configuration.

Although the first solar flare ever observed was detected in white light by Carrington (1859) and independently by Hodgson, only large flares are detectable in the continuum with normal patrols. Most flares are visible only by their line emission, and most flare images are obtained in Hα. Fig. 11.1 shows such a flare near the disk center in Hα along with the distant brightenings it produced. Disk flares almost always show two areas of emission on either side of the magnetic inversion line (Fig. 11.2), because energy released anywhere in a tube of force will rapidly heat the surface at its two footpoints where it meets the surface. When many lines of force are involved two ribbons of emission appear. Fig. 11.1 shows distant areas brightening simultaneously, showing that the exciting agent, most likely electrons, travels fast, at least 50 000 km/sec. In great flares the strands rapidly elongate ($\approx$100 km/sec) on either side of the neutral line and separate at 5–20 km/sec while the loop prominences connecting them rise higher and higher in the corona and a long soft X-ray (SXR) event takes place. If one ribbon is near a sunspot, it will be small and bright, because many flux lines converge there; the ribbons will not cross the spot since the other side involves flux lines connected away from the flare. In the late stages (Fig 11.2) the strands evolve into two thin lines formed by the intersection of a thin shell of hot coronal material with the surface.

Since reconnection means that two tubes of force interchange their end-points (Fig. 3.2), one expects four areas to brighten, and in larger flares these usually can be picked out. Occasionally, for reasons unknown, we see flares with three ribbons.

Often (Fig. 9.6) the filament rises tens of minutes before the flare; it may get exceptionally dark, blue-shifted or broadened in Hα. Then the flare breaks out with brilliant Hα emission and the filament blows away. This is called the flash phase; hard X-ray (HXR), Hα and microwave flux all increase sharply at once (Zirin 1978a). It is thought that the flare energy is concentrated in energetic electrons and possibly nucleons which produce the secondary effects.

In other cases the filament is not ejected, but breaks up with con-

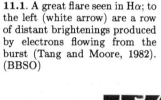

11.1. A great flare seen in Hα; to the left (white arrow) are a row of distant brightenings produced by electrons flowing from the burst (Tang and Moore, 1982). (BBSO)

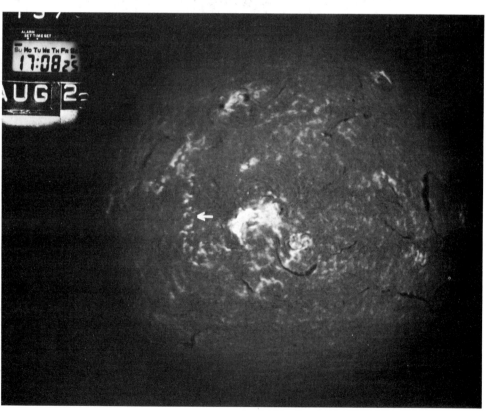

11.2. The great "sea horse" flare of Aug. 7, 1972, taken at Hα − 0.5 Å (Zirin and Tanaka 1973). In the first frames the flare is a group of bright patches, which merge at flare maximum (lower left) in a brilliant whole. The monochromatic flare image continues to increase in area, but the total line emission decreases as the emission line width narrows. After the maximum the double ribbons separate and elongate, the space between them being filled with loop prominences condensing from the hot coronal cloud. Although the fibril structure along the neutral line was initially parallel to the inversion line, the loops are now perpendicular to it. See also Figs. 10.20–21, and 11.9–10. (BBSO – Zirin and Tanaka 1973)

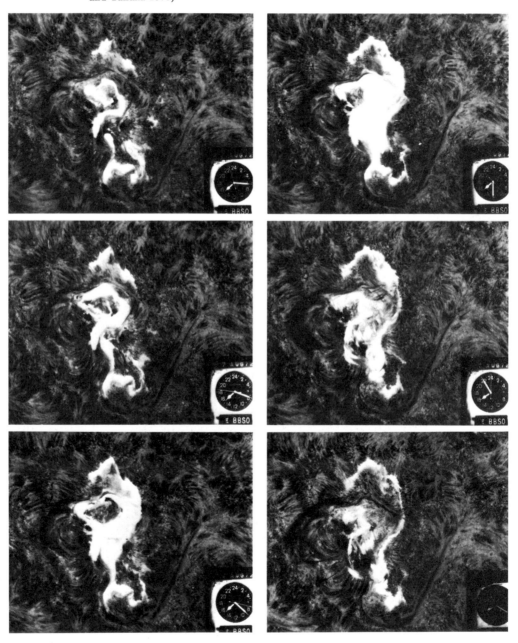

siderable twisting and turbulence at the start of the flare. A filament superposed on a plage or sunspot will surely erupt in a flare, because the photospheric field is vertical and the filament field, horizontal. The magnetic field must turn sharply, and, by Eq. (3.20d), be maintained by a current. The force-free field jumps to a lower-energy potential field. Fig. 11.4 shows a flare with the typical dark lane left behind by the erupting filament. Much of sheared horizontal structure remains; Tang (1986) has documented cases where the filament does not blow away.

The $H\alpha$ emission consists of three parts: bright kernels (Fig. 11.3) where the lines are broad and the intensity, up to three times the photospheric continuum; an extensive area of narrower (≈ 1 Å) emission not directly involved in the main energy release; and bright post-flare loops (Fig. 9.14, 11.5) connecting the two ribbons. The total $H\alpha$ emission from each of the first two is about the same, 10^{27} erg/sec in a big flare (Zirin and Tanaka 1973); the higher Balmer lines amount to three or four times that amount, and Lyα is equal to $H\alpha$. The $H\alpha$ emission from the loops is much less.

At the limb we may see an AR filament as a bright blob above the limb which violently erupts (Fig. 9.6). A loop display follows, but the ribbons are not visible because they are on the surface. In X-rays the loops are brightest at their tops; in $H\alpha$ material is seen to condense at the tops and rain down ("coronal rain"). The loop tops, where electron density is

11.3. $H\alpha$ spectrogram of the Aug. 7 flare between the first and second frames of Fig. 11.2, showing a region 10 Å wide. The widest kernel of emission was 10 Å FWHM, and the second, about 4 Å, but the rest of the $H\alpha$ emission was only a few Å wide. About half of the $H\alpha$ emission comes from the kernels. The background $H\alpha$ absorption line is about 1 Å wide. (BBSO – Zirin and Tanaka 1973)

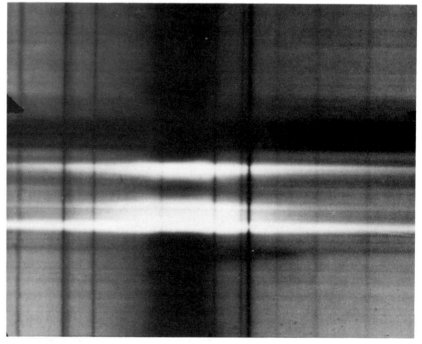

highest in defiance of hydrostatic equilibrium, coincide with the first step of cooling of the hot coronal cloud which emits intense SXR.

The ejecta include sprays (Fig. 9.12), surges, streams of electrons and nuclei and shock waves. One of the most spectacular phenomenon is the Moreton wave (Athay and Moreton 1961), an MHD shock which flies out of big flares at 1000 km/sec and may be observed crossing much of the disk (Fig. 11.8). These waves propagate through the interplanetary medium to the earth and produce intense type II radio bursts.

When flux emergence is rapid, active region plages will remain near flare brightness, and can be distinguished from flares only by their lack of impulsiveness or high-energy effects. Some flares, such as filament eruptions, are slow, but quite large. A reasonable definition is: "A flare is a transient increase in Hα brightness to at least two times chromospheric intensity, usually impulsive, accompanied by some increase in the X-ray or radio flux from the Sun."

Flares may be detected in many ways. Instruments monitor the Sun in Hα, X-rays and radio waves, and data were published in the various monitoring journals. But optical monitoring is limited by weather, and the X-ray monitoring is principally in soft X-rays. The microwave monitoring is most complete, but limited to larger events and presently provides no positional information.

Flares are ranked in importance by optical, X-ray or radio flux. The optical system approved by Commission 10 of the IAU in 1966 uses area (in degrees of heliocentric latitude), as given in Table 11.1.

A suffix (f, n or b) is added if the brightness (determined by visual estimate) is faint, normal, or bright. The area is supposed to be corrected for projection, but height effects make published areas of flares more than 65° from central meridian passage (CMP) quite inaccurate. The area is used because it can easily be measured; there is a fair correlation between area and other effects, but it disregards the impulsiveness and intensity of the event. There is no agreed measure of true importance; presumably that is related to total energy, but we are not sure what is the best measure of

Table 11.1. Flare Classification by Area.

Area (square deg)	Area in $10^{-6}A_\odot$	Class	Typical Flux at 5000 MHZ(sfu)	Typical SXR Class
≤ 2.0	≤ 200	S	5	C2
2.1–5.1	200–500	1	30	M3
5.2–12.4	500–1200	2	300	X1
12.5–24.7	1200–2400	3	3000	X5
>24.7	>2400	4	30000	X9

that. There often are impulsive or bright flares of small area with strong X-ray and radio effects which can be recognized in films, but are accorded small importance because their area is low. Their total energy may not be large, but the energy per second may be high. By contrast a spotless filament eruption may cover a large area, give strong SXR and have a large total energy, but such events are usually weak in HXR or microwave emission. Occasionally large events occur just behind the limb with strong X-ray and radio emissions, but the surface Hα effects are not seen at all. An ideal flare index would weight all these effects, but does not presently exist.

Smith and Smith (1963) estimated the number of flares per year at sunspot maximum as 20 of Classes 3 and 4, 200 of Class 2 and 2000 of Class 1. There probably are 20 000 Class S (subflares), excluding ephemeral region (ER) flares, which are not easily detected on patrols. Since the energy of a flare in each class is roughly ten times that of one in the next lower class, the total energy in each class is roughly the same. There are many other transient events: small flares in ephemeral regions, small flares at the poles, ejections in quiet regions, erupting quiescent prominences, Ellerman bombs and so forth.

In recent years there has been great improvement in the flare patrols all over the globe, and most of the important events are picked up; but there are still periods of poor coverage, particularly winter in the northern hemisphere. While most flare patrols are mediocre and cannot detect small events or those near the limb, the area measurements from different observatories are beginning to agree, at least to within 50%.

The Sun is regularly monitored in radio waves, which are not affected by clouds. Single frequency fluxes are given in solar flux units (1 sfu $= 10^{-22}$ w/m^2/Hz $= 10^4$ jansky). At frequencies above 2000 MHz, almost every event above 1 sfu is a flare. For impulsive flares, the area importance classification may be roughly related to the microwave flux by:

$$\text{Imp} = \log_{10} S(\text{sfu}) - 0.5 \qquad (11.1)$$

where S is the flux at 5 GHz (6 cm.). Since the flux is more accurately measured than the area, this is a good measure, but only for impulsive events. The range of flux is from 0.001 to 30 000 sfu at 5GHz.

In the meter range there are many bursts of different types, and not all come from flares; but type II bursts (Sec.11.7) are large and well correlated with big flares producing Moreton waves. All impulsive X-ray bursts come from flares. Occasionally a sizable radio or X-ray event is reported without corresponding Hα emission. This is invariably due to poor coverage.

A popular flare classification, often replacing the optical area class, is based on the 1–8 Å SXR flux monitored by the GOES spacecraft. The flares are designated by Cn, Mn or Xn, where the first letter is determined by whether the flux is greater than 10^{-6}, 10^{-5}, or 10^{-4} w/m^2 respectively and the integer n gives the flux for each power of 10. Thus M3 means a flux of 3×10^{-5}w/m^2 at the Earth. There is no such system for HXR bursts. Flares can also be detected by indirect ionospheric effects, of which there are a great number, resulting from X-rays of different wavelength affecting different parts of the ionosphere.

Flares show different time and energy profiles. Tanaka (1987) defines three classes:

A. *Hot thermal*, with $T = 3 - 4 \times 10^{7\,\circ}$, but limited and soft HXR emission from a compact source and little radio emission.
B. *Impulsive*, with spiky HXR and microwave emission from footpoints and low corona.
C. *Gradual-hard*, a long-enduring (> 30 min), large event with gradual peaks, a hard spectrum and a strong X-ray-microwave source high in the corona.

Tanaka points out that type B are frequently associated with a filament in a highly sheared neutral line. He shows that one well-observed case of type C shows slow optical evolution beginning with widely spaced flare ribbons, implying reconnection above the surface. Energetic particles

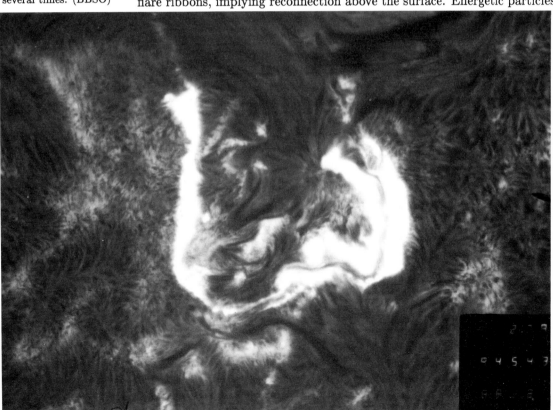

11.4. An intense flare in Hα involving the explosion of a filament in a δ spot. A p spot on the right was pushing through an f spot. The dark inversion line where the filament was can still be seen, separating the ribbons. The filament reformed and exploded several times. (BBSO)

(SEP) reaching the Earth are preferentially produced by flares accompanied by CME and Moreton waves; most gamma-ray line flares, on the other hand, direct their particles inward and do not produce proton storms at the Earth. But the largest flares produce almost all these effects.

11.2. The appearance of flares in the hydrogen lines

Of all the spectrum lines accessible from the ground, the solar atmosphere is best seen in Hα. Because we observe in the middle of the big dark Hα absorption line, brightenings are easily seen; because there is lots of hydrogen and it can be ionized only once, absorption phenomena such as surges, loops and sprays are visible, even at high temperatures. The high opacity in this line makes these phenomena visible at the limb. The fibrils which mark the transverse magnetic fields in the atmosphere are visible with Lyot filters only in Hα. In the center of Hα, however, flares are often saturated; everything is bright. The most important kernels of emission are better seen in the Hα wing, HeI D3, or the higher Balmer lines.

The Hα movies permit us to understand the relation of the flare to the context of local magnetic fields, filaments, sunspots, etc. From an information point of view, a multichannel record of the X-ray or microwave flux provides about a thousand bits per second; Hα cinematography every 10 sec provides about 4 million bps and a videotape, about 75 million (more frames, fewer bits.

Solar cinematography was pioneered by R. R. McMath (McMath 1937) using the spectroheliograph; Moreton and Ramsey (1960) at the Lockheed Solar Observatory brought it to a new level by using birefringent filters, frequent exposures and California seeing. Flare cinematography is most conveniently pursued with a 35 mm camera converted to pulse operation (some of the cameras involved were originally developed for Walt Disney) and subsequent reduction printing to 16 mm. If frames are taken every 15 sec, a 360-fold acceleration is produced upon projection. Stop-motion flickerless projectors are particularly valuable for studying the films. So strongly entrenched is the static view of solar activity that many observatories do not have even an ordinary projector, let alone take movies.

Videotape is an interesting alternative to film, offering high time resolution (32 frames/sec) and linearity but inferior spatial resolution and dynamic range. The videotape is reusable, linear and can be computer processed. But it is difficult to accelerate videotapes, and watching events in real time can be agonizingly slow. Considering the cost of film processing and reduction printing, film may be twice as expensive as video, and the capital costs are far higher. Most observatories don't have video systems either.*

Fig. 11.2 shows typical "great flare" behavior in Hα in the Aug 1972

* Movies or videos of flares are available at cost from Caltech Solar Astronomy

δ group. A filament eruption about 15:00 UT was followed by brightening at 15:05, reaching flare brightness in several kernels on either side of the neutral line at 15:15 UT. Emission increased and spread to a maximum 10 minutes later. After the maximum the two ribbons separated and bright loop prominences appeared. The broad profiles of Fig. 11.3 were confined to the kernels, while the rest of the Hα-emitting region had profiles only 1–2Å wide. Fig. 11.4 shows a more explosive type B flare which reached maximum in less than a minute. It occurred in a δ spot where a p umbra pushed continually into an f umbra, the separating filament exploding in homologous flares several times. In almost every case the high-energy processes rise in synchronism with the Hα kernels, while the thermal phase follows the extended narrow-band Hα emission.

Limb flares give us a view of the flare in plan. Fig. 11.5 shows a fine case of post-maximum development for a great flare (peak microwave 12000 sfu at 8800 MHz) observed in the wing of Hα, where saturation effects are lower. Our earliest frame (not shown) at 21:05 UT (flare maximum) shows intense Hα in the penumbral area of this δ spot, where the shear is maximum. The frame at 21:10 UT shows bright loops forming over the neutral line and connecting the two ribbons. In the frame at 21:29 we see emission at the loop tops while the rest are in absorption. In the last frame the ribbons have faded, but they still would be visible in centerline. Fig. 11.6 shows a series of small flares behind the limb along with their X-ray emission. While the SXR emission from the unseen coronal cloud is almost continuous, the HXR spikes coincide with each appearance of a brilliant cloud above the limb flare above the limb. The HXR must come from the elevated cloud. By contrast, Hα emission from disk flares last long after the end of the HXR event, because the chromosphere is always there to re-radiate the energy of the postflare coronal cloud.

Fig. 11.7 shows clearly the connection between the post-flare loops and footpoints. In this flare (Zirin 1979) the white-light footpoints were identical with the Hα footpoints in Fig. 11.7. The transition from the distorted and twisted fields of the pre-flare configuration is remarkable. The post-flare loops are regular at all heights, although Figs. 11.5–6 show their angle changing with height. But it is still not possible to determine whether these regular loops were always present and only became visible when filled with post-flare material, or actually untwisted from more complex arrangements.

A number of circumstances tell us that flares are likely. Persistency is most important; if a region has had many flares, it will have more. A rapidly growing δ spot covered with Hα emission, a p spot moving into an f spot, a big spot without penumbra, all presage flares. But only two definite precursors are known: darkening of a filament and gradual brightening of Hα accompanied by a rise in radio or SXR emission.

Smith and Smith (1963) discuss "homologous flares", events with

11.5. An excellent view of a flare near the limb in Hα − 0.5 Å, showing the development of loop prominences. The first frame is after the flare peak. Because the background is much brighter in the line wing, only the most intense features appear; only rarely are loops bright (21:10, 21:21 UT) in the Hα wing. They connect bright strands overlying the shear regions in the spot penumbrae. As they develop, the loops become darker, but the tops are always brightest and densest. Like those in Fig. 11.2, these loops rotate as they grow. In all loops the material rains down in the two branches; for this reason we see only the near branch in last two frames as we are observing in the blue wing. (BBSO)

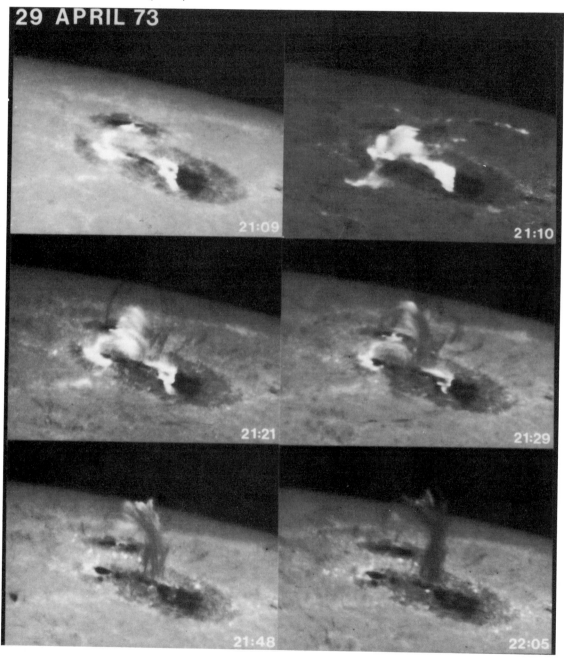

almost identical location and form occurring repeatedly in the same region (Fig. 11.4). These are due to shear built up by continuous flux emergence and sunspot motion. Successive flares occur as the energy builds up and is released in successive events a few hours apart. The similarity in form reflects the general magnetic layout of the region, the places where energetic electrons reach the chromosphere and produce Hα brightening. With higher resolution one finds that the main kernels of energy release always differ somewhat.

Since we do not see the cooler regions of secondary illumination, all the spectral lines of limb flares are very broad (up to 10 Å). In big events the erupting filament appears as a bright blob rising above the limb, slowly rising, brightening and exploding outward. If the explosion is violent, the impulsive X-rays occur at the time of explosion; if it is slow, SXR emission will occur when Hα emission does. Sometimes a brilliant loop appears and fades without explosion.

Intense flares observed in Hα centerline show (Fig. 11.9) a halo of diffuse emission about 20 000 km across first noticed by Zirin and Tanaka (1973) in the August 2, 1972 flare (Fig. 11.9). The halo is not scattering in the optics or film, because it does not quite match the peak Hα emission, nor is it symmetric about it. It is probably Hα scattered by the surrounding chromosphere; the irregularity may be due to absorbing clouds under the flare. Somov (1978) has discussed the possibility that the halo is due to scattered X-radiation, but the intensity would be less than scattered Hα. If we could understand it, the halo could be a help in determining the flare height. It is observed in most brilliant events and usually accompanied by continuum emission. Fig. 11.10 shows the continuum emission from the flare of Fig. 11.9. The saturated Hα is replaced by tiny bright flashes along the neutral line, smaller than 1000 km and closely connected with HXR spikes.

Most flares show a red asymmetry of about 0.5 Å in Hα. Probably this occurs because material explodes from a source above the surface; the upward moving material gives blue shifted SXR lines from the hot flare plasma, while the downward moving material excites chromospheric material it pushes downward. The Hα films reveal a wealth of ejecta, and in the largest events we see the Moreton wave (Fig. 11.8), a shock wave traveling across the surface at 1000km/sec. The Hα emission from the wave is due to trapped electrons; in the line wings the wave is seen because of the Doppler shift the wave induces in the chromosphere.

Henoux *et al.* (1983a) have detected polarized Hα emission in the thermal phase of a flare; they interpret the polarization, which is always directed toward the disk center, as the result of conductive heating from above, which establishes a preferred direction. Similar effects occur in XUV and SXR emission (Sec. 11.4).

In the other lines of the Balmer series flares appear as we might

11.6. A limb flare with associated X-ray emission (Zirin, 1978). In the HXR record at lower left (OGO-5 records courtesy Dr. Ken Frost) the bottom two records show the hard spike event that appeared when the bright arch at right peeked above the limb; it is reasonable to assume the X-rays come from the bright arch, which is probably the "real flare". (BBSO – GSFC)

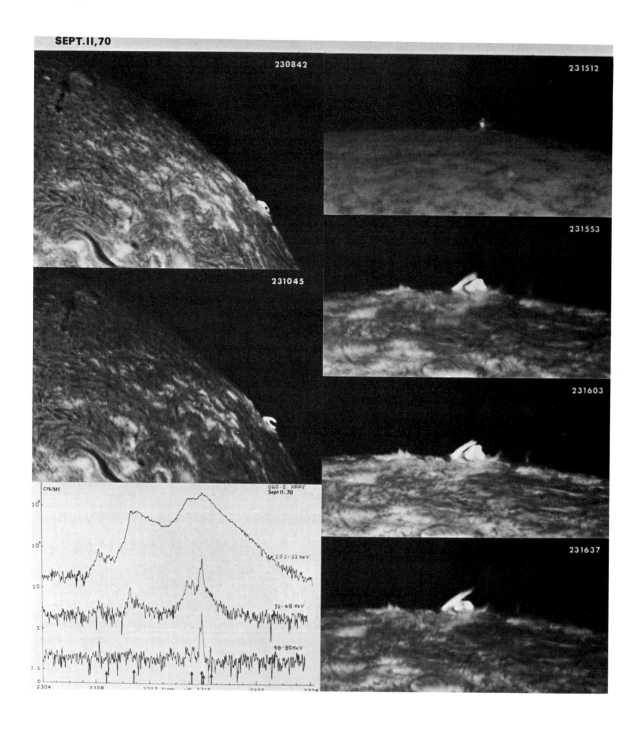

expect; the area of brightening is less, but the contrast is higher. The peak intensity in Hβ and Hγ is sometimes greater than that in Hα (Zirin *et al.* 1982). The effect is also found in stellar flares, where the strongest Balmer line is usually Hγ or δ. In these lines the optically deep line emission approaches the Planck function. Because the brightness temperature of the flare footpoints is greater than the 6000° photosphere, the peak of the Planck function shifts to the blue, so the higher Balmer lines are brighter than Hα.

An interesting contradiction exists in the ratio of the Lyα and Hα intensities. Since most of the electrons are in the ground state, the lower lying Lyman line will be more easily and frequently excited; in any event we should expect every Hα photon emission to be followed by a Lyα emission, and of course there is five times as much energy in a Lyα photon. In fact, this does not happen; although data are incomplete, it appears that the two lines are of roughly equal intensity (i.e. there are five Hα photons for every Lyα photon). The same surprising result is found in quasars and Seyfert galaxies (Zirin 1980). The effect must be due to collisional de-excitation of $n = 2$ in the long path of the Lyα photons. These lines are optically quite thick, and their relative intensities are determined by the Planck function. The observed ratio is in fact explained by a Planck function at 17 000°; the Hα flare intensity would then be about ten times the continuum instead of three as observed. That is not a large discrepancy, but one would expect

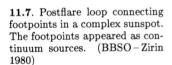

11.7. Postflare loop connecting footpoints in a complex sunspot. The footpoints appeared as continuum sources. (BBSO – Zirin 1980)

the temperature and consequently the Lyα : Hα ratio to vary much more than they do. These lines play an important part in the energy loss and may be limited by the overall energetics.

Flares observed in optically deep resonance lines such as CaII K, MgII λ2795 and the HeII λ304 look very much like Hα, except that the contrast is higher because they are on the steep side of the Planck curve. the only Lyα flare picture I know of is one taken in the decay phase by Foing and Bonnet (1984) from a rocket; it looks like Hα.

11.3. The appearance of flares in lines of other elements

The universal birefringent filter (UBF), as well as special single-line birefringent filters, have made possible pictures of flares in other lines, such as HeI D3, NaI D, and MgIb and CaII K. The appearance of flares in the lines of each ion depends on the ionization and excitation potentials and the abundance of the element. Optically thick lines like Ca K and HeII 304 display the same widespread emission as Hα, while optically thin lines such as Na D and He D3 show only kernels. The high excitation required to form D3 makes it end to show atmospheric phenomena, while Na reflects the low chromosphere.

Because only linear detectors are available for that wavelength, images in the strong λ10830 line of HeI can presently be recorded only with the spectroheliograph, which limits time and space resolution. Harvey and Recely (1984) have observed remarkable λ10830 enhancements following filament eruptions with the Kitt Peak spectroheliograph, but I am unaware of any flare images. The only high-resolution helium observations are in the D3 line (5876 Å), which is the strongest He line in the visible spectrum. The line appears in absorption in surges, eruptive prominences, and weaker flares, and in emission in more intense flares. It is a gold mine of information on the flare. Data on many He lines observed in the visible are given by Lites *et al.* (1986). Figs. 11.11 and 11.14 show examples of the extraordinary intensity of D3 in flares, reaching three times the continuum. In Fig. 11.11 we see the D3 emission starting as a number of small flashes along the inversion line, just like Fig. 11.10, then two bright strands form in the middle and develop into the two flare ribbons. Late in the flare bright D3 loops appear across the inversion line. The D3 emission may be compared with the Hα images of the same flare in Fig. 11.13.

D3 can only be observed against the disk if density or temperature are high. The normal excitation by coronal ultraviolet produces an optical depth of just a few per cent in D3, and we can only see it at the limb. The normal excitation of those triplets relative to the metastable $2\ ^3S$ level is only that of the photosphere with dilution factor 1/2; with the Boltzmann formula (Eq. 4.26) for 6000° and dilution factor 1/2, the ratio $2\ ^3P : 2\ ^3S$ is normally 0.165. Since λ 10830 has a radial optical depth of only 0.1, that of D3 is about 0.02. Thus surges and sprays appear in absorption either because the local X-ray flux is enhanced, raising the triplet population according to Eqs. (7.30-7.31); or because the temperature rises above 15 000°, where collisional excitation of triplets begins to occur.

When we see D3 (or any line) against the disk, we know from Kirchhoff's laws that the intrinsic emission of that gas must exceed the background. This requires that the temperature is greater than the background, that the density is high enough so that upward collisions are more frequent than photospheric excitations, and that the optical path be sufficient for $T_{ex}\tau$ to exceed 6000° and we see emission.

The D3 emission in the events illustrated is three times brighter than the photosphere, corresponding to a brightness temperature 8200°. Detailed calculations show that it is quite unlikely that such emission could occur from a source at that low temperature; the line must be optically deep, which means that there must be a great deal of triplet helium, or $T > 15 000°$. We can place a firm lower limit on the density as follows:

In Sec. 9.3 we pointed out that H lines will go into emission when collisional excitation exceeds photospheric excitation and the kinetic temperature exceeds 6000°. He is harder to excite and less abundant. For He we require $T > 15 000°$ and enough density to give $T_{ex}\tau > 6000°$. When the excitation of D3 is due to the radiation field of the photosphere with a dilution factor $1/2$, we have a steady-state relation

11.8. A Moreton wave in the Aug. 28, 1966 flare in blue wing, centerline and red wing. The blue wing shows bright-dark emission, corresponding to down-up motion; the red wing shows the opposite. In the line center a bright wave front is seen. (SPO)

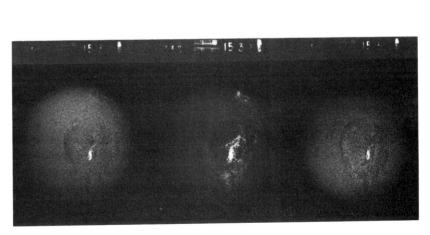

$$N(3\ ^3D)A_{UL} = N(2\ ^3P)B_{LU} \tag{11.2}$$

and from the Boltzmann formula,

$$\frac{N(3\ ^3D)}{N(2\ ^3P)} = \frac{1}{2}\frac{15}{9}10^{-5040\times2.11/6000} = 1.4 \times 10^{-2}. \tag{11.3}$$

Since $A_{UL} = 3 \times 10^7$/sec, $B_{LU} = 4 \times 10^5$/sec. For the excitation of D3 to be higher, the collision rate C_{LU} must exceed this value. Allen's approximation Eq. (4.64) with T=20 000°, $f \approx 0.5$ and $\chi = 2.11$ and $P \approx 0.1$ gives $C_{LU} = 7.66 \times 10^{-8}N_e$. For density $N_e = 5 \times 10^{12}$ upward collisions balance photospheric excitation, doubling the excitation rate, and for higher densities the emission increases correspondingly. The high density permits the gas to radiate with a source function closer to the Planck function for the corresponding temperature. Thus the transition from emission to absorption in a flare or surge occurs as the density drops below 10^{12}cm^{-3}. For Hα we may further judge the density by whether the loops are in emission in the wings. Why can't we use this rule to get the density of the chromospheric flare? Because the optical depth is so great that other effects may occur, while loops or D3 footpoints are easier to understand.

This pattern is observed in most events; compact, low-lying D3 sources are bright, while diffuse ejecta appear in weak emission or absorption. The transient flashes of the impulsive phase are replaced by the main ribbons as the hot thermal plasma is formed (Feldman *et al.* 1982) and conduction down to the level at which $N > 5 \times 10^{12}$ occurs. Here the effect is not a density increase but heating of dense layers; the bright D3 emission from a limb flare should be low. On some occasions (Zirin and Neidig 1981) a *third* bright kernel appears a few minutes later, quite separate from the first two and lasting longer than either; it must represent a reconnection to the thermal source. In addition to the emission general D3 absorption excited by the SXR source at the loop tops appears. The D3 flashes at footpoints must be caused by particle streams, and the main D3 sources by conduction from the hot thermal source. Na D or Mgb – in fact any Fraunhofer line – show the impulsive-phase kernels to some extent, although it is not known if the small flashes appear.

In all these descriptions we assume that the energy is coming down from above, because: (1) the surface density is so high that energetic particles would lose their energy before they are accelerated to high energies; (2) the microwave maps show the source at the loop tops; (3) whatever magnetic changes that occur seem to happen above the photosphere.

In great events post-flare loops are seen in D3 against the disk; from the fact that they start out in emission and end in absorption we may deduce that their density is above 5×10^{12}. Tanaka *et al.* (1983) found that when the D3 loops appear an initially hot plasma (5×10^7 deg) cools abruptly to only 3×10^7 deg. While the hot thermal plasma appears as the initial power-law electron distribution thermalizes, the emission measure continues to increase after the HXR burst ends, implying that energy release continues. One sometimes sees (Zirin 1980) a dark D3 shell in the explosive part of the flare, followed by two bright ribbons at the base.

The D3 emission gives us an excellent picture of how the overall flare develops: First, the irregular transient flashes of the impulsive phase suggest a tearing mode instability in various flux loops in the sheared region. This leads to a main phase reconnection between the two main poles which are geometrically closest. Since these footpoints do not ordinarily coincide with the initial flashes, the energy source must be new and not a remnant of the impulsive phase. Further energy release in the main phase occurs by ribbon spreading or reconnection to a third kernel. In both cases the magnetic energy of additional fields is released.

Because the flare energy is input at very high temperatures, we would like to observe the highest excitation lines with the highest resolution. But most high-excitation lines are in the UV where resolution is low, and the coronal lines are only emitted in the thermal post-flare regimes. The $\lambda 4686$ line of ionized helium is the highest excitation flare emission observable in the visible spectrum and has been observed in two intense flares (Zirin and Hirayama 1985; Zirin 1986). Fig. 11.14 compares the

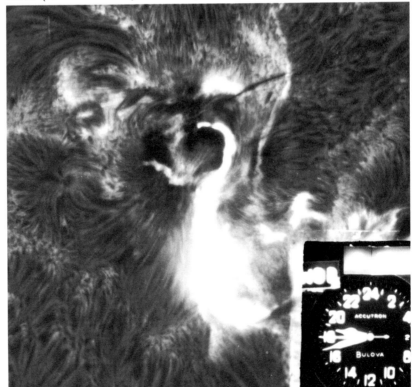

11.9. The Aug. 2 1972 flare (Zirin and Tanaka 1973), showing the $H\alpha$ halo. This flare occurred in the Aug 1972 region (Figs. 10.20–1, 11.2). It was unusually impulsive but not great; it evolved over a few hours into a huge type A thermal event. (BBSO)

11.10. Continuum flashes in the Aug.2, 1972 flare. These short lived (5–10 sec) bright points along the neutral line corresponded to spikes of HXR emission, and probably are due to energy release in individual flux loops. In the lower left and middle right we see twinning of the emission on either side of the magnetic inversion. (BBSO – Zirin and Tanaka 1973)

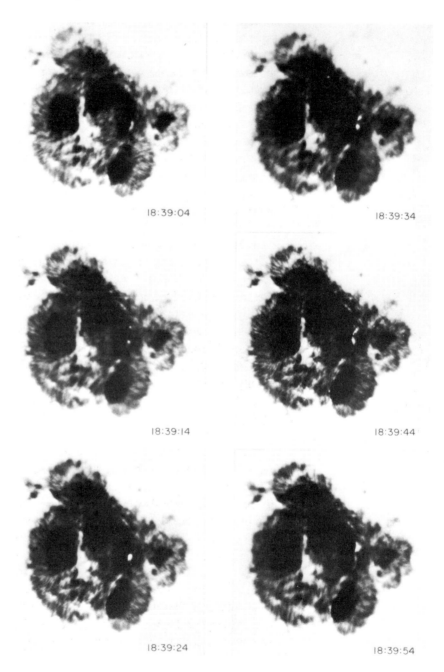

$\lambda4686$ emission from an intense flare with D3 and the continuum. $\lambda4686$ grows with the brightest D3 kernels in the impulsive phase but weakens with the continuum. Evidently the thermal phase is too cool for $\lambda4686$ excitation. The relatively high-brightness of the helium lines directly above the cool sunspot suggests they come from a dense hot region somewhat above the photosphere.

Na D and Mgb are low excitation lines emitted when the low chromosphere is heated, usually at footpoints of flux loops passing through the flare. Their behavior is similar to D3, but they are much weaker and only exhibit surface brightening; we have seen no effects of atmospheric phenomena such as surges in these lines. Flares may be observed in any Fraunhofer line; one simply needs good seeing and the right filter. All seem to show the pattern of the lines mentioned above: footpoints only.

Flare images in lines of highly ionized iron (Fig. 11.15) have been obtained by the NRL overlappograph (Widing 1975, Cheng 1977). These spectra are discussed in Sec. 11.8. The high ionization lines appear as tiny cores at the loop tops, while the low-ionization lines resemble Hα, showing that the height scale is not great. The behavior of these high-ionization lines tells us little about the impulsive stage of the flare; as Kane and Donnelly (1971) showed, the XUV and visible flare lines in the chromosphere brighten in the flash phase, and the coronal lines belong to the thermal phase.

I am not aware of any good flare pictures in the range 500–1500 Å, although excellent Skylab spectra exist. Some data of this type were obtained by the Solar Maximum Mission and may appear shortly. Thus far only the overlappograms show the XUV flare images. Unfortunately, although it reveals the truth, the overlappograph has been discarded by NASA for more quantitative but less informative techniques.

White-light flares mean just that, flares observed in the continuum. Although the first recorded observation of a flare was in the continuum, events large enough to record with the usual low-grade patrols are fairly rare, so the term "white-light flare" (WLF) has come to mean a really intense event. If one uses a filter that isolates a region several angstroms wide filled with absorption lines, the background is reduced and continuum emission may be detected from almost all flares class 2 or brighter. And when we go to super-high-resolution techniques, we find that every flare is a white-light flare. Because the continuum is enhanced in the blue and and the background depressed by the many lines, monitoring is best done below 4000 Å. I find it convenient to monitor the continuum with a 20 Å filter centered at 3862 Å, a blue region free of hydrogen emission. At both Sacramento Peak and Big Bear high-speed filter wheels are used to get the spectral dependence of the continuum.

The continuum morphology reflects the fact that we see the photosphere, perhaps 500-1000 km deeper than the D3 source. Only the most

11.11. Developments leading to the June 6 1982 flare; left: D3; right: magnetic fields. The spot A, which appeared alongside B near the limb (There was a great flare on June 3), moved rapidly westward, leaving behind a sheared δ configuration reflected by elongated Hα structure in Fig. 11.13a and a new spot (C). The velocity of spot A is plotted at upper right; after the great flare the motion continued, but more slowly. The new spot C connected to spot B. A number of flares occurred at the leading edge of spot A as it moved through opposite polarity; one appears in (c). (BBSO-Tanaka and Zirin 1985)

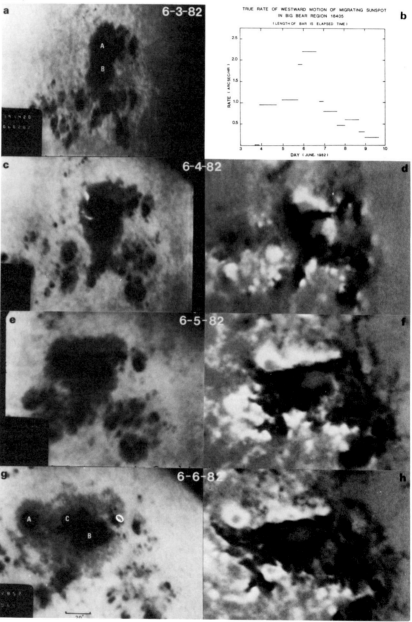

intense flare components will penetrate that deeply. The continuum thus appears like the impulsive part of the D3 event. Transient flashes a few arc sec in diameter (Fig. 11.10), roughly simultaneous with HXR spikes, appear at various points along the neutral line (Zirin and Tanaka 1973). The individual flashes do not move, but fade and are replaced by others. In the main flare phase brighter, fixed footpoints paired with opposite polarity are seen, usually in the penumbrae of the big delta spots, and an even brighter third point may appear in the thermal phase. Because these are the brightest continuum sources they are the kind most frequently appearing in published WLF pictures (Svestka 1976). In most cases the WLF emission disappears with the end of the impulsive phase.

Another kind of WLF emission consists of a slowly moving bright wave (Machado and Rust 1974, Zirin and Tanaka 1973) coincident with the brightest part of the spreading Hα double ribbons. These must result from the thermal conduction that excites the ribbons, and their existence shows that conducted energy can reach the surface if there is enough time and energy.

Understanding the spectrum of the WLF (Fig. 11.18) is a real challenge. In the visible the continuum is flat down to 4600 Å, where it begins to rise (Machado and Rust 1974, Zirin and Neidig 1981) to a strong Balmer jump. Zirin and Neidig found the ratio of continuum at the Balmer continuum (3500 Å), $\lambda3862$ and $\lambda4275$ to be 10:5:1, respectively, and the peak intensity at $\lambda3862$ to be 2.5 times the photosphere (without subtraction) in a class 2 flare. Other intense flares give the same results. Measurements from flare spectra all give much lower enhancements, of the order of 10%. Slonim and Korobova (1975) examined 17 flares and estimated that the integrated continuum intensity per cm^3 was 100 times that in Hα. Even modest flares appear in continuum emission at 1600 Å.

A similar increase in the blue continuum is seen in stellar flares (Kunkel 1970; Mochnacki and Zirin 1980). In one case observed by the author the continuum starts to rise around 4000 Å and is still rising at 3200 Å (the Balmer continuum is supposed to fall as we go to shorter wavelengths). The blue continuum is well marked in energetic objects such as quasars and Seyfert galaxies where there is a "3000 Å bump". Recently Neidig (1983) has argued that the blue continuum, at least in solar flares, is a visibility effect due to the sharp drop in solar luminosity (because of the many lines below 4000 Å), and he obtained a flat continuum after subtracting the background. But this was for a spectrogram where the enhancements were not great; Neidig shows that a substantial opacity enhancement is required for the sharp increases observed with two-dimensional detectors.

Could the blue continuum be due to emission in the many lines that crowd the blue part of the spectrum? Flare spectra (Grossi Gallegos *et al.* 1971, Neidig 1983) show that the many metallic lines below 4000 Å are inadequate to produce the continuum. But, as we mentioned, none of the recorded spectra shows more than 10% continuum enhancement, so the spectrum of the flare kernels was not recorded; the slit was not exactly in the right place. We badly need a flare spectrogram which really gets the kernel and gives the same continuum intensities that have been measured with filters.

Theoretical attempts to explain the blue continuum have all failed. All the known opacity sources fall off to the blue. The Paschen continuum opacity, which must contribute in the visible, falls off with the cube of the frequency and should be much weaker than the Balmer continuum, while the blue continuum isn't. The H⁻ opacity also falls off with frequency. But if we take images of the chromosphere with broad band filters below 4000 Å we see network emission, at least near the limb. So the many lines make us see higher in the atmosphere in the blue, and if the atmosphere at that height is heated more than the photosphere, we will see the blue continuum. Other attempts to explain the WLF by synchrotron, inverse Compton or other exotic emissions fail when subjected to realistic modeling.

In stellar flares the flare dominates the stellar spectrum in the blue, so we don't have to worry about getting the slit in the right place. Because we don't know the true size of the flare we can adjust it to explain the continuum by a hot spot radiating as a black body at 8–10 000°(Mochnacki and Zirin 1980). The peak of the continuum is below 3000 Å and the blue continuum follows naturally. In the Sun this won't work because we see the flare and know its surface area. The temperature required to produce the required spectral behavior would give a continuum brightness increase much greater than observed. This difficulty could be removed by the usual wretched device of introducing a filling factor; if the continuum kernels filled only a small fraction of the apparent area, one could fit the data. Maybe it really is that way. Good photometric continuum data in the 2000–3000 Å region could test this model and further delineate the spectral characteristics. But the VLA has resolution of 0.2 arc sec and reveals little structure below 1 arc sec in microwave sources. Another possibility is that the blue continuum is due to merging of higher members of the Balmer series. Zarro and Zirin (1986) showed that this indeed happened in two flares in YZ CMi, but there are several arguments against it in the WLF.

Another difficulty in understanding WLFs is how the energy reaches levels dense enough for continuum emission. The energy must reach the depth at which the density is 10^{17} atoms/cm³. But Eq. (3.8) gives a range of only 2 km for 100 keV electrons at this density (if fully ionized), and the scale height is 100 km. Since the atoms are mostly neutral (cross section $\approx 10^{-16}$cm² per atom) the situation is even worse. So the flare electrons cannot reach the level of continuum production. The FeI Kα line is observed in the impulsive phase of flares; it might be due to electron impact on the photosphere but more likely is due to photoionization by HXR emission from the flare.

11.12. D3 images of the June 6, 1982 flare overlaid with 25–35 keV HXR contours from Hinotori (Tanaka and Zirin 1985). Images (a) and (b) show multiple rapid flashes similar to Fig. 11.9; by (c) the double strand formed and a steep increase and hardening of HXR emission took place, saturating the detector. At the peak (d) the ribbons were fully formed. Postflare loops appeared in the last two frames; these are rare in D3 because the density must exceed 5×10^{12}. The peak emission at this time was centered on the loops. During the rise the HXR emission was always centered between the main kernels, even before they appeared. The HXR spatial resolution was inadequate to determine if the emission came from the kernels or between them. Sunspot development leading to this flare appear in Fig. 11.9. (BBSO – TAO)

6 JUNE 1982

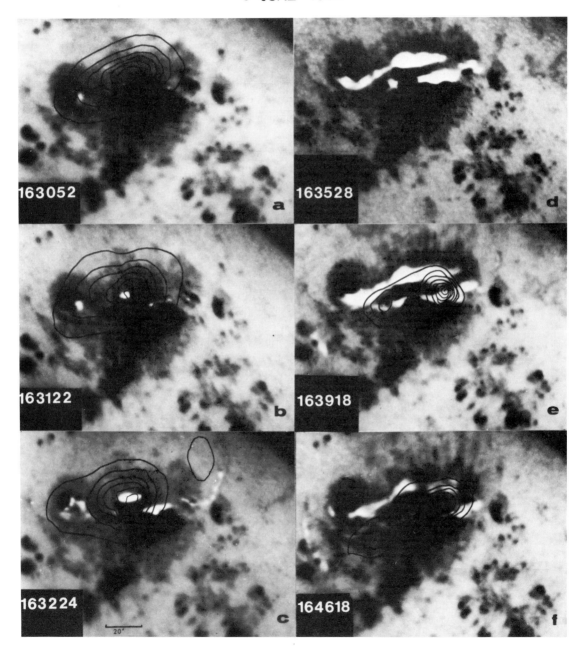

Svestka (1970) showed that electrons could not penetrate below the $n = 10^{14}$ level but that protons and other heavy particles could. This penetration was dramatically confirmed by the observation of nuclear gamma-ray lines from flares (Sec. 11.5). So it was thought that protons accelerated in the flare could produce the continuum. But those who calculate the proton flux steadfastly maintain that they don't have enough energy. Since students of protons probably want protons to be important, we must believe them when they say they aren't.

Could a two-stage mechanism work? We know that the continuum footpoints are similar to those seen in D3, except that the impulsive phase is more important in the former. The simultaneity of X-ray and D3 emission tells us that electrons reach at least the D3 level, where the density is 10^{13}, ten scale heights (1100 km) above the photosphere, and probably lower. From there a conduction front could move downward at the electron sound velocity, about 500 km/sec and reach the observed height of WLF emission. We would then expect a one- or two-second delay between X-rays and continuum; observations are inadequate. One might see the event at the limb; again there are no data. To be perfectly honest about it, most of the solar observatories are in the wrong sites and have the wrong equipment to see such things. Relaying the energy downward by radiation

Table 11.2. Space and time distribution of flare emissions.

Emission	Position	Time behavior	Source of emission or excit
Hα	Chromosphere everywhere	Rise with HXR, Fall with SXR	Beamed electrons, thermal conduction
HeI D3	Footpoints, Low chromosphere weak loops	Rise with HXR, Peak and fall with SXR	Beamed electrons, Conduction from high temp, recombination
HXR	Footpoints, some loop tops	Impulsive	Hard electron bremsstrahlung
SXR	Loop tops	Slow, peak minutes after HXR	Bremsstrahlung from thermal electrons
FeXXIII –XXVI	Loop tops	Post-impulsive	Excitation by high temperature electrons
FeXII –XVIII	Low corona	Decay	Recombination, 2×10^6 deg
CIV, OIV	Footpoints, chromosphere	Like D3?	Conduction from high temp
Microwave	Loop tops	Impulsive	Synchrotron by electrons >100 keV

doesn't work because the photospheric continuum has as much energy as the flare emissions, and radiation would produce diffuse kernels because it isn't focussed.

The timing and spatial distribution of the various flare emissions in a normal flare are summarized in Table 11.2. It should be noted that the SXR peak varies from one minute after the HXR for 10–20 keV to five or ten minutes for 3 keV. Further, as we have noted, there is another class of event with major X-ray and microwave emission quite late in the flare.

11.4. Flare electrons

Most of the electromagnetic effects detected from flares are produced by energetic electrons. Their energy distribution may follow a power law resulting from impulsive acceleration processes, or a Maxwellian distribution either resulting from the relaxation of the power-law distribution or produced directly in the flare. In addition to the optical effects discussed in the preceding section, the electrons produce the following easily recognized emissions:

- continuum X-rays (by bremsstrahlung);
- X-ray lines (by excitation of inner shells of ions)
- microwave emission (by synchrotron and free-free emission)
- meter wave emission (by plasma oscillations – type III bursts)
- distant Hα brightenings (by type III-RS bursts)
- direct detection at the Earth.

It is entirely possible that the electrons carry most of the energy initially imparted to the flare plasma, even that which later appears in macroscopic motion. The fraction of the energy in nuclei is not known. The electrons are probably responsible for the picturesque zoo of zebra, fiber and tadpole burst structure in radio noise storms (Sec. 10.4).

Four components of the flare electrons have been recognized:

1. Non-thermal electrons with a power-law distribution occuring in bursts or spikes in the impulsive phase; they can be recognized from the spiky profile of X-ray or radio emission they produce. Their spectral distribution may range from E^{-2} to E^{-9} and they are effective microwave emitters.

2. A hard non-thermal cloud of electrons occurring late in the flare high above the surface. Tanaka (1986) finds these to occur in flares without filament eruption. The X-ray spectrum of this component cannot be fitted by a thermal distribution (Ohki *et al.* 1982) and intense microwave emission is observed. The power-law exponent is 2.5–4.

3. A hot thermal cloud of electrons at 25–50 million deg, discovered by Feldman *et al.* (1980) in X-ray line spectra of FeXXV and later detected by Lin *et al.* (1981) from high-resolution X-ray continuum data. The

hot cloud appears at loop tops in the Skylab images slightly after the impulsive event, producing the most intense chromospheric emission.

4. A post-flare thermal cloud with T dropping from 20 to 3 million deg during the cooling of component (3).

At the onset of an impulsive flare, X-ray, microwave, Hα and continuum rise abruptly as electrons are accelerated. Even if there are several spikes, the X-ray and microwave fluxes track closely except that the microwave peaks are delayed by about one second. Hα and continuum emission brighten in different places for each spike. Each peak appears to be a separate acceleration process. In great flares the soft thermal component may last for many hours, during which the two ribbons separate and a loop display occurs.

Lin and Hudson (1971) proposed that the major energy input to the flare was in the power-law electrons producing the HXR emission, colliding with ambient electrons to form the thermal plasma that is the SXR source. In Fig. 11.17 we see the relation between the time behavior of HXR and SXR emission and the $\lambda 3862$ continuum. The hardest X-rays are synchronous with the continuum. The time derivative of emission in the Fe lines at 5 keV, which may be identified with the SXR flux, matches exactly the HXR time profiles. From this Tanaka (Tanaka *et al.* 1982; Tanaka 1983) concluded that the energy input for the SXR emission is the power-law electrons of the HXR burst. The exact relation between SXR continuum and line emission is not established. The morphology of the D3 and continuum emission indicates a different physical location for the SXR source so it is not clear how the energy is transferred. We have mentioned that the SXR emission continues so long that additional energy input is required. The subsequent increase in the emission of lines at lower ionization may be attributed to cooling from the hot material. Fig. 11.18 shows the overall X-ray spectrum. The SXR spectrum is fit by a thermal exponential distribution, the HXR by a power law. At the highest energy the spectrum is flatter still because of the contribution of nuclear γ-ray lines.

It has been argued by some that the hot thermal source is chromospheric material heated by the impulsive component which rises into the corona (the "evaporation" model). The support for this model is twofold: first, strong blue-shifts of FeXXV lines corresponding to upward flows of up to 600 km/sec have been observed (Acton *et al.* 1982; Antonucci *et al.* 1982; Tanaka and Zirin 1985); second, there is much more coronal material in the thermal plasma than was present before the flare. But no one has explained how you can produce a 50 000 000° plasma by heating the photosphere with particles, particularly electrons. Further, the energy of the non-thermal electrons barely equals that of the thermal plasma, so there is nothing left to compensate for other losses, such as the optical emission and heat conducted away. All the measurements of blue-shifted FeXXV

11.13. The same flare in Hα, with SXR (6–13 keV) contours (Tanaka and Zirin 1985). The postflare emission is centered on the loops. The spots are covered by Hα structure before and during the flare, afterwards the spots are visible as the horizontal sheared fields are broken and can no longer support material. (BBSO – TAO)

6 JUNE 82

show it occurring during the flash phase, the period of Hα expansion, usually before the maximum of the WLF, when the "evaporation" would be
expected to take place. On the other hand, there is plenty of chromospheric material that could be evaporated before the heating reached the
photosphere.

I prefer the explanation of the thermal cloud as coming from the
large amount of Hα absorbing material usually overlying the flare site, or
the filament that occupies it. This material often blows away during the
flare process (but rarely with the 400–600 km/sec speed observed in the
SXR lines) and might be heated directly by the HXR electrons with fewer
losses.

The X-ray flux from a power law electron distribution can be calculated in two ways. If the particles remain trapped in the acceleration
region we can calculate the resulting emission as though their distribution is unchanging. This is the "thin-target" model introduced by Kane
and Anderson (1970). The other possibility is that the electrons leave
the low-density acceleration region and precipitate in a denser region like
the photosphere, where they lose their energy immediately. This is the
"thick-target" model introduced by Hudson (1972). Since the high-energy
electrons lose only a fraction of their energy in each X-ray emission and are
being slowed down by other collisions, the calculation must account for all
the X-rays each electron will produce as it moves through the energy spectrum. The thick-target model thus represents the X-ray emission due to
the instantaneous flux of accelerated particles while the thin-target model
gives the emission resulting from a distribution which is the integral over
time of the acceleration process.

For simplicity we start with the thin-target model and follow the
treatment of Brown (1971, 1975), who was the first to solve the problem
of analytic deconvolution of the X-ray spectrum. The photon flux $I(\epsilon, t)$
measured at the Earth is due to the interaction of an electron distribution
$F(E, r, t)$ with a distribution $n(r, t)$ of field particles through a cross-section
$Q(\epsilon, E)$:

$$I(\epsilon, t) = \frac{1}{4\pi R^2} \int_V n(\mathbf{r}, t) \int_\epsilon^\infty F(E, \mathbf{r}, t) Q(\epsilon, E) dE \, dV \qquad (11.4)$$

where ϵ is the energy of the X-rays observed, and E is the energy of the
particles producing them. For simplicity, averaged values of n and $F(E, t)$
are introduced, so the formula becomes:

$$I(\epsilon, t) = \frac{\bar{n}V}{4\pi R^2} \int_\epsilon^\infty \overline{F}(E, t) Q(\epsilon, E) dE \qquad (11.5)$$

where $Q(\epsilon, E)$ is the Bethe–Heitler formula, given by Brown, which reduces
to normal free-free emission for lower energies. Brown shows Eq. (11.5) to

be a form of Abel's integral equation. We fit the observed photon spectrum by a power law:

$$I(\epsilon) = A\epsilon^{-\gamma} \quad \text{photons/cm}^2/\text{s/keV}. \qquad (11.6)$$

To avoid an infrared catastrophe the fit is cut off at some lower energy and normally used only above 25 keV. Brown found the solution to be:

$$\overline{F}(E,t) = \frac{6.7 \times 10^{50}}{\overline{n}V}(\gamma - 1)^2 B\left(\gamma - \frac{1}{2}, \frac{1}{2}\right) AE^{-\gamma+1}, \qquad (11.7)$$

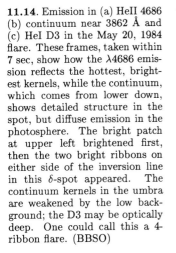

11.14. Emission in (a) HeII 4686 (b) continuum near 3862 Å and (c) HeI D3 in the May 20, 1984 flare. These frames, taken within 7 sec, show how the λ4686 emission reflects the hottest, brightest kernels, while the continuum, which comes from lower down, shows detailed structure in the spot, but diffuse emission in the photosphere. The bright patch at upper left brightened first, then the two bright ribbons on either side of the inversion line in this δ-spot appeared. The continuum kernels in the umbra are weakened by the low background; the D3 may be optically deep. One could call this a 4-ribbon flare. (BBSO)

where B is the beta function, a combination of Γ functions:

$$B(z,w) = \frac{\Gamma(w)\Gamma(z)}{\Gamma(z+w)}. \qquad (11.8)$$

The Γ function, in case you forgot, is

$$\begin{aligned} \Gamma(w+1) &= w!, \quad \text{if } w \text{ is an integer} \\ &= w\Gamma(w), \quad \text{if } n > 0, \end{aligned} \qquad (11.9)$$

and

$$\Gamma(1/2) = \sqrt{\pi}. \qquad (11.10)$$

The resulting energy dependence δ of the source spectrum $F(E,t)$ is one power harder than the observed photon spectrum. This analysis is valid so long as the electrons are not slowed by collisions with a cooler ambient medium; the cross-section for collisions with free electrons is about

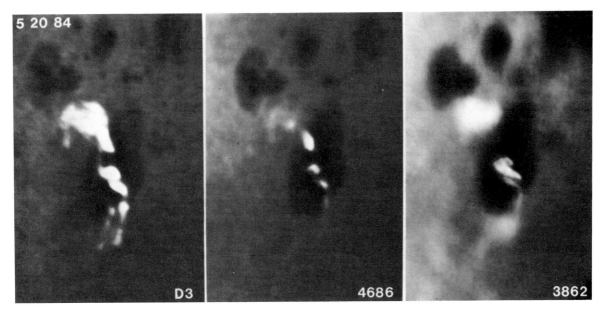

5 20 84

D3   4686   3862

$$Q_{ee} = \frac{2\pi e^4}{E^2}\Lambda_{ee} \approx \frac{3.3 \times 10^{-36}}{E^2 \, (\text{keV})} \quad \text{cm}^2. \tag{11.11}$$

so the number of background electrons along the path must be $< 10^{21}$ for 25 keV electrons.

In the thick-target model we seek the injection spectrum $\Im(E_0)$, where E_0 was the energy each electron had after acceleration. The solution is:

$$\Im(E_0) \, (\text{elec/sec/keV}) = 2.0 \times 10^{33} (\gamma - 1)^2 B\left(\gamma - \frac{1}{2}, \frac{1}{2}\right) A E^{-\gamma - 1}. \tag{11.12}$$

The deduced injection spectrum is one power *softer* than the measured photon spectrum, hence two powers softer than that deduced from the thin-target model. This is because the more energetic electrons contribute many photons as they slow down, so the assumed number of low-energy electrons injected is reduced.

The class 1 flare of May 18, 1972 was observed by the ESRO TD1A satellite. At its peak Hoyng (1975) found the photon spectrum above 25 keV to be fit by $\gamma = 4.6$ with $A = 5 \times 10^6$. Inserting this in Eq. (11.12), we find the thick-target flux to be 1.35×10^{35} electrons/sec above 20 keV. The flare lasted 30 sec so a total of 3×10^{36} electrons were produced. By contrast, the thin-target emission, using $N = 10^{12}$ is 7×10^{35}, about 5 times greater. So the thin-target emission is more efficient than thick-target, but of course the choice is not up to us.

Although there has been considerable controversy over thin- *vs* thick-target models, recent data favor the latter. The hard X-ray images obtained with the modulation collimator on Hinotori (Tsuneta 1983) as well as by the Solar Maximum Mission (Hoyng *et al.* 1983) show X-ray brightening at the footpoints of flare loops. Kane (1981) found a flare near the E limb of the Sun which was detected by an Earth-orbiting spacecraft but could not be detected by a Venus orbiter which was at a different heliographic longitude, further west. Since he knew the optical position accurately, he could conclude that the source height was below 4000 km; otherwise the Venus orbiter would have detected it above the limb. At this height the density is so great that thick-target processes must occur. Evidence on SXR from loop tops gives densities greater than 10^{12} cm^{-3}, which means the thick-target process can take place at the loop tops also. But Hinotori (Takakura *et al.* 1983, 1984) also observed HXR sources from 20 000 to 90 000 km above the limb in type C flares; models suggest densities around 3×10^{10} in those events, but there is no hard evidence. The high HXR sources do not radiate Hα or condense.

For the thermal flares of type A we can use standard free-free emission formulae; a simple approximation is given by Crannell *et al.* (1978):

$$I(E) = 1.3 \times 10^3 n_e^2 V E^{-1.4} T^{-0.1} \exp(-E/T) \quad \text{phot/cm}^2/\text{sec/keV}$$

$$(11.13)$$

where E and T are in keV.

Numerous workers are unhappy with the beamed power law model that is the most popular explanation for the HXR emission. They (Crannell *et al.* 1978, Mätzler *et al.* 1978) argue that the beams may be unstable, and that much of the electron energy is dissipated in collisions with the ambient electrons. They present evidence for adiabatic compression in HXR bursts. But the thermal models cannot explain the almost instantaneous lighting up of distant small footpoints, the fluxes above 100keV, or the high-frequency radio emission, which requires many electrons above 100 keV. Because it is hard to have a beam of electrons going one way without producing a big charge imbalance, a whole literature has grown up on return currents, diffuse currents induced by the charge imbalance caused by the electron beam.

The energy resolution of the normal X-ray detectors, scintillators or proportional counters, is not good. The energy spectrum of the particles falls off steeply, so that the X-rays detected in the 20-40 keV range, for example, may be due to a high flux from 10 to 20 keV, because the detector picks up photons from outside the nominal range and the spectrum falls off so rapidly that hardly any of the counts come from the upper part of the range. Worse, if the flux is too high, two 10 keV events close together may be recorded as a single 20 keV photon. This problem of pulse pile-up plagues all observations of X-rays from big flares, and the total electron energy from giant flares is a complete mystery because no one wants to fly an insensitive detector.

11.15. Skylab overlappograms of a class 2 flare showing the higher members of the HeII Lyman series, resonance lines of FeXV and XVI and the lithium-like resonance line of FeXXIV. The latter line is much brighter than the lower ionization Fe lines, and its emission is confined to the inversion line (loop tops). The HeII lines here are more extensive than λ4686 might be, because they are connected with the ground state; the higher HeII lines and continuum come from the loop tops. Later the lines of FeXVI and lower strengthen greatly, while HeII lines and continuum weaken. (NRL)

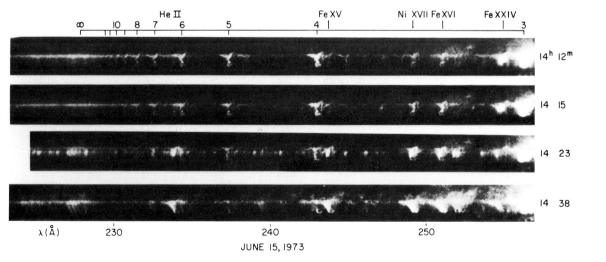

Lin *et al.* obtained a high-resolution X-ray energy spectrum of a flare burst with a germanium detector of high energy resolution and low pulse pile-up which revealed a thermal spectrum corresponding to a temperature of 35 million degrees. This flare was not a particularly impulsive one, but it produced a hot thermal cloud. At higher energies, the thermal spectrum merges into the power law. Most observations with coarser detectors are dominated (except for very impulsive events) by the thermal emission up to about 25 keV.

HXR bursts have been observed with photon energies up to 10 Mev. In the range above 400 keV the spectrum is strongly affected by gamma-ray lines (Sec. 11.6). Rieger *et al.* (1983) found that the flares with photon energies above 10 MeV all occurred near the solar limb; since then two events at 40° central distance have been recorded. Since big flares occurred all over the Sun in that period, this remarkable observation argues strongly that the hardest photons are emitted preferentially in the horizontal direction. When emitted, the γ rays are strongly directed along the electron velocity vector; if the electrons are spiralling about vertical field lines one would see horizontal beaming. This supports a thick foil model where these γ rays are emitted near the surface where the field is most nearly vertical. This process rules out thermal electrons, which would have neither the energy nor the directivity, as well as twisted fields, which would have no directivity.

In Sec. 9.3 we discussed deriving the electron density in the post-flare cloud from electron scattering. Densities up to 6×10^{11} are measured. At such densities other continuum emission processes, such as recombination, are unimportant by comparison. In one truly remarkable event, Harvey (unpublished) visually observed a flare in white light above the limb. The intensity must have been above 0.1 of the disk intensity and the emission could not have been due to electron scattering. The flare (June 21, 1980) produced one of the largest recorded nucleon events as well as a flux of neutrons.

There is growing, but not yet completely convincing, evidence for polarization in various flare emissions. Mandelstam *et al.* (1975) discovered polarization in the SXR emission from flares; this result is surprising because one thinks of the HXR source as beamed, and hence likely to be polarized, but the thermal emission should be isotropic or it isn't thermal. Henoux *et al.* (1983) found polarization in the 1437 Å SI line and by Henoux and Semel (1982) of polarization in Hα, both in the thermal phase of the flare. It is hard to decide about these results. The SXR source is high in the corona, while the other polarized sources are at footpoints, clearly heated by energy flowing down from above, which might establish the preferred direction required for polarization. But the thermal phase of most flares is much simpler geometrically than the impulsive event and that may result in polarization.

Observations of flares in microwaves are a powerful tool for the following reasons:

- With large modern interferometers the high-energy component of flares may be observed from the ground with extremely high spatial and temporal resolution and good frequency resolution.

- Big, sensitive dishes may be used, which can observe every day, while spacecraft are not always available. The energy resolution for microwaves can be less than one per cent, while that for most X-ray detectors is close to fifty per cent.

- We can derive considerable physical information on local magnetic fields through polarization and spectral properties of the microwaves.

- Synchrotron emission occurs whenever electrons spiral in a magnetic field and hence maps the acceleration region. X-rays are produced by bremsstrahlung when the electrons lose their energy, so we may not see the region of acceleration.

- Microwaves are preferentially produced by the highest energy electrons, so the antenna arrays can produce reasonable images of the high-energy electron distribution. Because their area is limited, X-ray detectors can only obtain good images below 40 keV.

11.16. A great white-light flare observed in the Balmer continuum by Neidig, along with the continuum distribution showing the sharp increase in brightness in the blue. High-speed BBSO observations of this event showed a pulsing emission point in the umbra. (SPO)

3610 Å

IONOSPHERIC DISTURBANCE CAUSED BY THIS FLARE NEGATED CRITICAL HF COMMUNICATION LINKS FOR 2 HRS

- X-ray sources are optically thin and relatively easy to interpret, while microwaves are subject to self-absorption, Razin effect and other maladies. Furthermore, microwave arrays cannot fly on spacecraft and thus are difficult to fund, while space agencies will pay handsomely for X-ray detectors.

Observations with the VLA may be made at 1.3, 2, 6 and 21 cm, but if more than one frequency is used the array must be split into subarrays. Because only the arrays with smaller spacings will give sufficiently clean images, the limiting resolution is a few arc seconds at the shortest wavelengths and the time constant is 3 sec. The Owens Valley solar interferometer, with about 5 arc sec resolution, observes 40 different frequencies in 1 sec. The Westerbork radio-synthesis array permits one-dimensional imaging with 5 arc sec resolution and 0.1 sec time resolution. The data reduction with these arrays is laborious and the images produced have limited dynamic range because of side lobes. But the burst sources are so strong that a good measure of the position and physical size is obtained.

Much can also be learned from observations with single dishes which give the time and spectral behavior of the microwaves. The microwave spikes are quite similar to the HXR, but lag by about 0.2 sec (Cornell *et al.* 1984) at 10 GHz behind the 25 keV X-rays. This probably reflects the time needed to accelerate the particles from 25 keV to the higher energies required to produce microwaves. But the intensities of the different spikes are not so well matched, because the microwave flux depends on the magnetic field and radiative transfer effects while the X-rays depend on the field particle density only.

The results of imaging microwave observations have been reviewed by Marsh and Hurford (1982). Observations by various groups have established that the microwave source at 2 cm and 6 cm is located at the top of low-lying loops passing through the flare neutral line. These loops can usually be identified as terminating in kernels of optical emission. When radio, X-ray and optical data can be compared (Hoyng *et al.* 1983; Takakura *et al.* 1983; Tanaka 1982; Tsuneta *et al.* 1983) the X-rays are found to come from the loop tops and footpoints and the microwave emission, from loop tops only. Loop tops have sufficient density for some bremsstrahlung to occur yet are suitable for acceleration and trapping of electrons. At footpoints the electrons are rapidly lost to the thick-target process and whatever microwaves may be emitted are absorbed by the overlying material.

Maps of microwave emission show the right- and left-hand circularly polarized sources slightly displaced, probably because on one side of the top of the arch we look along, on the other side against, the magnetic field. If we model this effect, we find a loop somewhat smaller than the main optical or X-ray kernel separation and much smaller than the separation of the impulsive footpoints. But of course if we had only a simple loop, there would not have been a flare there. In any event the observations strongly suggest that the primary flare-energy release occurs a few thousand km above the surface at the neutral line. The microwave source is small – about 2 arc sec at 2 cm, where it is optically thin and about 10 arc sec at 6 cm, where it is optically thick. This tells us that overlying absorption is not important, otherwise the sources would be bigger at the high frequencies where absorption is less. Usually only one source is seen, but there are a few examples of weaker secondary sources in the impulsive phase.

Schwinger (1949) showed that for high-energy electrons the gyroresonance emission, normally confined to the first few harmonics, increases strongly in the high harmonics. Since the latter are closely spaced, the magnetic field is not uniform and there is Doppler broadening, the high harmonics merge into a continuum. This emission was first observed in the electron synchrotron, hence the name. The radiation is preferentially in the direction of electron motion and highly polarized, but is smeared by the spiralling of the electrons and the field fluctuations. The spectrum is complicated, depending on pitch angle of the electron, position of the observer, energy distribution, etc. But simple approximations reveal the general pattern.

11.17. Time variation of various emissions from the June 6, 1982 (Figs. 11.12–3) flare plotted by Tanaka (Tanaka and Zirin 1985) against time. The top curve is the derivative of all the Fe line X-ray emissions, some of which are plotted in the next 4 curves; it bears a striking resemblance to the X-ray flux in 67–340 keV. From this correspondence one can conclude that the energy input to the soft X-ray emission is the non-thermal electrons that produce the HXR emission. However we only know time correspondence; the total energy in HXR is model dependent, although the energies are in the right order. The 30–67.5 keV channels were saturated; otherwise they might resemble the higher HXR energies.

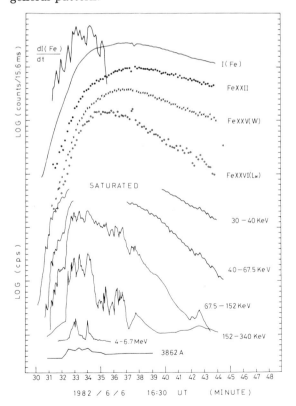

11.18. Flare spectra from SXR to γ-rays from Hinotori. (a): FeXXV–XXVI. (b): FeXXIII–XXV, high dispersion. (c): Flare spectrum from 1–100 keV showing the transition from thermal SXR to power-law HXR emission. FLM was a low-energy flare monitor. (d): The higher energy spectrum showing the additional contribution by "SGR" or solar γ-ray lines and possibly flatter continuum. (Tanaka 1986)

The microwave spectrum is a fascinating combination of the properties of synchrotron emission, radiation transfer and wave propagation in magnetic fields. These are described by Takakura (1967), Ramaty and Petrosian (1972) and others. Simplified equations for the mildly relativistic case have been published by Dulk and Marsh (1982). Although the formulas are complex the general idea is not. At low frequencies (below 5 GHz) the typical microwave burst is optically deep. No matter what magic emission device we have, the resulting emission cannot exceed the Planck function, which in this case is the Rayleigh–Jeans law ν^2. If the electron distribution is a power law there is a different effective temperature T_b at each frequency and a slightly different, roughly $\nu^{2.5}$ dependence should be found. But in real life the burst source is non-uniform, optical depth

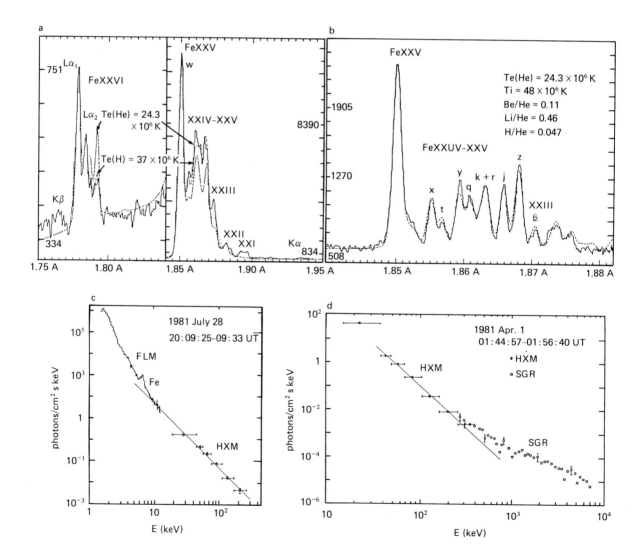

and magnetic field vary, and typically $S_\nu \propto \nu$, one power less steep than the Rayleigh–Jeans law, which one rarely sees. Above a frequency ν_{SA} the source is generally optically thin and the emission falls sharply, reflecting the particle energy distribution folded with the synchrotron emission formulae.

The Rayleigh-Jeans form for the flux from T_b is:

$$S_\nu = \frac{2\pi k T_b \nu^2}{c^2}\left[1 - e^{-\kappa_\nu L}\right]\Omega \qquad \mathrm{erg/cm^2/sec/Hz} \qquad (11.14)$$

where Ω is the solid angle of the source and κ, the absorption coefficient. The flux will rise to a maximum S_m (Slish, 1963) at frequency

$$\nu_{\mathrm{SA}} = 10^{12.5}\left(\frac{S_m}{\Omega}\right)^{2/5} B_\perp^{1/5}, \qquad (11.15)$$

where $B_\perp$ is the magnetic field perpendicular to the line of sight. Above this frequency the source is optically thin and the flux falls off as the product of electron energy distribution and emissivity. Slish's formula is valid for a relativistic source. For a non-relativistic source viewed at 45° to the magnetic field, Ramaty and Petrosian found approximately:

$$\frac{\nu_{\mathrm{SA}}}{\nu_B} = 10^6\left(\frac{S_m}{\Omega B^2}\right)^{2/5} \qquad (11.16)$$

Dulk and Marsh (1982) give a number of approximations for the microwave spectrum emitted by thermal, power law and relativistic particle distributions. Because the last do not follow a Maxwellian distribution, the slope of the rising curve follows $\nu^{2.5}$ for relativistic particles and $\nu^{2.9}$ for a power law. On the high-frequency side the flux falls off as ν^8 for a thermal source and ν^3 for a power law of $\delta = 3$. Spectra of real bursts often show self-absorption peaks, but little further resemblance to the theoretical curve. The optical depth varies from point to point, so the burst is a superposition of a number of different emission curves with different ν_{SA}. A spectrum of a real and reasonably typical burst (Hurford *et al.* 1984) is shown in Fig. 11.19. The curves rise as ν^2 and fall as $\approx \nu^{-1.4}$. This is much flatter than permitted by any of the models and might be due to the presence of kernels of higher peak frequency. But once one passes the peaks of those kernels, it is hard to see how such a flat falloff is possible. A remarkable feature of the spectra with the frequency-agile receivers is the stability of the spectrum. As the burst rises by as much as 100-fold in intensity, the spectral shape is preserved, as though the spectrum is due to a collection of narrow-banded kernels that rise and fall in harmony. Although little detailed analysis of the spectra has been carried out, we find a wide range of spectra, some rising continuously with no peak below 15 GHz, others falling through the entire spectrum, even some with double peaks.

Other factors may influence the microwave spectrum. The Razin effect (Eq. 8.16) produces a sharp cutoff at frequencies below $f_B = 20n/B_\perp$. For high enough densities, free-free absorption (Eq. 7.3) may be important; Ramaty and Petrosian showed it could be responsible for the flat spectrum of many bursts. In some cases (Zirin *et al.* 1971) the burst source is so low in the atmosphere that no emission escapes at frequencies below a cutoff determined by the free-free absorption. Unfortunately there are still too many free parameters for convincing matching of the spectra to electron distributions obtained from the hard X-ray spectrum; the new detailed spectra will restrict the models somewhat.

We can get a picture of what is needed to produce the observed flux from Eq. (11.13), which gives

$$S = 7.22 \times 10^{-29} T_b \mathrm{A}(\mathrm{arc\ sec}^2) \qquad \mathrm{sfu}. \qquad (11.17)$$

Thus a burst of 500 sfu at 5 GHz with area 100 arc sec^2 ($\Omega = 2.5 \times 10^{-9}$) has $T_b \approx 2.8 \times 10^9$ deg, but since the particle distribution is a power law and not thermal, the concept of temperature is not meaningful. It is not possible to explain the observed spectrum or flux with thermal distributions, which fall off much too fast at high energies.

Calculations of the emitted microwave power by Takakura and

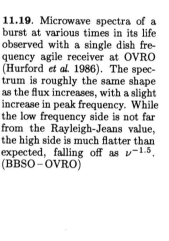

11.19. Microwave spectra of a burst at various times in its life observed with a single dish frequency agile receiver at OVRO (Hurford *et al.* 1986). The spectrum is roughly the same shape as the flux increases, with a slight increase in peak frequency. While the low frequency side is not far from the Rayleigh-Jeans value, the high side is much flatter than expected, falling off as $\nu^{-1.5}$. (BBSO – OVRO)

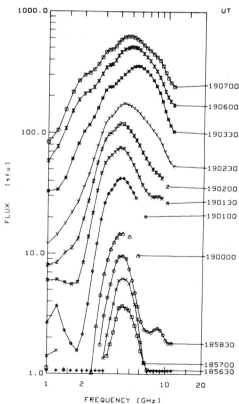

Scalise (1970) showed that a small fraction of the electrons required to produce the observed X-ray flux was required for the radio burst. Various models to explain this were advanced, but the most convincing solution was given by Gary (1985) who resolved the contradiction by introducing modern data. He showed that if the field strength is below 1000 gauss and the electron energy distribution is derived from a thick-target model, one gets roughly equal numbers of electrons in the X-ray and microwave sources. He found this result to hold for ten well-observed flares. Since the microwaves come from all regions with proper magnetic fields and the X-rays, only from regions where the electrons are decelerating, some difference is inevitable.

Not all flare electrons remain in the flare region. type III (or fast-drift) meter-wave bursts are produced by a stream of electrons moving through the corona with 1/3 the velocity of light (or an energy of 40 keV), which often reaches the Earth. And we saw in Fig. 11.1 how the streams may return to the surface and produce distant brightenings.

11.5. Solar energetic particles

Certain flares produce great clouds of energetic particles (SEP) which fill the solar system. SEP also produce nuclear γ-ray lines at the Sun. The SEP reaching the Earth are filtered by the effects of interplanetary propagation, but those producing γ rays are not. The total energy in SEP may approach that emitted in optical wavelengths. Ramaty *et al.* (1980) plot results of many observations which show that the flux of electrons between 0.2 and 1 MeV is about 100 times that of protons at 10 MeV. Since the velocities are roughly the same, the total energy in each band may be the same. There are virtually no data on protons below 0.2 MeV.

Except for the very largest events there is little coincidence between the greatest X-ray or γ-ray line events and the greatest nucleon producers. Kahler *et al.* (1984) showed that almost all SEP events were associated with coronal mass ejections (CME). The charge on the SEP corresponds (Luhn *et al.* 1984) to ions common at 1–2 million deg, such as FeXIV. Mullan and Waldron (1986) show that the ionization is not uniform, but it certainly is not very high. Thus there is a real possibility that the SEP are accelerated in the CME – Moreton wave – type II burst, and second-stage acceleration may actually occur in these cases. That means there are at least two quite different nucleon acceleration processes. Obviously it would be hard to bring the CME particles back down to the surface to produce γ-ray lines. Cane *et al.* 1986 selected flares lasting more than one hour at above 10% of their maximum 1–8 Å flux and showed that these were far more likely to produce SEP events than impulsive events. These "long-duration" events are typically associated with filament eruptions and CME.

McGuire *et al.* (1981) find the proton spectrum in interplanetary space may be fit by a form proposed by Ramaty *et al.* (1979) using a Bessel function K_2 of order 2 (Abramowitz and Stegun 1965):

$$\frac{dJ}{dE}(E) \approx \beta K_2[2(3\beta/\alpha T)^{1/2}] \tag{11.18}$$

where β is an abundance parameter and αT is a fit parameter composed of an acceleration rate α and an escape time T.

Particles from flares near the west limb appear with little delay because they travel freely along the Archimedean spiral. The most energetic nuclei (greater than 100 MeV) reach the Earth about 20 min after the optical photons; then the slower ones catch up. After the primary pulse dies down (and it may last several days), a new pulse of low-energy (≈ 5 MeV) nuclei arrives with the shock front from the flare; presumably they are trapped in it, perhaps they are accelerated by it as well.

Particles from flares in the eastern hemisphere of the Sun reach the Earth with some delay and a more gradual build up; but these particles arrive on the same spiral trajectories as the west-limb events. The particles must diffuse through the magnetic envelope of the Sun and escape along the outward-leading field lines.

McDonald and Desai (1971) detected co-rotating SEP events at 27 day intervals after great flares, matching the central meridian passage of the place on the Sun where the original event happened, but occurring long after the sunspots and solar manifestations had disappeared. More recent evidence (Christon and Simpson 1979, McDonald *et al.* 1976) suggest these streams are accelerated far from the Sun along the leading edges of high-velocity solar wind streams in which they are trapped. The energetic nuclei reach the Earth from the normal spiral trajectories.

The relative abundances of the solar energetic nuclei agree with coronal abundances with a few exceptions. As noted earlier, they are poor in high-FIP elements, as is the corona (Breneman and Stone 1985). This supports the case for acceleration of coronal particles. Relative to galactic cosmic rays SEP are rich in He and poor in C (Cook, Stone and Vogt 1980). Scholer *et al.* (1979) found co-rotating events to be rich in He and C relative to O. The He abundance in the co-rotating particle stream was 0.08 relative to H compared with 0.02–0.05 in normal flare events or the solar wind. Hirshberg (1972) found helium enhancements up to 0.20 in the solar wind after flares; it is not certain if these were direct or co-rotating events. Hirshberg *et al.* (1974) found the He abundance to vary by a factor of two in and out of high-velocity solar wind streams.

The most remarkable nucleon enrichments are the He^3-rich events (Hurford *et al.* 1975). In these relatively small flares the He^3/He^4 ratio ranges from 0.2 to 8 at 2.9–12.7 MeV/nucleon compared to a normal value of $10^{-2} - 10^{-3}$. All also show enhancement of nuclei with $Z > 6$ (van Hollebeke 1979), but deuterium or tritium are not enhanced. Because the events are small and the delay between the flare and the arrival of nucleons is longer than the interval between flares, it is difficult to identify the He^3 events with known solar flares, because there are usually several candidate flares occurring in the 2–3 day propagation period.

Recently a step toward solving this problem was made by Reames *et al.* (1985), who found the He^3 events arrived at the Earth about six hrs after impulsive bursts of electrons in the 2–100 keV range. Kahler *et al.* (1987) took these events and found the electron events could be identified with type III bursts observed at kilometric wavelengths near the Earth. These were identified with $H\alpha$ flares with start times within one min. The six-hour window is small enough that a unique flare identification could be made, except for a few cases with no optical counterpart. The flares are unremarkable and not even of a particular type, except small. Kahler *et al.* suggest that the electron spectrum gives a minimum path length to $0.2R_\odot$ above the surface, and suggest that the acceleration must take place at that height.

Luhn *et al.* (1986) found that the ionization state of the nucleons arriving with the He^3 nuclei is high, corresponding to 10^7 deg. Ions like FeXXIII would only exist in the loop-top SXR source. The clear association with direct electron streams found by Kahler *et al.* supports the idea that the He^3 rich material is the true stuff of the flare, and we only see these events where the SEP are produced in the flare. If the association of big SEP events with CME is correct, those events are coronal material, accelerated by the flare wave.

The most popular explanation of the He^3 rich events is Fisk's (1978) model. Fisk notes that electrostatic ion–cyclotron waves can be excited in the corona at the proton and He^4 cyclotron frequencies; since only He^3 has a cyclotron frequency between these nuclei, it is preferentially heated and injected into the still unknown particle acceleration process. Fisk's model also gives preferential acceleration of heavy particles, and Fe nuclei are overabundant in He^3-rich flares. Further, this model would fit well with the picture of acceleration in the flare, where magnetic fields are strong. But there are many Fe-rich flares that do not show He^3 enrichment. If Fisk's argument is correct, it implies that the injection process is the key selector, and the acceleration works on everything. In the big SEP events the subsequent acceleration in the corona swamps the He^3 overabundance generated in the flare itself.

High-energy flare particles are generally thought to be accelerated by variants of two processes suggested by Fermi (1954). First-order Fermi acceleration takes place when a magnetic shock front collides with trapped particles. The particles are in a kind of magnetic bottle between the moving magnetic front and the rest of the trap, and they undergo adiabatic compression. Blandford and Ostriker (1980) showed that a strong shock takes particles of momentum p and spreads them into a power law distribution p^{-q}, where q depends on the compression and is > 4 for adiabatic shocks. The energy of all these particles is much greater than they started with, and the final distribution is the sum of these power laws. Whether we can distinguish between the power law and Ramaty's Bessel function I do not know.

Second-order Fermi acceleration comes from the stochastic collision of particles with magnetic field elements, resulting in an energy equipartition of the single particle with the larger magnetic elements. The idea is that the acceleration of particles to MeV energies in 1-sec can take place if there is strong turbulence on the scales of $10^6 - 10^7$ cm. This may be true in flares, and Ramaty (1986) argues that the fit by the K_2 Bessel function supports this model. But the rate of escape for different energies must be the same for this distribution to result.

Other mechanisms are associated with strong electric fields, current sheets, and shock waves. Usually they are proposed with a great deal of arm waving but without observational tests, so we cannot judge their validity. The subject is reviewed by Ramaty and Forman (1987).

Solar neutrons were first detected by Chupp *et al.* (1982) in the γ-ray records of the June 21, 1980 white-light limb flare. The neutrons arrive at the Earth minutes after the γ-rays photons. They could be distinguished from late-phase photons by their interaction with the detector. Neutrons are not filtered by the interplanetary magnetic fields, but come straight to Earth. Thus what we see at the Earth is close to what was produced at the Sun. Ramaty *et al.* (1983) have used those observations to extend the data on charged particle energies to 100 MeV.

11.6. Gamma-ray lines

Lingenfelter and Ramaty (1967) predicted that the powerful flux of energetic nuclei produced in the flare should collide with surface nuclei and produce various γ-ray lines. A γ-ray spectrometer built by Prof. Chupp of the University of New Hampshire was flown on OSO-7, but to save power it was turned on only when high solar activity was predicted. Since the official predictions are usually contrary to reality, and no one had thought of using them as contrarian predictors (as is commonly done for the stock market), no data had been obtained up till August 1972. Chupp, who was on sabbatical in Germany, turned on the radio and heard news of a large flare. He immediately had the experiment turned on (it was, of course, still turned off) and was rewarded by observations (Chupp *et al.* 1973, 1975) of γ-ray lines in the great flares of August 4 and 7, 1972. In these flares four lines were observed, at 0.51 Mev (positron–electron annihilation), 2.22 MeV (neutron–proton capture), 4.4 and 6.2 MeV (excitation of nuclear levels in C^{12} and O^{16}, respectively). These are the strongest lines predicted. The processes involved and the observations are summarized by

Ramaty (1986). Calculations by Ramaty *et al.* (1979) also show a number of weaker lines which form a pseudo-continuum. Since the initial detection γ-ray lines have been observed in many events; a typical spectrum is shown in Fig. 11.20.

The strongest γ-ray line (2.223 MeV) results from the radiative capture of neutrons on protons, forming deuterium, with the emission of a γ-ray. The neutrons are produced in spallation reactions of the flare protons with heavier nuclei and come off with energies of the order of 10 MeV. They travel until they decay naturally (932 sec) or are slowed down and captured. There is a delay of up to 100 sec between the appearance of hard electrons (as evidenced by HXR and microwave bursts) and the onset of 2.2 MeV emission. This indicates a somewhat higher density than modeled by Wang and Ramaty (1974). Because the neutrons must thermalize before capture, the process takes place in the low photosphere and the line is limb-darkened, in contrast with the limb-brightened 10 MeV photons. For hard nucleon spectra, most neutrons are produced by protons incident on α-particles; for steep spectra, with few high-energy protons, the neutrons come from α-particles incident on heavy nuclei. Ramaty concludes that the γ-ray lines are predominantly due to thick-target interaction.

The positron–electron annihilation which produces the 0.5 MeV line requires the slowing down of positrons from radioactive nuclei produced in the flare, a coulomb effect which is regulated by Eq. (3.8). Obviously the delay would give some idea of ambient density, but the time dependence of the 0.5 MeV line has not been well determined. The ratio of positron to neutron yield is between 0.2 and 0.4, depending on whether the source is a thick- or thin-target interaction, and most of the positrons are rapidly slowed down till they produce positronium, which annihilates in 10^{-8} sec. to give two 0.511 MeV γ-rays.

The 0.511 MeV line is not seen in all flares. Limb darkening explains some of this effect, but it is not clear. It is hard to tell from the literature what fraction of γ-ray flares show the 0.5 MeV line; to date it appears to have been detected in only a few. In many events it is swamped by other nuclear lines in that region.

The lines of C and O at 4 and 6 MeV should consist of two components, a narrow line due to emission by ambient C and O nuclei excited by flare protons and a Doppler-broadened one due to excitation of high-energy nuclei colliding with the photospheric protons. The SMM data show some lines broadened in this way. By comparing the energy in the 4–8 MeV range with the 2.2 MeV line, Ramaty *et al.* (1977) were able to set limits on the spectrum of the flare nucleons. The exponent in a power law spectrum above 10 MeV must be less than two; by contrast the energy dependence of interplanetary nucleons is roughly E^{-5}. The energy deposition in the Aug 4, 1972 flare was about 10^{27} erg/sec, an amount adequate to produce the white-light flare.

Ramaty *et al.* (1980) found that the continuum γ-ray spectrum indicates a heavy-element enrichment. The flux above 2 MeV is high relative to the extrapolation of the lower-energy electron bremsstrahlung spectrum. They explain the excess by the many weak nuclear lines that would be produced if the heavy-element abundance was enhanced. This model could be tested with microwave spectra, which give us an independent measurement of the electron spectrum.

For some time it was thought that the WLF was due to protons impinging on the surface; now it is felt (Hudson and Dwivedi 1982) that the energy in particles is insufficient (but opinions have been known to change from time to time). The γ-ray lines show that the nucleons do reach the surface and the timing of WLF and γ-ray emission matches well, allowing for a 50 sec delay in the 2.2 MeV line. The prompt lines as well as the high-energy continuum appear to be simultaneous with the white-light emission and Kα excitation (Culhane *et al.* 1981, Tanaka 1983).

There has been a lot of fuss about "second-stage acceleration," an idea introduced by Ramaty because the γ-rays in the Aug 4 1972 event were late compared to a not-too-accurately-timed microwave burst. The idea was that the main nucleon acceleration took place in the big outward moving shock; it never was explained how the particles got back to the photosphere. All the SMM and Hinotori events show simultaneity of gamma-ray lines, HXR, and microwaves (allowing for the built-in delay in the nuclear lines, so the Aug 4 burst probably was wrongly timed and the evidence for second-stage acceleration γ-ray line flares has dwindled. But it has reappeared as a real effect in other cases. First, we now see that SEP

11.20. Smoothed γ-ray spectrum of a large flare accumulated over 32 min with the continuum background subtracted. The features marked are: (1) 6.129 MeV: O^{16}; (2) 4.438 MeV: C^{12}; (3) 2.123 MeV: N^{14} and 2.223 MeV: $H^1(n,\gamma)$ H^2; (4) 1.634 MeV: Ne^{20}; (5) 1.24-1.37 MeV: complex of lines from Fe^{56} and Mg^{24} and other nuclides; (6) 0.8-0.9 MeV: Fe^{56} and others; (7) 0.431 and 0.478 MeV: Be^7 and Li^7 resulting from $\alpha - \alpha$ interactions and 0.511 MeV: positron-electron annihilation. (SMM γ-ray Spectrometer Team – U of New Hampshire, Max Planck Institute and NRL. Courtesy Prof. E. L. Chupp, Principal Investigator)

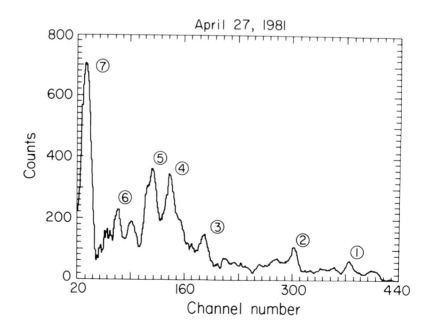

events are related to CME and the process Ramaty had in mind, but γ-ray line and WLF flares are not. Second, there is the new class of flares with late, long-lived X-ray and microwave events, which must be accelerated by something quite late in the flare. We now have three classes of flares: the big γ-ray events, which accelerate nucleons near the surface simultaneous with electrons; the long-duration CME events producing SEP events; and the He^3-rich flares, which may be a dwarf subset of the other two.

11.7. Waves, ejections and meter-wave bursts

When an energy release as large as a flare occurs it is bound to make waves, which is exactly what happens. Both the gas and magnetic pressure in the event produce substantial ejecta, involving the flare material itself and nearby structures. These are among the most spectacular solar phenomena and have the most profound terrestrial effects. We see ejected material, shock waves and streams of high-energy particles.

The primary source of ejected material is the flare filament. Usually it rises gradually 20 minutes or so before the flare onset, whereupon it is violently expelled from the surface. Because the filament is still cool, the lifting is a magnetic phenomenon, either breaking the bonds that hold the filament down or expanding fields involved in the flare. The flare phase is more violent; the slowly rising filament accelerates from 10 km/sec to 100 km/sec, simultaneous with sharp increase in $H\alpha$ and X-ray emission. One would conclude that the same process is responsible for all; this suggests that a magnetic process occurs, in which electrons are accelerated and magnetic pressure expels the material. Velocities up to 2000 km/sec have been observed. These events are usually followed by a shock wave and CME. The ejecta are irregular and disorganized and termed a spray. Their trajectories are roughly straight with weak spiralling around the invisible field lines.

Often slower eruption takes place with no impulsive phase. The filament becomes a great loop (Fig 11.14) which erupts outward, typically at 50–200 km/sec. The eruption becomes elongated as it rises and there is continued acceleration (Schmahl and Hildner 1977). The instability appears to be in the erupting loop field itself, accounting for the continuous acceleration. Because the eruption is slow the particles lose energy as fast as they are accelerated and only relatively soft thermal particle spectra are observed. These eruptions usually occur in active regions; they appear to result from changes in the underlying magnetic configuration, usually flux emergence. The movement is invariably outward. This makes sense; the material is trapped in a magnetic medium, and downward motion would require energy to compress the underlying magnetic fields. Fig. 9.17 is a good example of the "spotless flare", which appears to have occurred because the old bipolar plage had weakened to where it no longer could hold the filament down. These flares are extensive and are almost always class 2 or 3. Despite their thermal electron spectra, the shock wave they generate in the corona can accelerate nucleons. They are followed by standard post-flare loops and are flares in every way except impulsiveness.

The fate of the field lines attaching the erupting events to the surface is intriguing. We know that lines of force are carried out by the solar wind, and here we see the field lines attaching the CME to the surface stretching out into space. There is no evidence that these lines reconnect behind the eruption; films late in the event show material draining back down along the legs of the arch. There are instances of coronal holes forming behind these eruptions, emphasizing that closed fields have been transformed into open ones. It might be that the reconnection behind the CME is a source of energy for the long-duration events.

Filaments in the quiet sun also erupt; the difference is that few or no energetic particles are produced and little Hα emission occurs, though CME may result. We can make a simple estimate of the energy involved in a prominence eruption. The brightest prominences have intensity 10^{-4} of the continuum, which gives 6×10^{20} atoms/cm^2. If the area of the prominence is 10^{19} cm^2 it weighs 10^{16} gm (10^{10} tons), and if every last bit of it is accelerated to 500 km/sec we get an energy of 10^{31} ergs, similar to the total optical output. All of the energy must come from magnetic buoyancy. In most cases the velocity is much less, or only part of the material is expelled.

The results of Widing *et al.* (1986) on an eruption observed in the UV (Chap. 9) show that the temperature of the ejected material is not high. The ejecta may even be cooled by adiabatic expansion as they rise. The same is true of surges, discussed below. The eruption is purely magnetic.

The major characteristic of the flare spray is the overwhelming of neighboring fields by the tremendous explosive power of the flare. By

11.21. A type II burst and its second harmonic (below), each twinned (unexplained). Usually a cluster of type III bursts are found at the onset, but they do not appear here. (CSIRO–Div. of Radiophysics)

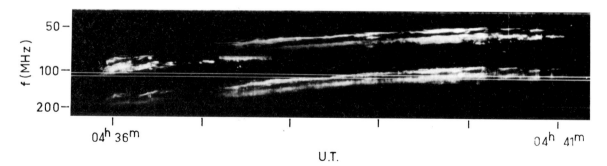

contrast, surges (Chap. 9) follow the existing magnetic patterns. The initial temperature is greater than the chromosphere, or else we would not see emission. The surge goes from emission to absorption as it moves out, but this does not mean it is cooling, since expansion alone would drop the density below the level at which emission should appear (Sec. 11.3). While it seems likely that the surge material expands and cools adiabatically as it rises, it is also possible that the material is heated somehow and merges into the corona. However the fact that surge spectra are generally of low excitation argues for cooling.

Surges typically occur near large spots with well-organized fields. Zirin and Lazareff (1975) found a series of surges at the front edge of a spot group to be highly productive of type III bursts, an indication that they flowed into open field lines but also that hard electrons were produced. But HXR bursts from surges are generally weak; in some cases if an HXR burst is in progress and a surge comes out, the HXR flux drops sharply, as though the surge material carries away the energy. The velocities usually remain constant for most of the lifetime of the surge. Surges rarely are associated with large flares.

Are surges due to magnetic ejection, or are they merely high-pressure flare gas expanding into open field lines? In some cases we see a small nozzle form between the dipole field and the overall field through which the material is injected. A typical surge can quite reasonably be explained by conversion of all the internal energy of a small flare into kinetic energy via a magnetic nozzle. A surge velocity of 100 km/sec corresponds to thermal energy at $500\,000°$; if the surge volume is 10^{28} cm^3, then a volume increase of 10^4 times takes place if the source flare has volume 10^{24}cm^3, a typical value. Then if $T \approx 10^7$ deg, there is plenty of energy.

11.22. The "Westward Ho" eruption: An erupting loop observed at the Univ. of Hawaii Mees Observatory was followed by a moving type IV source at 80 MHz probably associated with a CME. (CSIRO–Div. of Radiophysics)

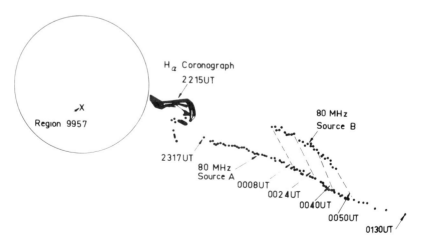

Some of these parameters have been modeled by Noci (1980).

Despite the attraction of the nozzle model, we saw in Fig. 9.13 that some ejections are threaded by field lines and have more complex behavior. Hurford and Tang (unpublished) found a surge which absorbed the emission from a radio source in an active region. The absorption was due to invisible material moving ahead of the visible surge, later appearing in absorption. The surge is not always simple.

Moreton waves (Moreton and Ramsey 1960; Athay and Moreton 1961) are shocks moving 1000 km/sec produced by large impulsive flares (microwave flux > 700 sfu). They appear (Fig. 11.7) as a bright front in Hα centerline or a dark front (corresponding to a wavelength shift) in the Hα wing. They propagate with constant velocity in an arc about the source, following the curve of the surface, sometimes for more than a solar radius. The wave normally propagates until it reaches a magnetic boundary of some sort (such as a filament or active region). Subtraction of images in red and blue wings of Hα show that a downward shift followed by an upward recovery produces the optical effect; the centerline brightening may be due to the same effect, because a shift in the absorption line increases the intensity of light passing through the Lyot filter, or it may be excitation by particles trapped in the front. The up and down motion produced by the wave passage causes a dramatic winking of filaments as they shift in and out of the band pass. Moreton waves are invariably accompanied by type II bursts.

Uchida (1974) explained the Moreton wave as a weak MHD shock wave propagating spherically from the flare site; the optical phenomenon is the intersection of this expanding sphere with the surface. He fitted actual trajectories of waves to the magnetic circumstances. Suggestions that Moreton waves may represent the locus of flare particles intersecting the surface are wrong; the phenomena are entirely different. The flare particles hitting the surface (Tang and Moore 1982) travel much faster (100 000 km/sec) and appear only in high-field areas of opposite polarity. The waves, by contrast, appear in the undisturbed chromosphere and produce an up and down motion which cannot be caused by precipitating particles. However, passing Moreton waves may leave brightening behind, probably due to energetic electrons trapped in the front.

The magnetic nature of the Moreton wave is shown by its disappearance at magnetic boundaries and propagation along the curve of the Sun's surface. The means of initiation is unknown. The wave becomes visible only after it travels some distance and the arch it forms becomes visible against the undisturbed chromosphere. In most cases it is seen to overtake spray material ejected earlier that is moving only a few hundred km/sec, so its generation must occur a minute or two after the initial explosion that ejects the spray. It appears as though the spreading flare brightness reaches a barrier and there initiates the wave; all the waves I have seen move in the direction of the main spread of Hα brightening and ejecta. This is in the flash phase, before the slow outward spread of the two Hα ribbons. The waves have been detected as far out as Jupiter. The famous Aug 4 1972 flare produced several interplanetary shocks. Curiously it showed no Moreton wave in the chromosphere, but a filament 300 000 km away was the source of a secondary Moreton wave; the initial wave did not intersect the surface but could affect the filament field higher up.

Flare waves and material ejection should be detectable in the characteristic radio frequencies emitted by the corona, and they are. In their famous early work Wild and his co-workers (Wild 1950a, b; Wild and Mc-Cready 1950) classified the meter-wave bursts by their distinctive spectral properties:

I: Circularly polarized narrow band ($\approx$ 5MHz) bursts lasting a second or less with complex spectral properties. Superposed on a continuous background, they make up the type I noise storm (Sec. 10.9).

II: Bursts (Fig. 11.21) of great intensity following large flares. A slow drift to lower frequencies of 1 MHz/sec indicates that they are moving outward at 1000–1500 km/sec. Harmonics at twice the frequency are often observed. The emission is thought to be produced by plasma waves excited by the Moreton wave as it travels outward.

III: The most common flare-associated bursts, observed from 5 to 600 MHz. They are characterized by a fast drift from low to high frequency, around 20 MHz/sec at 100 MHz. Connected with streams of electrons around 40 keV, the frequency drift corresponds to such a stream moving 100 000 km/sec and successively exciting lower local plasma frequencies.

IV: Broad-band continuum radiation lasting for hours after a flare, moving outward with velocities from several hundred to 1000 km/sec. Originally thought to be a noise storm initiated by the flare, type IV was shown by Boischot (1957) to be a different phenomenon. Both moving and stationary type IV bursts are now recognized. The moving type IV's last around half an hour, and are partly circularly polarized; the stationary type IV's last a day or more, and may in fact be identical to the continuum portion of noise storms. The broad-band nature of the type IV emission suggests synchrotron emission as its source. In this case the energetic electrons would have to be trapped for a fairly long period of time by a magnetic bottle of some sort.

V: Broad-band continuum radiation following a type III burst.

Wild concluded that the downward drift of frequency in types II and III was due to outward movement of a source out through the corona, exciting emission at the local plasma frequency. Interferometer measurements and other data confirmed this idea. The sources travel all the way to the Earth, emitting very long (kilometric) wave emission at the interplanetary plasma frequency, and bursts of 40 keV electrons were detected by a spacecraft near the Earth as they arrived. Both type II and type III bursts are detected near the Earth (Malitson *et al.* 1973).

The combination of imaging meter-wave systems and images of the outer corona from SMM has made possible identification of the loci of type III bursts with coronal condensations. Trottet *et al.* (1982) have identified a double source of type III emission with white-light coronal condensations along two separate lines of force coming out of the flare region. Their conclusion is that the stream of flare electrons travels along transient streamers of enhanced density. There are problems with the propagation of these streams. When a stream of electrons passes through a plasma, the velocity distribution has two peaks: one for the ambient electrons, one for the stream. This is subject to a coherent process called two-stream instability, which rapidly reduces the distribution to a single Maxwellian distribution. Various explanations have been set forth as to why the streams succeed in reaching the Earth, but they are not conclusive.

In large flares (Tang and Moore 1982) distant points brighten in Hα within a few seconds of the flash phase (Fig. 11.1). Since the distance may be 100 000 km or more, the agent must be fast electrons; Tang and Moore in fact found that type III-RS (reverse slope) bursts were emitted in these events, the reverse slope indicating that electrons were moving *downward*. Coronal X-ray pictures have shown long loops connecting the main flare to distant points. The velocities determined from films are lower limits because of the time resolution; the velocity is probably $\approx$ 100 000 km/sec.

While there may be clusters of type III bursts associated with a flare, the major ones are usually well correlated with the onset of the optical event. For microwave bursts the relation is generally closer. The exceptions turn out to be flares with a slow beginning followed by a well-defined flash phase; in these cases the beginning of the bursts is usually connected with the start of the flash phase. Of course the type C flares show a late microwave and type IV burst. Because the type II disturbance must reach the plasma level at which we can see it, it usually appears a few minutes late, but we can extrapolate it back to the surface and find a good match.

Type II and IV bursts are associated with the largest flares. Wild's conclusion that the type II bursts were travelling outward through the corona at 1000 km/sec was confirmed when the Culgoora radioheliograph formed images in meter waves of the travelling disturbances. The discovery of the Moreton waves revealed the source. Finally Dulk *et al.* (1976) showed that the type II burst from a level closely followed the white-light transient or CME, beginning just after the leading edge passed and ending after the transient had gone by.

About half the type II bursts show harmonic structure with two parallel bursts, the fundamental and second harmonic. Often these harmonics are themselves twin, with closely parallel bands of emission in the frequency–time diagram. The twinning has been explained in various ways: mixing with the gyrofrequency, Doppler shifts, or parallel streams (Krüger 1979). Two-dimensional pictures of type II bursts from Culgoora show a roughly spherical emission front corresponding in form and velocity to Uchida's (1974) model of the Moreton wave. The exact mechanism of excitation of the plasma oscillations is not certain. The type II bursts have been observed (Malitson *et al.* 1973) at 30 KHz near the Earth coincident with shock-front arrival. The brightness temperature of the type II source reaches 10^{12} deg, evidence that they are produced by a coherent mechanism, presumably plasma oscillation. The detailed type II burst structure is even more complicated. Sometimes a complex "herringbone" structure of type III bursts is excited by the passage of the type II front.

Type IV bursts are defined as any long-lived flare-associated continuum events, usually following the passage of the type II burst. Aside from the microwave bursts which we discussed earlier, they may occur at meter and dekameter wavelengths and be moving or stationary. They are all thought to be due to synchrotron emission, since that is the only way we know that a continuum can be produced. The stationary type IV bursts are the long wavelength extension of the microwave burst; the source does not move, although it is much higher than the microwave burst source. Presumably the flare electrons penetrate to greater heights and produce the observed emission.

Even more spectacular are the moving type IV events (Fig. 11.22), which are observed to move out to several solar radii, the emission lasting several hours. In some cases an arch is seen, a bright source near the top and two additional oppositely circular polarized sources at the intersection of the arch with the plasma level (Wild 1969). The acceleration and trapping of the energetic electrons in the moving type IV is quite remarkable; the particles remain in erupting arches of regular form for hours. Stewart *et al.* (1978) have shown that they are in fact trapped near the outward-moving Hα material behind the CME.

Radio noise storms were discussed in Sec. 10.9. There is some evidence that they are flare triggered, and they are certainly associated with the most active regions. Lantos *et al.* (1981) found a close time association between a soft X-ray event and a meter-wave noise storm, and events associated with flares were found by Wild and Zirin (1956). But it is not really established if the noise storms take place in the absence of flares.

11.8. Flare spectra

We have already discussed flare spectra in the microwave and HXR region. In this section we consider the ultraviolet, SXR and optical regions.

The main observational limitation to flare spectra is the problem of pointing the spectrograph at the flare at the right time and place. This is particularly difficult in the visible, where the flare emission is weak compared to the photospheric background. In the ultraviolet the flare emission is so strong it is all we see in most of the lines below 400 Å but of course we still must point to the right place.

Fig. 11.15 shows the overlappogram images at four times during the June 15 1973 flare observed by Skylab; the H-like higher lines of HeII merge into the HeII recombination continuum at 228 Å. The early phases are dominated by chromospheric lines and those from the hottest plasma. The former must be excited by precipitating electrons and the latter are associated with the hot thermal plasma. Later in the flare we see extended images in the permitted transitions of coronal ions such as FeXIV, XV, XVI, reflecting the cooling coronal loops. These last long after the higher HeII lines and continuum have faded. The FeXXIII and XXIV cores sit on the loop tops, evidence that at least the SXR source and maybe the energy release takes place along the sheared neutral line. These ions radiate mainly in the X-ray range, but there are screening doublets of the type $2s \rightarrow 2p$ with relatively long wavelength. In fact Feldman *et al.* (1974a) have identified a forbidden line (analogous to the coronal lines in the visible) of FeXXI at 1354 Å. The continuum emission in these spectra has not been analyzed but probably is due to H and He recombination at the loop tops.

The evolution of the UV lines in this flare is quite characteristic. The hot core lines last about five minutes, during which the lower stages of ionization of iron and other elements (e.g. FeXIV–XVI, Ca XVII) gradually strengthen and dominate in the decay phase. The lower ionization lines cover the entire flare region but are missing, cooked out, in the core.

Widing and Dere (1977) found that the lines from highly stripped ions in this flare followed the time behavior of the 0–3 Å SXR flux, increasing in intensity for several minutes after the end of the microwave impulsive burst. The core of the hot plasma lines coincided with a gap in the lines of lower ionization and although there was little Doppler shift, there was continued shift in position as higher loops were activated, a phenomenon we know well from limb observations. Widing and Dere used emission measure and observed area to deduce the density in the loop cores as 2×10^{11}. From the FeXXIII/XXIV ratio they found the temperature to be around 1.3×10^7 deg, the normal temperature at which those ions are observed.

Beautiful spectra of the same flare in the 1100–1600 Å range were obtained by the NRL slit spectrograph on Skylab. They show (Brueckner 1975) very broad Lyα emission, broad chromospheric lines of higher ionization, and many narrow lines from the low chromosphere. We also find the forbidden line of FeXXI at 1354.2 Å and, late in the decay phase, the forbidden line of FeXI, as well as a few unidentified lines of similar nature. No continuum appears in these spectra, probably because the slit was set across the neutral line, missing the footpoints. All the resonance lines are greatly enhanced. The HeII 1640 (Hα) line is weak.

Cheng (1977) analyzed the spectra of this flare and found upward motion of 80–100 km/sec in FeXXIII-XXIV and turbulence of about 150 km/sec, which was fed into the plasma and produced much of the continued heating. From line ratios he arrived at a temperature of 14×10^6 deg. Cheng *et al.* (1979) used the broadening of FeXXI λ1354 to obtain a temperature of 10^7 deg for the peak of FeXXI in the ionization curve, supporting calculations of Jordan (1969) and Jacobs *et al.* (1977), but contradicting the results of Summers (1974).

As mentioned in Sec. 8.4, the SXR line emission permits spectroscopic analysis of plasmas up to 50 000 000°. The ionization equilibrium gives temperatures and density diagnostics exist if we are lucky and the plasma has the right density for permitted and forbidden lines to change their ratios.

Each important ion has a cluster of satellite lines produced in the presence of outer shell electrons which slightly changes the energy of the transition. There is a whole series of lines between 1.77 and 1.94 Å associated with $2p \rightarrow 1s$ transitions following inner-shell ionization in all the ions of Fe. The structure of this spectrum was first revealed by high-resolution spectra by Mandelstam's group (Grineva *et al.* 1973). The spectrum in the 1.75–1.85 Å region (Figs. 11.18, 11.23) consists of the Lyα line of FeXXVI and dielectronic recombination satellites, the Kβ line due to $3 \rightarrow 1$ jumps in neutral Fe after K-shell photoionization, and a range of lines due to FeXX–XXVI. The dominant line w is analogous to λ584 of HeI, $2\,^1P \rightarrow 1\,^1S$. The electron temperature is determined by fitting to the theoretical curves (Tanaka 1986), using either the line ratio $w : j$ or the fits for FeXXVI are used. Tanaka finds that the ionization temperature from the FeXXVI : XXV ratio usually exceeds the electron temperatures considerably, no doubt because of the effect of cascade as the plasma cools. We can interpret the time profiles of these lines in terms of thermalization of the hard electron input but the exact mechanism by which an FeXXVI ion is created is unclear. We never see the ionization increasing at the flare onset, FeXXVI just appears and cools. Further, it is not known if the ions are high-speed Fe nuclei (which would give lines about 0.004 Å wide for 50 keV) or field Fe ions stripped by collisions, which would give narrower lines. Details of these and other techniques along with line lists are given by Feldman (1981) and Feldman and Doschek (1977).

Fig. 11.23 shows a remarkable series of spectra of a large flare obtained by Hinotori (Tanaka *et al.* 1982; Tanaka and Zirin 1985). Tanaka has compared the Hinotori data with the calculations of Vainstein and Safronova (1978) and Dubau *et al.* (1981) and obtained beautiful fits showing the temperature dropping from 38 million deg at the beginning of the thermal event to 10 million deg 20 min later. All the present observations of ratios of this type show that the ions are in ionization equilibrium. However, Tanaka finds the ionization temperature of FeXXVI systematically higher than that of FeXXV, no doubt because of selective radiation from the hotter plasma. Doschek *et al.* (1981) computed the ratios for a series of He-like ions up to FeXXV observed in flares. They find the OVII lines, which vary in ratio for densities between 10^9 and 10^{13}, give peak densities of 2×10^{12} cm^{-3} in the flare. For the Fe lines the density might be higher.

A remarkable observation of the FeI Kα line, caused by inner-shell ionization of Fe, is reported by Culhane *et al.* (1981) with the SMM crystal spectrometer. They show how the Kα line has the same behavior as the HXR, a short spike during which the non-thermal electrons are precipitating on neutral atoms in the photosphere, and a later peak, twice as high, during the thermal phase, when the hot thermal electrons again excite the line. It appears to be confirmation that impulsive electrons can reach the photospheric levels at which Fe is neutral. Calculations (Hudson and Dwivedi 1982) show that electrons should penetrate only to heights where the density is $< 10^{14}$ cm^{-3}, where the Fe should be singly ionized. Something we don't understand must allow the electrons to penetrate deeper and excite the Kα along with white-light and D3 emission. Tanaka points out that the FeXXVI emission may be excited by flare X-ray photons incident on the surface.

The FeXXVI line spectra obtained by Hinotori (Tanaka *et al.* 1982) are illustrated in Figs. 11.18a and 11.23. Besides the two lines of the resonance doublet, there is a series of satellites due to dielectronic recombination from doubly excited FeXXV states. Fig. 11.23 shows the FeXXVI emission during the impulsive phase of the June 6 1982 flare, along with the Kβ $3p \rightarrow 1s$ transition, and the relative brightening of the satellites as the density increases. The Hinotori also revealed rapid variations in lines of highly ionized atoms at the times of hard X-ray spikes; temperature changes of the order of 5 million deg in seven seconds were observed.

Like He (Fig. 5.5), the $1s^2$ shell of the Fe ion has a series of permitted and forbidden lines. But the effective nuclear charge is so great that the spin-orbit splitting is huge and the singlets and triplets are mixed, so relative intensities of intersystem lines like $2\ ^3P \rightarrow 1\ ^1S$ and totally forbidden lines such as $2\ ^3S \rightarrow 1\ ^1S$ are unaffected by density. As long as the collision rates are small relative to radiative transitions this is the case; but as the density increases we reach a point where de-exciting collisions are important relative to the slow "forbidden" line, and it becomes relatively weaker. Later the same happens for the intersystem line. For the highly ionized atoms discussed above, this density is much higher than found in flares. Doschek *et al.* (1980) calculated that in FeXX and FeXXI there are ratios that discriminate in density between $N_e = 3 \times 10^{12}$ and 10^{13}, but they could observe no change in the ratio, so the flare densities must have been lower. They found the ratio did change in Tokomak plasmas which are much denser.

11.23. Time evolution of Hinotori SXR spectra of the June 6, 1982 flare (Tanaka and Zirin 1985). Low dispersion left, high dispersion right. The blue shifted rising material is seen as an excess above the theoretical curve at upper right. At upper left we see initial twinning of the FeXXVI line. The lines of lower ionization grow relatively stronger as the plasma cools; coupled with decreased Doppler motion, that brings them out more clearly in the last frame. (TAO)

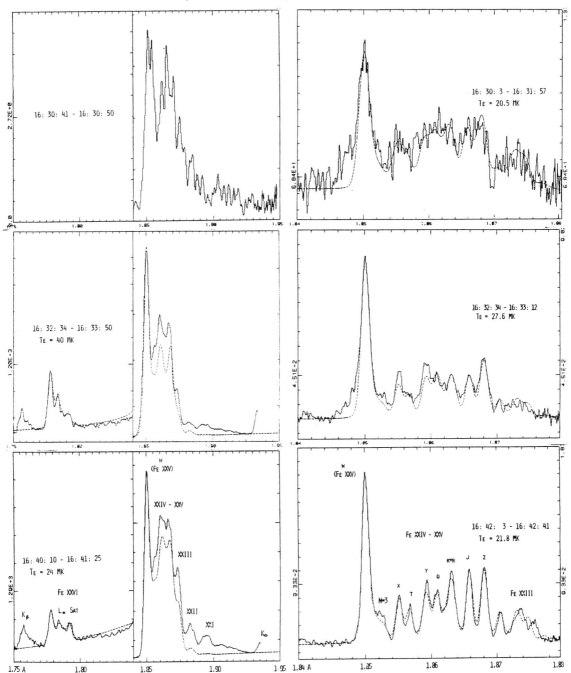

Flare spectra in the visible range have been studied for a long time. They are still useful because our new knowledge of flares enables us to ask the right questions, and many of the spectra were never analyzed properly. The high spatial and spectroscopic resolution available in the visible is unique.

It is very difficult to obtain good flare spectra, since we don't know when or where they will occur, and one must move the slit to the kernels. Those who obtained good spectra waited long to get them. Flares also have a distressing habit of occurring in the penumbra and other places of uncertain intensity and line profile, so it is difficult to remove the background. Except for the Balmer lines the real site of energy release is invisible, and the best we can do is to look at the spectra of footpoints and deduce the properties of the exciting flare accordingly.

The properties of the spectrum vary considerably depending on where in the flare we look. Big flares are bright enough to show the weaker lines and last long enough to be photographed, so there is a tendency for the events studied to be the big ones. Only in the most intense kernels of emission are representative spectra of the real footpoints recorded, with broad H lines, He and continuum emission; outside the kernels the H lines are fairly narrow, as can be seen from the limited extent of flare emission in pictures made in the wing of Hα. For many purposes the echelle spectra of a 3b flare obtained by Jefferies *et al.* (1959) are the best compilation. Higher-resolution spectra have been obtained by Hiei (Dinh 1980), by Neidig (1983) for the flare in Fig. 11.16 and by Donati Falchi *et al.* (1985). None of these show the strong continuum emission observed in two-dimensional observations. Neidig notes that his slit was 3 arc sec away from the brightest kernel. The spectrum consists of intense broad H, He and resonance lines (such as CaII H and K, Na D) which represent the heated chromosphere, and many narrow lines in the cores of Fraunhofer lines. Neidig's spectrogram is useful because it shows not only the blue continuum but also the narrow low-excitation metallic lines. These lines are less than 0.1 Å wide; there is no motion at the low heights where they are formed because the excitation is either by fluorescence of flare emission or conduction from hot thermal plasma above.

The H lines can be very broad; Zirin and Tanaka (1973) found Hα half widths of 10 Å in a great flare (Fig. 11.3). Neidig (1982) found 5 Å in H15. These widths correspond to velocities of 500 km/sec or energies of one keV per nucleon. This is occasionally referred to as "turbulence", but one way or another it is the energy of the H atoms, the dominant species, and therefore may be thought of as equivalent to a temperature of millions of degrees. Most of the Balmer emitting region has much narrower profiles resulting from simple heating (Zirin and Tanaka 1973). The kernels of broad emission are associated with the elevated bright regions such as in Fig. 11.6 as well as footpoints. The bombardment by flare particles produces an extraordinary hot, dense plasma in the chromosphere.

One interesting facet of the Hα emission is the frequent presence of red shifts. They indicate that material is moving downward rapidly in the Hα emitting region. It suggests that the extremely broad Hα profiles come from footpoints where energetic heating is taking place and a shock wave is moving downwards, ionizing and heating the material below.

Svestka (1965) pointed out that the Balmer lines in flares should reach a minimum half width at a certain value of n. The early members have a high optical depth which produces a large half width by curve of growth effects; this decreases with n. But the Stark broadening increases as n^4 and eventually dominates for high n, so there is a value of n (depending on temperature, density and optical depth) for which the half width is a minimum. Neidig's half width of 5.2 Å for H$_{15}$ corresponds to $N_e = 6.0 \times 10^{13}$. Further correction for curve of growth broadening led Neidig to a density of 5.2×10^{13} in the Hα emitting region of the flare. For the number of atoms in the ground state of Hα he found 3.4×10^{16} cm^{-2}, leading to an optical depth $\tau = 10^4$ in the center of Hα. The temperatures indicated by these line widths are 4 million deg. When just the higher Balmer lines are observed, one can simply use the Inglis–Teller formula (Eq. 5.17) to obtain the electron density; the results suggest densities of 10^{12} cm^{-3}. How can H emission come from such hot plasma? Hydrogen can be ionized only once, so there is always some neutral H around. This means that a temperature of millions of degrees exists at the Hα level, only a few thousand km above the photosphere. Small wonder we see white-light emission from the photosphere below.

In Sec. 11.3 we showed how the HeI D3 line goes into emission against the disk for densities above 5×10^{12}. Similar arguments apply to any spectrum line, and for most permitted lines the conclusions are similar. For the λ4686 4$\rightarrow$3 transition in ionized He, the highest excitation line in the visible, Zirin and Hirayama (1985) found a temperature of 30 000°– 50 000° and density 3×10^{12}. A similar minimum density in the electron acceleration region is deduced by Ramaty *et al.* (1983) from 10 MeV photon emission from flares.

The high temperature associated with the H line broadening is a real skeleton in the closet of solar flare research, since most models use far lower temperatures. In limb flares, for example, the metallic lines disappear completely from limb flare spectra, and only broad lines of H, He, and HeII are seen (Zirin 1964). Since the He line widths do not exceed 2 Å it is likely that the H widths are indeed thermal, and the H emission may come from dense plasma at million-degree temperatures.

Svestka (1976) and others proposed a filamentary model of flares, based on the idea that the occupation number of the lower level of Hα is so large that the physical thickness must be low, a few km, or the line would be too bright. This analysis assumed that (1) the H-emitting region is at 10 000 deg (how the million-degree profiles were produced was not explained), and (2) the flare does not emit in the Balmer continuum. Since these assumptions (justified by the data at their disposal at that time) have been shown to be in error, that particular problem has gone away.

Other problems of interpreting flare spectra have not gone away. The spectrum is fairly well understood in the SXR region, where we know what the temperature is. The He ratios were reconciled by Zirin and Hirayama (1985) and, with a little fudging, fit the UV and visible He lines. But we are unable to explain the peculiarities of H line emission as explained here and Sec. 11.3. There is no model for the hydrogen lines which has the broad profiles and high energies observed, no self-absorption, equal $H\alpha$ and $Ly\alpha$ intensities, and $H\gamma$ stronger than all lines. I believe these problems will be solved only by models with very high temperature (>1 million deg) and high density.

The interpretation of spectra has been somewhat diverted by the proliferation of computer codes. While one might expect these to be useful, because they can keep track of all the important parameters, they have, in my opinion, enjoyed little success. The many parameters make it possible to always make some kind of fit, even with wrong physics, while the problem of getting the code to run obscures the physics of the problem by focussing on the mechanics. A general rule of thumb is that the error in a calculation equals the number of atomic levels included; a 50-level calculation might be expected to be in error by a factor 50, a 2-level calculation by no more than 2. Once we have the physics right, we don't need many levels; if the physics is wrong, no number of levels will help.

11.9. How do flares occur?

The understanding of why flares occur has come from the study of many $H\alpha$ and magnetic observations of active regions; the critical problem is now *how* they take place.

In 1968 I began to operate, for the first time, a high-resolution $H\alpha$ telescope. Because the telescope field included only part of the Sun, I tried to pick the region most likely to flare. Having read the earlier literature on flares as well as my own foolishness, I pointed at the biggest sunspot group I could find. I was dismayed to find that while I had observed no flares, large flares were occurring elsewhere on the Sun. One had to actually observe the Sun day after day to learn where to point the telescope. I soon found that while large spot groups do have more and bigger flares than little ones, groups with simple magnetic structure, no matter how large, had relatively little activity, as did all αp groups with a spot and following plage. But if the magnetic axis was rotated from the customary E–W line, or if there were steep magnetic gradients in the region, considerable activity might be expected. Even more flares occurred in groups where bright $H\alpha$ emission and arch filaments told us new flux was coming up inside the group. In the end, flare activity could be tied to one crucial factor: magnetic field change, usually involving flux emergence but sometimes flux decay. These changes cannot be accommodated by the normal magnetic diffusion rate (Sec 3.3), so magnetic stresses build up until an instability occurs and the magnetic field changes to the lowest energy configuration.

While observers knew that bright Hα, which is a mark of emerging flux, was connected with high activity, the idea that it was flux emergence was first recognized by Martres *et al.* (1968), who termed the growing field "*structure magnetique evolutive*". As we learned about emerging flux regions and obtained better observations the picture became clear. Liggett and Zirin (1985) have shown that there is strong tendency for magnetic flux to erupt in existing active regions. The new flux must either cancel the existing fields or push them aside. Since diffusion is quite slow, only a little reconnection takes place (marked by bright Hα) but the old flux is pushed aside, strong gradients being established at the edges of the emerging field. As the dipole emerges the p spot usually moves forward at a rapid rate, squeezing the flux ahead of it. The new dipole expands in other directions, too, replacing the old field. Fig. 10.18 shows such a dipole emerging in an AR. The AR was one of only three active regions on the Sun at that time, covering not more than a few percent of the active latitudes, yet one of the two groups to emerge that week came up right in the middle of it, and the other, not far away. The p spot rapidly pushed forward into an extensive area of f plage near the main spot, and a series of flares ensued.

As new flux pushes into old flux of the opposite sign, there is immediate reconnection, but as the material is pushed out of the way, the field lines are drawn out along the line perpendicular to the motion, and a neutral line filament forms separating the two opposite polarities. Now lower energy magnetic connection to material directly across the neutral line is possible (Fig. 3.2), and a series of flares will occur.

None of the foregoing happens if the moving spot pushes into flux of the same sign. Further, the compressing and shearing of fields generally occurs only with moving sunspots; plage fields do not appear to have the cohesiveness to push material aside and make steep gradients. As time goes by, the sheared field line produced by the original motion and marked by the filament is replaced by fibrils directly arching across the inversion line; at that point the magnetic fields are connected in the lowest possible energy state and, unless further flux emergence takes place, the flares are over.

A striking example of the effects of spot motion is seen in Fig. 11.11. An island δ spot composed of two large spots appeared on June 3 1982, having emerged on the far side of the Sun. There was a big flare on June 3, after which the upper spot, of p polarity (there is some doubt of the polarity; the spot had f polarity for that hemisphere, but it is entirely possible that this group, at 9° N, was really from the other hemisphere and that the spot was really *p* polarity) moved rapidly forward, producing a series of flares along its leading edge. Finally on June 6 a great flare occurred (Figs. 11.12–13), resulting in loops connecting directly across the inversion line from the big spot to new spots growing behind the moving spot. Since the big spot was previously connected to the moving spot, we know the field lines have reconnected. We have observed this ejection of a spot from a δ configuration on two occasions.

I have already mentioned that spotless flares (Fig. 9.17) occur upon the eruption of large filaments imbedded in extensive plage left over from an old active region where the spots have decayed. As the old fields decay, the filament becomes unstable and erupts, producing brightening over an extensive area. These flares usually exhibit a thermal electron spectrum but are soft in particle energy. The contradiction in magnetic fields that leads to the flare can be inferred from the fact that the fibril structure along the neutral line runs along the filament axis, but the post-flare loops connect directly across the neutral line. If a filament far from plages erupts, it will usually not produce much flare brightening, possibly because the magnetic fields involved are too weak.

The following list (Zirin and Liggett 1987) of circumstances leading to big flares may be useful:

- δ spots, preferably those of types 1 and 2.
- Umbrae obscured by Hα emission.
- Bright Hα, which marks flux emergence.
- New flux erupting on the leading side of the penumbra of a dominant *p* spot.
- A filament crossing a δ spot.
- All these effects are associated with greatly sheared magnetic configurations.

These conditions are all symptoms; the cause is invariably sunspot eruption, growth or motion which pushes one spot into another of opposite magnetic polarity, producing shear.

Although this describes what is seen it begs the question of how the magnetic anomalies occur. Why does new flux erupt in the midst of old? What is the relation between old and new flux tubes? This problem is particularly striking in the case of very active spot groups, such as that of July 1974, which appear on the disk in complex, twisted form and then are highly active as the field straightens itself out. What twisted them up? Why did the straightening out occur on the surface? Are there sub-surface flares? It appears as though the sunspot structures break loose from their old ties when they hit the surface and begin a new life with different rules. As we noted in Chap. 10, the most active δ spots emerge as such, clearly a sub-surface production.

While the picture of what magnetic features lead to flares is fairly clear, the actual mechanism is far from certain because of the difficulty of handling MHD problems. Many MHD flare models have been published, each claiming to explain the process. But there are few, if any, predictive models. The natural state of affairs is often too complicated to model. A good example is the Petschek (1964) mechanism, which shows how nearby oppositely directed fields may reconnect. But the fields along the shear line are not oppositely directed; they are parallel. So that model is irrelevant. The tearing-mode instability seems to describe the reconnection of fields by the development of short-cut connections, as in the pattern of Fig. 11.12 (Low 1985). Flare models which include the real circumstances of a sheared parallel field with a filament are needed. For example, the extensive analysis of Hood and Priest (1980) and Hood (1983) addresses a symmetrical arcade, more typical of the post-flare environment. These authors show the importance of line-tying, the fact that all of the lines of force are anchored in the photosphere. Line-tying helps stabilize the fields, which must vary within the fixed anchors. But of course the anchors change and are not well understood. The system must give an explosive energy release when we want it. An MHD configuration which is highly unstable will never store enough energy to produce flares.

Since I am not very expert in these matters I advise the reader to read here the observational evidence, then read one of the various MHD books, and see if he can put them together.

Laboratory work with Tokomaks, devices for confining a hot, dense plasma in the hope of producing fusion, has given useful insights for solar flare theory, although the two have many differences. The tearing-mode instability is marked by a series of closed magnetic islands that appear in the middle of the toroidal configuration. The inner lines of force in the toroid have a lower energy connection to the outer ones, and, worse still, to the walls. The islands increase the resistivity and decrease the scale so further reconnection takes place. The islands could well be the sources of the bright flashes observed in the impulsive flare, as well as the kernels of radio emission. The way in which we step from the irregular flashes of the impulsive event to the more sedate and organized main phase is also reflected in these laboratory events, where the islands produce enough resistivity and small-scale structure for reconnection to go forward.

The mechanisms of particle acceleration are another challenge. Although it is generally felt that Fermi acceleration is the key process, no detailed description has been given. The remarkable fact is that in a few minutes all the magnetic energy is converted into acceleration of fast electrons and protons. It is surprising that the particles are not slowed down by collisions with field particles during that process. The relative proportions of fast protons and electrons are also not well understood.

Generals and government officials often are struck with a desire to predict solar flares, and astronomers often oblige these desires with false promises. Other astronomers, who have not bothered to study the question, will say it is impossible. The truth is somewhere in between. If one follows the points made in this chapter it is easy to determine that a high probability of a solar flare exists. If there have been many flares, persistency will work; there probably will be more. If we have a big δ spot with umbrae covered by Hα emission, there will be flares. If the Hα emission dies down, there won't. If a filament on a sheared neutral line shows a big blue shift, there will be flares. If a new EFR comes up rapidly in the middle of an old AR, its p spot pushing into the f spot of the old group, there will be flares. When is not so easy to tell, although a study of past experience could yield probabilities per day. It is not a foolish subject; being forced to predict flares should help us understand them.

We see that all we have to do now is figure out why new flux erupts in the middle of old, why the p spot pushes forward, how particles get accelerated, and finally, how the tearing mode (or whatever instability sets off the flare) actually takes place.

References

Ramaty, R. 1986. *Physics of the Sun*, ed. Sturrock, Ch. 14. Dordrecht: Reidel.

Forman, M. A. *et al.* 1986. *Physics of the Sun*, ed. Sturrock, Ch. 13. Dordrecht: Reidel.

Neidig, D. F. 1986. *The Lower Atmosphere of Solar Flares* Sunspot: National Solar Obs.

Svestka, Z. 1976. *Solar Flares* Dordrecht: Reidel.

Tanaka, K. 1986. *Publ. Astr. Soc. Japan.* **38**, 225.

Tanaka, K. 1987. *Publ.Astron.Soc.Japan.* **39**, 1.

12

Questions

The purpose of a text is not only to educate, but also to spur research by raising questions worthy of further investigation. There are questions on almost every page of this text, quite simply because I don't know the answers; in this chapter I shall remind the reader of the more important or intriguing ones.

Of course, everyone wants to work on **Big Questions**. That is why cosmology is so popular. Big Questions, however, have a way of defying quick solution, so we must chip away at them by answering little questions, which are also pretty hard. Because of instrumental developments, better observation and understanding, solar physics is in a position to break some big chips off the Big Questions. So we will spend a bit of time with them.

The solar interior Bethe (1986) recently proposed that the MSW mechanism permits electron neutrinos to change to muon neutrinos in the high density of the solar interior and avoid detection. If this idea is correct, we can stop worrying about the neutrino deficit and be more confident in our understanding of the solar interior. The development of helioseismology now permits us to explore the nature of that interior in more detail. There are still small discrepancies between predicted and observed frequencies of solar oscillatory modes, but we may hope these will be resolved by further work (Sec. 6.2).

Does the Sun rotate as a solid body? Recent helioseismology data suggest that it does and further improvements in this field may settle that question (Sec. 6.4–5).

Does differential rotation (DR) exist throughout the Sun? This is a more difficult challenge for helioseismology because of the high resolution required. If we know the internal distribution of DR, we may understand its cause. Some models of the sunspot cycle, such as that of Babcock (1961), require the differential rotation to intensify the fields and drive the cycle, in which case the source of the DR is the source of the sunspot cycle. Various proposals explaining the DR in terms of pole–equator temperature differences have foundered on our inability to test them adequately. Particularly frustrating is the contrast between measured differential rotation in sunspots, magnetic fields and coronal holes. Surely, in view of the close correspondence between surface magnetic morphology and atmospheric structure, these cannot differ by much or the correspondence would be destroyed. How, then, can the measured results be so different? What is the true differential rotation? The sunspot and spectroscopic determinations are fairly consistent, but the coronal holes are probably the best way to study this large-scale phenomenon (Sec. 6.4–5).

The sunspot cycle The 22-year cycle, the complex but regular pattern of the Hale–Nicholson law, the butterfly diagram and the reversal of the polar field form the most complex and challenging of the big questions. We have discussed the Babcock–Leighton model in Chap. 10; it is not a bad phenomenological representation of the general pattern. But it depends on Joy's law, the differential rotation and a magnetic field diffusion rate three times that observed, so obviously it is far from perfect. Further, if the general solar field is reversed by kinematic effects at the surface, we still must explain how these fields move into the interior, where they must again be amplified. The little surface field elements are supposed to random-walk freely about the surface and finally become the internal solar field. If they were still attached below, I assure you they would not walk so randomly. There also are problems with periods like the Maunder minimum, when there is no longer any field to regenerate. The large-scale nature of details of solar activity – spots may erupt all over the Sun at once, or the activity may show a hemispheric pattern – suggests to me that we are seeing the surface effects of a large-scale global field, unaffected by surface events (Sec. 10.1).

Heating the chromosphere and corona We know that the temperature begins to increase somewhere above the photosphere, but we do not know the exact extent to which the chromosphere is heated; except for the transition to the corona, most of it is no hotter than the photosphere. How the transition region and the corona got so hot nobody knows. The old ideas of acoustic heating have foundered on the observation that heating is localized in regions of enhanced magnetic fields. There is a further contradiction in the location of the peak intensities of transition region lines a few thousand km above the photosphere, well below the level of the spicules, which are supposed to play a big role in the transition. This contradiction is strengthened by the absence of spicules above plages, a fact I have pointed out over and over (starting on p. 239 of the 1966 edition) with little effect. If spicules account for the atmospheric heating, one would not expect transition region and corona to be brightest where they are absent.

These are brightest where reconnection takes place (EFR's and active regions), so it is likely that the heating is due to reconnection phenomena, which are now known to occur all over the Sun (Sec. 7.8).

12.2. Little questions

We discuss here not only questions that are obviously unexplained, but also those where popularly held views in fact are contradicted by observations. Some are considered mere details, but in fact the physics is just as fundamental, although the religious significance may be less. These are all questions that bother me, that I feel we should be able to make progress on, or at least be attacking.

The supergranulation and the network We think that the SG is the offspring of the hydrogen convection zone and maybe it is. The SG lifetime is unknown and that of the network is thought to be 24 hrs, but that value is suspiciously close to the interval between observations. But we know that enhanced-network cells are much longer lived. So there is not a one-to-one match. The idea that the supergranulation creates the magnetic network by dragging fields to the edges should not work, because IN fields of both signs are dragged to the boundaries, so there is no net gain. Yet we see the SG flow moving the IN flux elements to the cell boundaries. It cannot be coincidence that the Sun is covered with a magnetic network of about the same scale as the SG, or that the fairly uniform fields of active regions quickly break up into the enhanced network. The effect exists, but how it works is unclear (Sec. 6.7).

The sunspot phenomenon During sunspot emergence the EFR clearly reflects the emergence of subsurface field. But once the sunspot group is formed, it behaves as though cut off from its origins. Spot fields diffuse away into enhanced network. Does that mean the subsurface fields also have spread out? Why do the p spots run forward (the Hale–Nicholson force)? Why are they larger than the f spots? What makes the great island δ spots? Are they an original product of the cycle, or distorted in the convective zone? If they live only a few weeks, their passage through the interior should take less, or they would reconnect. If the δ spot sinks below the surface, what happens to that flux? Is it regenerated? Are there sub-surface flares, or is the surface a unique site of reconnection (Sec. 10.4)? Where does the missing energy go that the spots block from the solar constant? The energy must be stored for a long time, perhaps months.

Sunspots as bosons A large fraction of solar flares are due to the propensity of sunspots to emerge in the same place as existing spots, just as children always like to play in the same place. This tendency must stem from the morphology of the subsurface field and the different groups erupting in the same center of activity must be related somehow. But on the surface the spots behave as though they are completely independent, blundering into one another. On a more global scale, the number of EFR's on the Sun is strongly non-random; the emergence of one new group increases the probability that another will emerge, often far away (Sec. 10.4).

Filaments The role of filaments in establishing magnetic boundaries is poorly understood. Whether or not a special process occurs in filaments to stabilize the magnetic inversion line with which they are always associated is not known. In the absence of some barrier one expects the connected elements of opposite field either to submerge under magnetic tension or to erupt because of magnetic buoyancy. We do not know how the material accumulates in the filament or what supports it, although we feel the horizontal field lines play some role. All the known models are unstable in some way. Another remarkable unexplained characteristic of filaments is that they always erupt outward and never fall down. This suggests that magnetic buoyancy pushes them against the restraint of the overlying fields. As a result many filaments have a sharp upper bound (Sec. 9.2).

Spicules These little jets are a mysterious phenomenon to which have been ascribed many effects, including heating the corona and producing the solar wind. We do not know what they are, or if they represent a net upflow or downflow. We do know that they occur only in unipolar elements of the chromospheric network; they are not present in plages (except on the fringes) or in weak intranetwork magnetic elements. Because they occur in unipolar elements, it is doubtful that they are related to flares, which invariably occur near a magnetic inversion. Thus the energy source in spicules must be hydrodynamic, the magnetic field performing the function of focussing the material and determining the trajectory. As noted above, the evidence that spicules play an important role in the high-temperature transition zone is shaky, because the transition appears to occur lower down. However, they appear quite clearly in monochromatic CIV images, so they must have some hot component. They are so poorly understood that we can make no definite claims (Sec. 7.2).

Solar flares The evidence is strong that magnetic reconnection provides the energy; however, the details of the acceleration mechanism are unknown, particularly when we consider the short time scales and prodigious energies involved. And the flares themselves are quite varied. How are the properties of the pre-flare configuration related to the different types of X-ray events described by Tanaka (1986)? What flares will produce a CME? Obviously there must be a filament, but there are plenty of flares with only small filaments, and sometimes the spray comes from a filament outside the main inversion line. We know that the big impulsive gamma-ray flares will come from the island δ regions; we should be able to pick up more precursors of these giant phenomena. Impulsive flares involve filament eruptions and often have a hot thermal component, while flares with long-lived non-thermal events do not have filaments on the magnetic inversion line. Apparently the high coronal densities produced by the filament eruption promote thermalization. What is the significance of the loop tops, where much of the energy release seems to occur? It is really hard to understand how density can peak at the loop tops.

What is the energy of great flares? Since the X-ray detectors are always saturated, the maximum energy is not known, but it can be projected as above 10^{34} ergs. The electron input to M class flares has been measured as 10^{32} ergs. If we add up the magnetic field energy over the flare area, it is usually less. Either a sizable amount is stored in the currents that stabilize the force-free sheared field, or the flare energy is less than we think (Chap. 11).

Reconnection We have discussed the reconnection of magnetic field lines many times. As noted in Chap. 3, this phenomenon should take place rather slowly. In fact, we have seen that it is common and occurs rather faster than we would expect. In the weak fields it seems to happen within hours and in flares it occurs in minutes; one would expect much longer diffusion times. The cause of fast reconnection is obscure and probably connected with the small scale of the solar fields. We need to learn a lot about plasmas before we can understand reconnection better.

Structure of background magnetic fields In Sec. 6.9 we went back and forth over the problem of whether all solar magnetic fields are clumped in strong elements. There is good evidence that plage fields are about 1000 gauss, but also good evidence that the weak fields are really weak. The weakness of microwave emission from ER's suggests that their fields are weak. There may be two classes of fields, the strong ones associated with the cycle and the weak ones generated by the SG. We need a better picture of the creation and destruction of the background field and how exactly the polar fields are formed and maintained. Is there a weak diffuse background field?

Element separation Over the years there has been discussion of the possible separation of elements in the solar atmosphere. One might expect the heavier ions to fall off more rapidly than the lighter. But there always appeared to be enough turbulence to keep the elements all mixed. Now it seems that elements of high first ionization potential (FIP) are underabundant in the solar energetic particles from flares, as well as in the chromosphere, corona and solar wind. But the high FIP elements, though abundant, are hard to excite and have few lines in the photospheric and chromospheric spectrum. There is a small chance that they are less abundant in the photosphere than we thought. If these elements are underabundant in the atmosphere, that is a momentous clue to the formation of the chromosphere and corona, a clue that one would not have expected from SEP analysis.

Here before our eyes is a real star, whose remarkable phenomena we can sample in every way if we will only take the trouble. The signal-to-noise ratio is high; we can explore the world of magnetic phenomena and convection in plasmas. We can watch particles being accelerated in flares and measure the output at the Earth. We can probe the interior of this star through the measurement of its oscillations, which can be carried out with exquisite precision. And with even a little telescope we can see the marvels of the sunspot cycle. I invite the reader to join the fun.

12.1. As the Sun sets behind the San Bernardino mountains about 4 km from the observatory, Caltech students signal goodbye to our readers and friends.

Appendix

Abbreviations

AFS arch filament system
AR active region
ARF active region filaments
BBSO Big Bear Solar Observatory
CME coronal mass ejection
CSIRO Commonwealth Scientific and Industrial Research Organization
DR dielectronic recombination
DR differential rotation
EFR Emerging Flux Region
em electromagnetic
ER ephemeral region
FF free-free
FIP first ionization potential
FTA Field transition arches
FTS Fourier Transform Spectrometer
FWHM full width at half-maximum
GSFC Goddard Space Flight Center
HAO High Altitude Observatory
HXR hard X-ray
IN intranetwork
KDP potassium dihydrogen phosphate
KPNO Kitt Peak National Obs., National Solar Obs.
LTE local thermodynamic equilibrium
MHD magnetohydrodynamic
MTF modulation transfer function

NRL Naval Research Laboratory
OSO Orbiting Solar Observatory
OVRO Owens Valley Radio Obs.
P–C prominence–corona (interface)
P–R photoionization–recombination
PdM Observatoire du Pic du Midi
QRF quiet region filaments
RP resolving power
RP running penumbral
RTV Rosner Tucker and Vaiana
SAT solar acoustic tomography
SEP solar energetic particles
sfu solar flux unit $= 10^4$ Jansky
SG supergranulation
SMM Solar Maximum Mission
SPO Sacramento Peak Obs., National Solar Obs.
SXR soft X-ray
TAO Tokyo Astronomical Observatory
UBF Universal Birefringent Filter
UV ultraviolet
UVSP UV spectrometer on SMM
VLA Very Large Array
VMG videomagnetograph
WLF white light flare
XUV extreme ultraviolet

Bibliography

Abel, W. C., *et al.* 1963. *Space Res.* **3**, 635.

Acton, L. W. and H. Zirin 1967. *Ap.J.* **148**, 501.

Acton, L. W. 1964. Ph. D. thesis, Univ. of Colorado.

Acton, L. W. *et al.* 1982. *Ap.J.* **263**, 409.

Adams, W. and A. Joy 1933. *Pub.A.S.P.* **45**, 301.

Adams, W. and F. Tang 1977. *Solar Phys.* **55**, 499.

Agnelli, G. *et al.* 1975. *Solar Phys.* **44**, 509.

Alfvén, H. and C. G. Falthammer 1963. *Cosmical Electrodynamics–Fundamental Principles, 2d ed.* Oxford: Oxford Univ. Pr.

Allen, C. W. 1974. *M.N.R.A.S.* **107**, 211.

Allen, C. W. 1981. *Astrophysical Quantities, 3d ed.* London: Athlone.

Aller, L. H. 1953. *Astrophysics.* New York: Ronald, p. 289.

Aller, L. H. and A. K. Pierce 1952. *Ap.J.* **116**, 176.

Alpher, R. 1950. *J.G.R.* **55**, 437.

Ando, H. and Y. Osaki 1977. *Publ.Astr.Soc.Japan.* **29**, 221.

Antonucci, E. *et al.* 1982. *Solar Phys.* **78**, 107.

Anzer, V. 1979. *IAU Colloq.* **44**, 322.

Athay, R. G. and D. H. Menzel 1956. *Ap.J.* **123**, 285.

Athay, R. G. 1976. *The Solar Chromosphere and Corona: Quiet Sun.* Dordrecht: Reidel.

Athay, R. G. and G. E. Moreton 1961. *Ap.J.* **133**, 935.

Athay, R. G. and R. N. Thomas 1956. *Ap.J.* **123**, 309.

Athay, R. G. *et al.* 1985a. *Ap.J.* **288**, 363.

Athay, R. G. *et al.* 1985b. *Ap.J.* **291**, 344.

Athay, R. G. *et al.* 1986. *Ap.J.* **303**, 877.

Avrett, E. H. *et al.* 1976. *Ap.J.* **20**, L199.

Ayres, T. R. and J. R. Linsky 1976. *Ap.J.* **205**, 874.

Ayres, T. R. and L. Testerman 1981. *Ap.J.* **245**, 1124.

Ayres, T. R. *et al.* 1986. *Ap.J.* **304**, 542.

Babcock, H. D. 1959. *Ap.J.* **130**, 364.

Babcock, H. D. and H. W. Babcock 1955. *Ap.J.* **121**, 349.

Babcock, H. W. 1953. *Ap.J.* **118**, 387.

Babcock, H. W. 1961. *Ap.J.* **133**, 572.

Babin, *et al.* 1974. *Izv.Krym.Astrofiz.Obs.* **55**, 3.

Badalyan, O. G. and N. S. Shilova 1984. *Solnechnye Dannye* **8**, 3.

Bahcall, J. N. 1965. *Science* **147**, 115.

Bahcall, J. N. *et al.* 1982. *Rev. Mod. Phys.* **54**, 767.

Bahcall, J. N. and R. Davis, Jr. 1982 in *Essays in Nuclear Astrophysics* ed. Barnes. Cambridge: Camb. Univ. Pr.

Bahcall, J. N. 1985. *Solar Physics* **100**, 53.

Bahcall, J. N. and R. A. Wolf 1968. *Ap.J.* **152**, 701.

Bahng, J. and M. Schwarzschild 1962. *Ap.J.* **134**, 312.

Baliunas, S. L. *et al.* 1984. *Proc. 3d Camb. Wkshp.* p.326.

Barbier, D. 1943. *Ann.d'Astrophys.* **6**, 113.

Bartoe, J. -D. and Brueckner, G. E. 1975. *J.O.S.A.* **6**, 13.

Baumbach, S. 1937. *Astron.Nachr.* **263**, 121.

Beckman, J. E. *et al.* 1973. *Solar Phys.* **31**, 319.

Beckers, J. M. 1963. *Ap.J.* **138**, 648.

Beckers, J. M. 1973. *J.Opt.Soc.Am.* **63**, 484.

Beckers, J. M. 1978. *Ap.J.* **224**, L143.

Beckers, J. M. and P. E. Tallant 1969. *Solar Phys.* **27**, 71.

Beckers, J. M. and R. W. Milkey 1975. *Solar Phys.* **43**, 289.

Behring, W. E. *et al.* 1976. *Ap.J.* **203**, 521.

Belkina, I. L. and N. P. Dyatel 1972. *Soviet Astron.A.J.* **16**, 476.

Bell, B. and H. Glaser 1954. *J.G.R.* **59**, 551.

Bell, B. and H. Glaser 1956. *J.G.R.* **61**, 179.

Bell, B. and H. Glaser 1957. *Smith.Contr.Astr.* **2**, 159.

Benz, A. O. and P. Zlobec 1982. *Solar Radio Storms.* Trieste: Oss. Astronomico.

Bethe, H. A. 1939. *Phys.Rev.* **5**, 434.

Bethe, H. A. 1986. *Phys.Rev.Lett.* **56**, 1305.

Bethe, H. and E. E. Salpeter 1957. *Hdb.d.Phys.* **35**, 88.

Bhatnagar, A. and K. Tanaka 1972. *Solar Phys.* **24**, 87.

Biermann, L. 1941. *Vierteljahresschr. Astronom. Gesellschaft.* **76**, 194.

Biermann, L. 1947. *Naturwiss.* **34**, 87.

Biermann, L. 1948. *Zs.Ap.* **2**, 161.

Biermann, L. 1951. *Zs.Ap.* **29**, 274.

Billings, D. E. 1966. *A Guide to the Solar Corona.* New York: Academic Pr.

Billings, D. E. and W. O. Roberts 1964. *Astrophysica Norvegica.* **9**, 147.

Blaha, M. 1969. *Astr.Ap.* **1**, 42.

Blandford, R.D. and J. E. Ostriker 1980. *Ap.J.* **237**, 793.

Bogorodsky, A. F. and N. A. Khinkulova 1950. *Bull. Comm. Sun.* No. 5-6.

Boischot, A. 1957. *Compt. Rend.* **244**, 1326.

Boischot, A. and B. Clavelier 1967. *Astrophys. Lett.* **1**, 7.

Bommier, V. and S. Sahal-Brechot 1978. *Astr.Ap.* **69**, 57.

Bonnet, R. M. 1982. *Astr.Ap.* **111**, 125.

Boreiko, R. T. and T. A. Clark 1986. *Astr. Ap.* **157**, 353.

Borrini, G. *et al.* 1982. *J.G.R.* **87**, 7370.

Borrini, G. *et al.* 1983. *Solar Phys.* **83**, 367.

Bowen, I. S. and B. Edlen 1939. *Nature.* **143**, 374.

Brandt, J. C. 1970. *Intr. to the Solar Wind.* San Francisco: W.H.Freeman.

Branscomb, L. M. and S. J. Smith 1985. *Phys.Rev.* **98**, 1028.

Brault, J. C. 1979. *Proc 1978 JOSO Workshop*, ed. Godoli. Oss. Mem. Arcetri, **106**, 33.

Brault, J. W. and R. Noyes 1983. *Ap.J.* **269**, L61.

Bray, R. J. and R. E. Loughhead 1964. *Sunspots.* London: Chapman and Hall.

Bray, R. J. and R. E. Loughhead 1974. *The Solar Chromosphere.* London: Chapman and Hall.

Bray, R. J. *et al.* 1984. *The Solar Granulation.* Cambridge: Camb. Univ. Pr..

Bray, R. J.*et al.* 1976. *Solar Phys.* **49**, 3.

Breneman, H. H. and E. C. Stone 1985. *Ap.J.* **299**, L57.

Brueckner, G. E. 1975. *IAU Symp.* **68**, 105.

Brueckner, G. E and J. -D. Bartoe 1983. *Ap.J.* **272**, 329.

Brookes, J. R. *et al.* 1976. *Nature.* **259**, 92.

Brown, J. C. 1971. *Solar Phys.* **18**, 489.

Brown, J. C. 1975. *IAU Symp.* **68**, 245.

Bruzek, A. 1967. *Solar Phys.* **2**, 451.

Bumba, V. 1965. *IAU Symp.* **22**, 192.

Bumba, V. 1976. *IAU Symp.* **71**, 47.

Bumba, V. and R. F. Howard 1965. *Ap.J.* **141**, 1502.

Burgess, A. 1964. *Ap.J.* **139**, 776.

Burlaga, L. F. 1979. *Space Sci. Rev.* **23**, 201.

Byard, P. L. and K. E. Kissell 1971. *Solar Phys.* **21**, 351.

Cane, H. V. *et al.* 1986. *Ap.J.* **301**, 448.

Carlier, A., *et al.* 1968. *Comptes Rendus.* **266**, 199.

Carrington, R. 1858. *M.N.R.A.S.* **19**, 1.

Carrington, R. 1859. *M.N.R.A.S.* **20**, 13.

Chandrasekhar, S. 1945. *Ap.J.* **102**, 223, 395.

Chandrasekhar, S. and F. Breen 1946. *Ap.J.* **104**, 430.

Chang, E. S. and R. W. Noyes 1983. *Ap.J.* **27**, L11.

Chapman, G. and A. P. Ingersoll 1972. *Ap.J.* **17**, 819.

Chapman, R. D. 1981. *The Universe at Ultraviolet Wavelengths.* NASA Conference Publ. 2171.

Chapman, S. 1931. *Proc.Phys.Soc.* **26**, 433.

Chapman, S. 1957. *Smithsonian Contrib. Astrophys.* **2**, 1.

Chapman, S. and J. Bartels 1940. *Geomagnetism.* Oxford: Oxford Univ. Pr.

Cheng, C. -C. 1977. *Solar Phys.* **55**, 413.

Cheng, C.–C. *et al.* 1976. *Ap.J.* **210**, 836.

Cheng, C.-C. *et al.* 1979. *Ap.J.* **233**, 736.

Chevalier, R. A. and D. L. Lambert 1969. *Solar Phys.* **10**, 115.

Chevalier, R. A. and D. L. Lambert 1970. *Solar Phys.* **11**, 243.

Chitre, S. M. and G. Shaviv 1967. *Solar Phys.* **2**, 150.

Chou, D.-Y. 1986. *Ap.J., submitted,* BBSO Prepr. #0249.

Chou, D.-Y. and H. Zirin 1987. *Ap.J., submitted,* BBSO Prepr. #0273.

Christiansen-Dalsgaard, J. *et al.* 1985. *Nature* **315**, 378.

Christon, S. P. and J. A. Simpson 1979. *Ap.J.* **227**, L49.

Chupp, E. L., *et al.* 1973. *Nature.* **241**, 333.

Chupp, E. L., *et al.* 1975. *IAU Symp.* **68**, 341.

Chupp, E. L. *et al.* 1982. *Ap.J.* **263**, L95.

Clerke, A. M. 1903. *A Popular History of Astronomy.* London: A. & C. Black, p. 159.

Cohen, L. *et al.* 1978. *Ap.J. Suppl.* **37**, 393.

Condon, E. U. and G. H. Shortley 1959. *The Theory of Atomic Spectra.* Cambridge: Camb. Univ. Pr.

Connes, P. 1970. *Ann.Rev.Astr.Ap.* **8**, 209.

Cook, J. W. *et al.* 1983. *Ap.J.* **270**, 89.

Cook, W. R. *et al.* 1980. *Ap.J.* **238**, L97.

Cornell, *et al.* 1984. *Ap.J.* **279**, 875.

Cragg, T. *et al.* 1963. *Ap.J.* **138**, 303.

Cram, L. E. *et al.* 1983. *Ap.J.* **267**, 442.

Crannell, C. J. *et al.* 1978. *Ap.J.* **223**, 620.

Crannell, C. J. *et al.* 1986. *Proc.S.P.I.E.* **571**, 142.

Culhane, J. L., *et al.* 1964. *Space Res.* **4**, 74.

Culhane, J. L., *et al.* 1981. *Ap.J.* **244**, L141.

Dainty, J. C. and R. Shaw 1974. *Image Science.* New York: Academic Pr.

Danielson, R. E. 1961a,b. *Ap.J.* **134**, 275, 289.

Davis, R. Jr. *et al.* 1968. *Phys.Rev.Lett.* **20**, 1205.

Deinzer, W. G. *et al.* 1984. *Astr.Ap.* **139**, 435.

DeMastus, H. L. *et al.* 1973. *Solar Phys.* **31**, 449.

Dere, K. P., J.-D. Bartoe, and G. E. Brueckner 1982. *Ap.J.* **259,**, 366.

Dere, K. P., J.-D. Bartoe, and G. E. Brueckner 1983. *Ap.J.* **287**, L65.

Deubner, F. L. 1975. *Astr.Ap.* **44**, 371.

DeVore, C. R. *et al.* 1985. *Aust.J.Phys.* **38**, 999.

Dicke, R. H. 1963. *Nature.* **202**, 432.

Dicke, R. H. 1976. *Solar Phys.* **47**, 475.

Dicke, R. H. *et al.* 1987. *Ap.J.* to appear.

Dicke, R. H. and H. M. Goldenberg 1967. *Phys. Rev. Letters.* **18**, 313.

Dinh, N. V. 1980. *P.A.S. Japan.* **32**, 495.

Donati Falchi, A. *et al.* 1985. *Astr.Ap.* **152**, 165.

Doschek, G. A. 1985. In *Autoionization.* ed A. Temkin New York: Plenum.

Doschek, G. A. *et al.* 1976. *Ap.J. Suppl.* **31**, 417.

Doschek, G. A. *et al.* 1977. *Ap.J.* **212**, 905.

Doschek, G. A. and U. Feldman 1978. *Astr.Ap.* **69**, 11.

Doschek *et al.* 1980. *Ap.J.* **239**, 725.

Doschek *et al.* 1981. *Ap.J.* **249**, 372.

Dubau, J. *et al.* 1981. *M.N.R.A.S.* **19**, 705.

Dulk, G. A. *et al.* 1976. *Solar Phys.* **49**, 369.

Dulk, G. A. and K. A. Marsh 1982. *Ap.J.* **259**, 350.

Dulk, G. A. and D. E. Gary 1982. *Bull.A.A.S.* **13**, 878.

Dunn, R. B. 1960. *Ap.J.* **130**, 972.

Dunn, R. B. 1960. Ph. D. thesis, Harvard Univ.

Dunn, R. B. 1968. *Ap.J. Suppl.* **15**, 275.

Dunn, R. B. 1985. *Solar Phys.* **100**, 1.

Dunn, R. B. and J. B. Zirker 1973. *Solar Phys.* **33**, 281.

Dunn R. B. *et al.* 1968. *Ap.J. Suppl.* **15**, 139.

Durney, B. R. and A. J. Hundhausen 1974. *J.G.R.* **79**, 3711.

Duvall, T. L. Jr. and J. W. Harvey 1982. *Nature.* **300**, 24.

Duvall, T. L. Jr. and J. W. Harvey 1984. *Nature.* **310**, 19.

Duvall, T. L. Jr. *et al.* 1984. *Nature.* **310**, 22.

Duvall, T. L. Jr. 1982. *Nature.* **300**, 242.

Eddy, J. A. 1976. *Science.* **192**, 1189.

Eddy, J. A. 1980. *The Ancient Sun.* New York: Pergamon.

Edlen, B. 1937. *Zs.Ph.* **104**, 407.

Edlen, B. 1942. *Zs.Ap.* **22**, 30.

Edlen, B. 1964. *Hdb.d.Phys.* **27**, 80.

Edlen, B. 1969. *Solar Phys.* **9**, 432.

Edmonds, A. R. 1957. *Angular Momentum in Quantum Mechanics.* Princeton: Princeton Univ. Pr.

Edmonds, F. N. 1960. *Ap.J.* **131**, 57.

Efanov, V. A. *et al.* 1980. *IAU Symp.* **86**, 141.

Elwert, G. 1952. *Z. Naturforschung.* **7A**, 432.

Engvold, O. 1972. *Solar Phys.* **23**, 346.

Evans, J. C. and Testerman, L. 1975. *Solar Phys.* **4**, 41.

Evans, J. W. 1949. *J.O.S.A.* **39**, 229.

Evans, J. W. 1957. *J.O.S.A.* **48**, 142.

Evans, J. W. 1953. *The Sun,* ed. Kuiper, Chicago: U of Chi. Pr.

Evans, J. W. *et al.* 1962. *Ap.J.* **136**, 682.

Evans, J. W. and R. Michard 1962. *Ap.J.* **136**, 487, 493.

Evans, J. W. *et al.* 1963. *Ann.d'Astrophys.* **26**, 368.

Feldman, U. 1981. *Physica Scripta.* **24**, 681.

Feldman, U. and G. A. Doschek 1977. *J.O.S.A* **67**, 726.

Feldman, U. *et al.* 1974a. *Ap.J.* **196**, 613.

Feldman, U. *et al.* 1974b. *Ap.J.* **199**, L67.

Feldman, U., G. A. Doschek, and J. T. Mariska 1979. *Ap.J.* **229**, 369.

Feldman, U., G. A. Doschek, and R. W. Kreplin 1980. *Ap.J.* **238**, 365.

Feldman, U. *et al.* 1980. *Ap.J.* **241**, 1175.

Feldman, U. *et al.* 1982. *Ap.J.* **260**, 885.

Feldman *et al.* 1983a. *Ap.J.* **271**, 832.

Feldman *et al.* 1983b. *Ap.J.* **273**, 822.

Fermi, E. 1954. *Ap.J.* **119**, 1.

Finn G. D. and D. A. Landman 1973. *Solar Phys.* **30**, 381.

Firor, J. W. and H. Zirin 1962. *Ap.J.* **13**, 122.

Fisher, R. R. *et al.* 1981. *Appl.Opt.* **20**, 1094.

Fisk, L. A. 1978. *Ap.J.* **224**, 1048.

Flower, D. R. 1977. *Astr.Ap.* **54**, 163.

Foing, B. and R. M. Bonnet 1984. *Ap.J.* **279**, 848.

Foukal, P. V. 1971. *Solar Phys.* **19**, 59.

Foukal, P. V. 1987. *J.G.R.* **92**, D801.

Foukal. P. V. and J. Lean 1986. *Ap.J.* **302**, 826.

Frazier, E. N. 1968. *Zs.f.Ap.* **68**, 345.

Frazier, E. N. and J. O. Stenflo 1972. *Solar Phys.* **27**, 330.

Friedlander, G. 1978. *Comments on Astrophysics.* **8**, 47.

Friehe, C. A. 1975. *J.O.S.A.* **65**, 1502.

Furst, E., W. Hirth, and P. Lantos 1979. *Solar Phys.* **63**, 257.

Gabriel, A. H. 1971. *Solar Phys.* **21**, 392.

Gabriel, A. H. 1972. *M.N.R.A.S.* **160**, 99.

Gabriel, A. H. and C. Jordan 1972. In *Case Stud. Atom. Coll. Phys. II,*
    ed. McDaniel and McDowell. Amsterdam: North Holland.

Gary, D. E. 1985. *Ap.J.* **297**, 799.

Gary, D. E. 1986. *Ap.J.* submitted.

Geiss, J. 1982. *Space Sci. Rev.* **33**, 201.

Geiss, J. and P. Bochsler 1984. In *Proc. Int. Conf. on Isotope Ratios in
    the Solar System.* Paris: in press.

Geltman, S. 1962. *Ap.J.* **136**, 935.

Geltman, S. 1965. *Ap.J.* **141**, 376.

Giovanelli, R. G. 1973. *I.A.U.Symp.* **56**, 83.

Glasco, H. P. and H. Zirin 1964. *Ap.J. Suppl.* **90**, 193.

Gold, T. and F. Hoyle 1960. *M.N.R.A.S.* **120**, 89.

Goldberg, L. 1939. *Ap.J.* **89**, 673.

Goldberg, L. 1957. *Ap.J.* **126**, 318.

Goldberg, L. *et al* 1960. *Ap.J.* **132**, 184.

Goldberg, L., E. Muller, and L. H. Aller 1961. *Ap.J. Suppl.* **4**, 1.

Goldreich, P. and D. A. Keeley 1977. *Ap.J.* **212**, 243.

Goldreich, P. and G. Schubert 1968. *Ap.J.* **154**, 1005.

Grec, G., E. Fossat, and M. Pomerantz 1980. *Nature.* **288**, 541.

Green, L. C., P. P. Rush and C. D. Chandler 1957. *Ap.J. Suppl.* **3**, 37.

Griem, H. 1964. *Plasma Spectroscopy.* New York: McGraw-Hill.

Grineva, Yu. I. *et al.* 1973. *Solar Phys.* **29**, 441.

Grossi Gallegos, H. *et al.* 1971. *Solar Phys.* **16**, 120.

Grotrian, W. 1939. *Naturwissenschaften.* **27**, 214.

Hagyard, M. J. *et al.* 1982. *Solar Phys.* **80**, 33.

Hale, G. E. 1912. *Pub. A.S.P.* **24**, 223.

Hale, G. E. *et al.* 1919. *Ap.J.* **49**, 153.

Hale, G. E. *et al.* 1925. *Ap.J.* **62**, 270.

Hale, G. E. and S. B. Nicholson 1938. *Carnegie Inst. Wash. Publ.* **49**, 8.

Hanle, W. 1924. *Z.Phys.* **29**, 93.

Harris, D. L. 1948. *Ap.J.* **108**, 112.

Harvey, J. W. and W. Livingston 1969. *Solar Phys.* **10**, 283.

Harvey, J. W. and N. R. Sheeley 1977. *Solar Phys.* **54**, 343.

Harvey, K. L. 1985. *Austr. J. Phys.* **38**, 875.

Harvey, K. L. and S. F. Martin 1973. *Solar Phys.* **32**, 389.

Harvey, K. L. *et al.* 1975. *Solar Phys.* **40**, 87.

Harvey, K. L. and F. Recely 1984. *Solar Phys.* **91**, 127.

Heasley, J. N. and R. W. Milkey 1976. *Ap.J.* **210**, 827.

Heasley, J. N. and R. W. Milkey 1978. *Ap.J.* **221**, 677.

Henoux, J. C., *et al.* 1983a. *Ap.J.* **265**, 1066.

Henoux, J. C. and M. Semel 1983b. *Astr.Ap.* **119**, 233.

Hermans, L. M. and C. Lindsey 1986. *Ap.J.* submitted.

Heroux, L., M. Cohen, and M. Malinowsky 1972. *Solar Phys.* **23**, 369.

Hill, H. A., 1980. *Highlights of Astronomy.* **5**, 449.

Hill, H. A. and R. T. Stebbins 1975. *Ap.J.* **200**, 471.

Hirayama, T. 1971. *Solar Phys.* **17**, 50.

Hirayama, T. 1979. *IAU Colloq.* **44**, 4.

Hirayama, T. 1985. *Solar Phys.* **100**, 415.

Hirayama, T. and M. Irie 1984. *Solar Phys.* **90**, 291.

Hirshberg, J. E. 1972. *Solar Wind II.* NASA SP-308 p. 582.

Hirshberg, J. *et al.* 1972. *Solar Phys.* **23**, 467.

Hirshberg, J. E., J. R. Asbridge, and D. E. Robbins 1974. *J.G.R.* **79**, 934.

Hoang-Binh, D. 1982. *Astr.Ap.* **112**, L3.

Hood, A. W. 1983. *Solar Phys.* **87**, 279..

Hood, A. W. and E. R. Priest 1980. *Solar Phys.* **66**, 113..

Horne, K. *et al.* 1981. *Ap.J.* **244**, 340.

House, L. L. 1964. *Ap.J.* **138**, 1323.

House, L. L. 1970a. *J.Q.R.S.T.* **10**, 909, 1171.

House, L. L. 1971. *J.Q.R.S.T.* **11**, 367.

Howard, R. A. *et al.* 1982. *Ap.J.* **263**, L101.

Howard, R. F. 1958. *Ap.J.* **127**, 108.

Howard R. and B. J. LaBonte 1980. *Ap.J.* **239**, L33.

Howard, R. F. *et al.* 1983. *Solar Phys.* **83**, 321.

Hoyng, P. 1975, Ph. D. thesis, Utrecht.

Hoyng, P. *et al.* 1983. *Ap.J.* **268**, 865.

Hudson, H. S. 1972. *Solar Phys.* **24**, 414.

Hudson, H. S. and B. N. Dwivedi 1982. *Solar Phys.* **76**, 45.

Hurford, G. J. *et al.* 1975. *Ap.J.* **201**, L95.

Hurford, G. J. *et al.* 1984. *Solar Phys.* **94**, 413.

Hurford, G. J. *et al.* 1985. in *Radio Stars*, ed. Hjellming and Gibson. Dordrecht: Reidel p. 379.

Hyder, C. L. 1967a. *Solar Phys.* **2**, 49.

Hyder, C. L. 1967b. *Solar Phys.* **2**, 267.

Inglis, D. R. and E. Teller 1939. *Ap.J.* **90**, 439.

Jackson, B. V. 1981. *Solar Phys.* **73**, 133.

Jacobs, V. L. *et al.* 1977. *Ap.J.* **211**, 605.

Jain, N. K. and U. Narain 1976. *Solar Phys.* **50**, 361.

James, Jesse C. 1966. *Ap.J.* **146**, 356.

James, Jesse C. 1970. *Solar Phys.* **12**, 143.

Janssens, T. 1970. *Solar Phys.* **11**, 223.

Jefferies, J. T., E. P. Smith, and H. J. Smith 1959. *Ap.J.* **129**, 146.

Jefferies, J. T. *et al.* 1971. *Solar Phys.* **16**, 103.

Jordan, C. 1964. *M.N.R.A.S.* **134**, 463.

Jordan, C. 1969. *M.N.R.A.S.* **142**, 501.

Jordan, C. 1971. *Solar Phys.* **21**, 381.

Jordan, C. 1979. *Phil.Trans.Roy.Soc.* **297**, 541.

Kahler, S. W. *et al.* 1978. *Solar Phys.* **57**, 429.

Kahler, S. W. *et al.* 1984. *J.G.R.* **89**, 9683.

Kahler, S. W. *et al.* 1987. *Solar Phys.* **107**, 385.

Kane, S. 1981. *Astrophys. Space Sci.* **75**, 163.

Kane, S. R. and K. A. Anderson 1970. *Ap.J.* **162**, 1003.

Kane, S. R. and R. F. Donnelly 1971. *Ap.J.* **164**, 151.

Kanno, M. 1979. *Publ.Astr.Soc.Japan.* **31**, 115.

Karpinsky V. N. 1977. *Solar Phys.* **54**, 25.

Karzas, W. J. and R. Latter 1961. *Ap.J. Suppl.* **5**, 167.

Keil, S. L. 1980. *Ap.J.* **237**, 1035.

Kiepenheuer, K. O. 1953. *The Sun.* ed. Kuiper. Chicago: Univ. of Chic. Pr., p.322.

Kinman, T. D. 1952. *M.N.R.A.S.* **112**, 425.

Kinman, T. D. 1953. *M.N.R.A.S.* **113**, 613.

Kippenhahn, R. and A Schlüter 1957. *Zs.Ap.* **43**, 36.

Kirchhoff, G. 1859, *Sitzungsber.Akad.Wiss.Berlin.* p. 783.

Knuth, D. E. 1984. *The TEXbook.* Reading: Addison-Wesley.

Kohl, J.L. *et al.* 1983. *Solar Wind V.* NASA Conference Publ. 2280, p.47.

Kopp, R. A. and G. W. Pneuman 1976. *Solar Phys.* **50**, 85.

Kosugi, T., M. *et al.* 1986. *Publ. Astron.Soc.Japan.* **38**, 1.

Kozlovsky, B.-Z. and H. Zirin 1968. *Solar Phys.* **5**, 50.

Kraft, R. P., G. W. Preston, and S. C. Wolff 1964. *Ap.J.* **140**, 235.

Krieger, A. S., A. F. Timothy, and E. C. Roelof 1973. *Solar Phys.* **32**, 505.

Kristenson, H. 1951. *Stockholm Obs.Ann.* **17**, No. 1.

Krüger, A. 1979. *Introduction to Solar Radio Astronomy and Radio Physics.* Dordrecht: Reidel.

Kuiper, G. P., ed. 1951. *The Sun*. Chicago: Univ. of Chicago.

Kulsrud, R. 1967. In *Proc. Int. Sch. "Enrico Fermi"*. New York: Academic Pr.

Kundu, M. 1959. *Ann.d'Ap.* **22**, 1.

Kundu, M. 1965. *Solar Radio Astronomy*. New York: Interscience.

Kunkel, W. E. 1970. *Ap.J.* **161**, 503.

Kunzel, H. 1960. *Astron. Nachr.* **285,**, 271..

Kuperus, M. and M. A. Raadu 1974. *Astr.Ap.* **31**, 189.

Kurochka, L. N. 1967. *Soviet Astronomy - A.J.* **11**, 290.

LaBonte, B. J. *et al.* 1973. *Bull. AAS.* **5**, 275.

LaBonte, B. J. 1979. *Solar Phys.* **61**, 283.

LaBonte, B. J. 1986. *Ap.J.* submitted.

Labrum, N. R. 1972. *Solar Phys.* **27**, 490.

Lambert, D. L. 1969. *M.N.R.A.S.* **142**, 71.

Landini, M. and B. C. Monsignori Fossi 1971. *Solar Phys.* **20**, 322.

Landman, D. A. 1984. *Ap.J.* **279**, 438.

Landman, D. A. 1985. *Ap.J.* **290**, 369.

Landman, D. A. and R. M. E. Illing 1976. *Astr. Ap.* **49**, 277.

Lantos, P. *et al.* 1981. *Astr.Ap.* **101**, 33.

Lantos, P. 1980. *IAU Symp.* **86**, 41.

Lee, R. H. *et al.* 1965. *Applied Optics.* **4**, 1081.

Leer, E. *et al.* 1982. *Space Sci.Rev.* **33**, 161..

Leibacher, J. and R. Stein 1981. In *The Sun as a Star*. NASA SP-450.

Leighton, R. B. 1959. *Ap.J.* **130**, 366.

Leighton, R. B. 1964. *Ap.J.* **140**, 1547.

Leighton, R. B. 1969. *Ap.J.* **156**, 1.

Leighton, R. B., R. W. Noyes, and G. W. Simon 1962. *Ap.J.* **13**, 471.

Lena, P. 1970. *Astr.Ap.* **4**, 202.

Leroy, J. L. 1977. *Astr.Ap.* **54**, 811.

Leroy, J. L. 1979. *IAU Colloq.* **44**, 56.

Libbrecht, K. G. 1984. Ph. D. thesis, Princeton.

Libbrecht, K. G. and H. Zirin 1986. *Ap.J.* **308**, 413.

Libbrecht, K. G. *et al.* 1986. *Nature* **323**, 235.

Liggett, M. A. and H. Zirin 1983. *Solar Phys.* **84**, 3.

Liggett, M. A. and H. Zirin 1984. *Solar Phys.* **91**, 259.

Liggett, M. A. and H. Zirin 1985. *Solar Phys.* **97**, 51.

Lin, R. P. *et al.* 1981. *Ap.J.* **251**, L109.

Lin, R. P. and H. S. Hudson 1971. *Solar Phys.* **17**, 412.

Lin, R. P. and H. S. Hudson 1976. *Solar Phys.* **50**, 153.

Lindsey, C. *et al.* 1982. *Ap.J.* **264**, L660.

Lindsey, C. *et al.* 1984. *Ap.J.* **281**, 862.

Lingenfelter, R. E. and R. Ramaty 1967. In *High Energy Nuclear Reactions in Astrophysics*. New York: W. A. Benjamin, p. 99.

Linsky, J. L. 1973a. *Solar Phys.* **28**, 409.

Linsky, J. L. 1973b. *Ap.J. Suppl.* **25**, 163.

Linsky, J. L. *et al.* 1976. *Ap.J.* **203**, 509.

Liszka, L. 1970. *Solar Phys.* **14**, 354.

Lites, B. W. 1973. *Solar Phys.* **32**, 283.

Lites, B. W. *et al.* 1986. In *The Lower Atmosphere of Solar Flares*. Sunspot: National Solar Obs. p. 101.

Livingston, W. *et al.* 1971. *Publ. Roy. Obs. Edinburgh.* **8**, 52.

Livshits, M. A. *et al.* 1976. *Solar Phys.* **49**, 315.

Low, B. C. 1985. *Solar Phys.* **100**, 309..

Luhn, A. *et al.* 1984. *Adv. Sp. Res.* **4**, 161.

Luhn, A. *et al.* 1986. *Proc. 19th Int. Cosmic Ray Conf.* **4**, 285.

Lust, R. and H. Zirin 1960. *Zs.Ap.* **49**, 8.

Lynch, D. K., J. M. Beckers, and R. B. Dunn 1973. *Solar Phys.* **30**, 63.

Lyot, B. 1930. *C.R.Acad.Sci.Paris.* **191**, 834.

Lyot, B. 1933. *C.R.Acad.Sci.Paris.* **197**, 1593.

Lyot, B. 1939. *M.N.R.A.S.* **99**, 586.

Machado, M. E. and D. M. Rust 1974. *Solar Phys.* **38**, 399.

MacQueen, R. M. *et al.* 1980. *Solar Phys.* **65**, 91.

Malinowsky, M. and L. Heroux 1973. *Ap.J.* **181**, 1009.

Maltby, P. 1975. *Solar Phys.* **43**, 91.

Malitson, H. H. *et al.* 1973. *Ap.J.* **14**, L111.

Malville, J. M. and J. A. Eddy 1967. *Ap.J.* **150**, 289.

Mandelstam, S. L. *et al.* 1975. *Nature.* **254**, 462.

Mariska, J. T. 1980. *Ap.J.* **235**, 268.

Marsh, K. A. 1978. *Solar Phys.* **59**, 105.

Marsh, K. A. *et al.* 1978. *Ap.J.* **224**, 1043.

Marsh, K. A. *et al.* 1981. *Astr.Ap.* **94**, 67.

Marsh, K. A. and G. A. Hurford 1982. *Ann.Rev.Astr.Ap.* **20**, 497.

Martin, S. F. 1973. *Solar Phys.* **31**, 3.

Martres, M.-J. *et al.* 1968. *Solar Phys.* **5**, 187.

Mätzler, C. *et al.* 1978. *Ap.J.* **223**, 1058.

Maunder, E. W. 1922. *M.N.R.A.S.* **82**, 534.

Mayfield, E. B. *et al.* 1969. *Sky and Tel.* **37**, 208.

Mayfield, E. B. *et al.* 1964. *Aerospace Corp Rept.* ATR-65(8102)-1.

McDonald, F. B. and U. D. Desai 1971. *J.G.R.* **76**, 808.

McDonald, F. B., ed. 1963. *Solar Proton Manual*. NASA TRR-169

McDonald, F. B. *et al.* 1976. *Ap.J.* **203**, L149.

McGuire, R. E. *et al.*1981. *17th Int. Cosmic Ray Conf.* **3**, 65.

McMath, R. R. 1937. *Publ.Mich.Obs.* **7**, 42.

Mead, J. M. *et al.* 1984. *Future of Ultraviolet Astronomy Based on Six Years of IUE Research*. NASA Conference Publ. 2349.

Menzel, D. H. 1937. *Ap.J.* **85**, 330.

Menzel, D. H. and J. W. Evans 1953. *Atti dell' Convegno Volta*. Rome: Academia Lincei, p. 119.

Merrill, W. 1956. *Lines of the Chemical Elements in Astronomical Spectra.*, C.I.W. Publ 610.

Meyer, J.-P. 1985a,b. *Ap.J.Supp.* **57**, 151, 173.

Mihalas, D. 1978. *Stellar Atmospheres.* San Francisco: W. H. Freeman.

Mikheyev, S. P. and A. Yu. Smirnov 1986. *Nuovo Cim.* **96**, 17.

Milkey, R. W. 1975. *Ap.J.* **199**, L131.

Mitchell, S. A. 1923. *Eclipses of the Sun.* New York: Columbia Univ. Pr.

Mochnacki, S. W. and H. Zirin 1980. *Ap.J.* **239**, L27.

Moore, C. E. 1959. *U.S.Natl.Bur.of Std.Tech.Note 36.*

Moore, R. 1981. *Ap.J.* **249**, 390.

Moore, R. *et al.* 1977. *Ap.J.* **218**, 286.

Moore, R. *et al.* 1980. *Solar Flares,* ed. P. A. Sturrock. Boulder: Colorado Assoc. Univ. Pr. p. 341.

Moreton, G. E. and H. E. Ramsey 1960. *Pub. A.S.P.* **72**, 357.

Mosher, J. M. 1976. *The Caltech Videomagnetograph.*, Caltech BBSO Prepr. #0159.

Mosher, J. M. 1977.Ph. D. thesis, California Institute of Technology.

Mosher, J. M. and T. P. Pope 1977. *Solar Phys.* **53**, 375.

Mullan, D. and W. L. Waldron 1986. *Ap.J.* **308**, L21.

Munro, R. H. *et al.* 1979. *Solar Phys.* **61**, 201.

Nakagawa, Y. *et al.* 1973. *Solar Phys.* **30**, 421.

Neidig, D. F. 1983. *Solar Phys.* **85**, 285.

Ness, N. F., C. S. Scearce, and J. B. Seek 1964. *J.G.R.* **69**, 3531.

Ness, N. F. and J. M. Wilcox 1964. *Phys. Rev. Letters* **13**, 461.

Neugebauer, M. 1983. *Solar Wind V.* NASA Conference Publ. 2280, p.135.

Newkirk, G. A. 1961. *Ap.J.* **133**, 983.

Newkirk, G. A. and M. Altschuler 1970. *Solar Phys.* **13**, 131.

Newton, H. W. and M. L. Nunn 1951. *M.N.R.A.S.* **111**, 413.

Nikolskaya, K. I. 1966. *Astr. Zh.* **43**, 936.

Noci, G. 1980. *IAU Symp.* **91**, 307.

Nolte, J. T. *et al.* 1976. *Solar Phys.* **46**, 303.

Norton, R. H. *et al.* 1967. *J.G.R.* **72**, 815.

Noyes, R. W. and W. Kalkofen 1970. *Solar Phys.* **1**, 120.

Noyes, R. W., *et al.* 1984. *Ap.J.* **279**, 763.

Ohki, K. *et al.* 1982. *Hinotori Symp.* p. 69.

Öhman, Y. 1938. *Nature.* **141**, 157,291.

Öhman, Y. *et al.* 1968. *Nobel Symp.* **9**, 95.

Orrall, F. Q. and E. J. Schmahl 1976. *Solar Phys.* **50**, 365.

Orrall, F. Q. and E. J. Schmahl 1980. *Ap.J.* **240**, 908.

Orrall, F. Q. *et al.* 1981. *Ap.J.* **247**, L135.

Osterbrock, D. E. 1960. *Ap.J.* **134**, 347.

Owocki, S. P. *et al.* 1982. *Solar Phys.* **78**, 315.

Pagel, B. E. J. 1964. *Ann.Rev.Astr.Ap.* **2**, 267.

Parke, S. J. and T. P. Walker 1986. *Phys. Rev. Lett.* **57**, 2322.

Parker, E. N. 1957. *Ap.J.Suppl.* **3**, 51.

Parker, E. N. 1958. *Ap.J.* **128**, 669.

Parker, E. N. 1979. *Ap.J.* **230**, 905.

Parker, E. N. 1963. *Interplanetary Dynamical Processes.* New York: Interscience.

Pecker, J. C. 1965. *The Solar Spectrum.*, ed. de Jager, Dordrecht: Reidel, p. 29.

Petschek, H. E. 1964. *AAS-NASA Symp.on the Physics of Solar Flares* NASA SP-50, p. 409.

Phillips, E., D. L. Judge, and R. Carlson 1982. *J.G.R.A.* **87**, 1433.

Pierce, A. K. 1968. *Ap.J.Suppl.* **150**, 1.

Pierce, A. K. and J. Waddell 1961. *Mem.Roy.Astro.Soc.* **68**, 89.

Pneuman, G. W. 1984. *Solar Phys.* **94**, 387.

Pneuman, G. W. and R. A. Kopp 1978. *Solar Phys.* **57**, 49.

Poland, A. I. 1978. *Solar Phys.* **57**, 141.

Poland, A. I. R. H. Munro 1976. *Ap.J.* **209**, 927.

Pottasch, S. R. 1963. *Ap.J.* **137**, 945.

Pottasch, S. R. 1964. *Space Science Rev.* **3**, 816.

Priest, E. R. 1982. *Solar Magnetohydrodynamics.* Dordrecht: Reidel.

Racah, G. 1942a. *Phys.Rev.* **61**, 186.

Racah, G. 1942b. *Phys.Rev.* **62**, 438.

Racah, G. 1943. *Phys.Rev.* **63**, 367.

Ramaty, R. 1969. *Ap.J.* **158**, 753.

Ramaty, R. 1986. In *Physics of the Sun,* ed. Sturrock. Dordrecht: Reidel.

Ramaty, R. and V. Petrosian 1972. *Ap.J.* **178**, 241.

Ramaty, R. *et al.* 1977. *Ap.J.* **214**, 617.

Ramaty, R. *et al.* 1979. *Ap.J. Suppl.* **40**, 487.

Ramaty, R. *et al.* 1980. *Solar Flares,* ed. Sturrock. Boulder: Colo. Assoc. Univ. Pr. p. 117.

Ramaty, R. *et al.* 1983. *Ap.J.* **273**, L41.

Ramaty, R. and M. A. Forman 1987. *Essays in Space Science*, ed. Ramaty *et al.* NASA Conf. Publ. 2464, p. 47.

Ramsey, H. 1971. *Solar Phys.* **21**, 54.

Reames, D. V. *et al.* 1985. *Ap.J.* **292**, 716.

Redman, R. O. 1942. *M.N.R.A.S.* **102**, 134.

Rhodes, E. J., R. K. Ulrich, and G. W. Simon 1977. *Ap.J.* **218**, 901.

Rieger, E. *et al.* 1983. *Proc. 18th Int Cosmic Ray Conf.* **10**, 338.

Roberts, W. O. 1945. *Ap.J.* **101**, 136.

Roelof E. C. *et al.* 1975. *Solar Phys.* **41**, 349.

Rohrlich, F. 1956. *Ap.J.* **123, 129**, 521, 441.

Rösch, J. 1957. *Astronomie.* **71**, 129.

Rosen, S. P. and J. M. Gelb 1986. *Phys.Rev.D.* **34**, 3499.

Rosner, R., W. H. Tucker and G. S. Vaiana 1978. *Ap.J.* **220**, 643.

Rotenberg, M. *et al.* 1959. *The 3-j and 6-j Symbols.* Cambridge: MIT Pr.

Rottman, G. J. *et al.* 1982. *Ap.J.* **260**, 326.

Roy, J.–R. and A. Michalitsanos 1974. *Solar Phys.* **35**, 47.

Rust, D. M. 1972. *Solar Phys.* **25**, 1972.

Rust, D. M. 1980. *Solar Flares*, ed. Sturrock. Boulder: Colo. Univ. Pr.

Rust, D. M. and V. Bar 1973. *Solar Phys.* **33**, 205.

Rust, D. M. and D. F. Webb 1977. *Solar Phys.* **54**, 403.

Saito, K., A. I. Poland, and R. H. Munro 1977. *Solar Phys.* **5**, 121.

Sakurai, T. and Y. Uchida 1977. *Solar Phys.* **52**, 397.

Sakurai, T. 1985. *Solar Phys.* **95**, 311.

Sakurai, T. 1981. *Solar Phys.* **69**, 343.

Schatzman, E. 1949. *Ann.d'Astrophys.* **12**, 203.

Scherrer, Ph. H. *et al.* 1977. *Solar Phys.* **52**, 3.

Schlüter, A. 1957. *I.A.U. Symp.* **4**, 362.

Schmahl, E. J. and E. Hildner 1977. *Solar Phys.* **5**, 473.

Scholer, M. *et al.* 1979. *Ap.J.* **227**, 323.

Schroter, E. H. 1985. *Solar Phys.* **100**, 141.

Schwabe, M. 1849. *Astronomische Nachr.* **21**, 234.

Schwabe, M. 1851, *Bern Mitt* p. 94.

Schwarzschild, K. 1906. *Gottinger Nachr.* p. 41.

Schwarzschild, M. 1948. *Ap.J.* **107**, 1.

Schwarzschild, M. 1959. *Structure and Evolution of the Stars.* Princeton: Princeton Univ. Pr.

Schwarzschild, M. 1960. *Ap.J.* **130**, 345.

Schwinger, J. 1949. *Phys. Rev.* **75**, 1912.

Sears, R. L. 1964. *Ap.J.* **140**, 477.

Seraph, H. E. *et al.* 1968. *Phil.Trans.Roy.Soc.* **264**, 77.

Serio, A., *et al.* 1978. *Solar Phys.* **59**, 65.

Severny, A. B. 1957. *A.J.U.S.S.R.* **34**, 328.

Severny, A. B. 1958. *Izv.Krym.Astrofiz.Obs.* **1**, 102.

Severny, A. B. *et al.* 1976. *Nature.* **259**, 87.

Severny, A. B. *et al.* 1980. *Highlights of Astronomy.* **V**, 87.

Sheeley, N. R. 1972. *Solar Phys.* **25**, 98.

Sheeley, N. R. *et al.* 1981. *EOS* **62**, 153.

Sheeley, N. R. *et al.* 1982. *Spac.Sci.Rev.* **33**, 219.

Sheeley, N. R. Jr. and J. W. Harvey 1978. *Solar Phys.* **59**, 159.

Sheeley, N. R. Jr. and J. W. Harvey 1980. *Solar Phys.* **70**, 237.

Sheeley, N. R. Jr. *et al.* 1985. *J.G.R.* **90**, 163.

Shelke, R. N. and M. C. Pande 1985. *Solar Phys.* **95**, 193.

Shklovsky, I. S. 1948. *A.J.U.S.S.R.* **25**, 145.

Shklovsky, J. S. 1958. *Soviet Astronomy.* **2**, 786.

Shortley, G. H. 1940. *Phys.Rev.* **57**, 225.

Simon, G. W. and R. B. Leighton 1964. *Ap.J.* **140**, 1120.

Simon, G. W. and J. B. Zirker 1974. *Solar Phys.* **3**, 331.

Simon, M. and H. Zirin 1969. *Solar Phys.* **9**, 317.

Skumanich, A. *et al.* 1984. *Ap.J.* **282**, 776.

Slish, V. I. 1963. *Nature.* **199**, 682.

Slonim, Yu. M. and Z. B. Korobova 1975. *Solar Phys.* **40**, 397.

Slottje, C. 1981. *Atlas of Fine Structure Dynamic Structure.* Dwingeloo, NFRA.

Smerd, S. F. 1950. *Aust.J.Sci.Res.* **A3**, 34.

Smith, H. J. and E. v. P. Smith 1963. *Solar Flares.* New York: Macmillan.

Smith, S. F. 1968. *I.A.U. Symp.* **35**, 267.

Smith, W. T. 1966. *Modern Optical Engineering.* New York: McGraw Hill.

Smithson, R. C. 1973. *Solar Phys.* **29**, 365.

Snodgrass, H. B. 1983. *Ap.J.* **270**, 288.

Somov, B. V. 1978. *Proc. Lebedev Inst.* **88**, 303.

Speer, R. J. *et al.* 1970. *Nature.* **226**, 249.

Spitzer, L. W. 1956. *Ap.J.* **124**, 20.

Spitzer, L. W. 1962. *Physics of Fully Ionized Gases, 2d ed.* New York: Interscience.

Spiegel, E. A. 1966. *Stellar Evolution*, ed. Stein and Cameron. New York: Plenum Pr p. 143.

Spruit, H. C. 1976. *Solar Phys.* **50**, 269.

Stenflo, J. O. 1973. *Solar Phys.* **32**, 41.

Stewart, R. T. *et al.* 1978. *IAU Colloq.* **44**, 315.

St. John, C. E. 1913. *Ap.J.* **37**, 322.

Stromgren, B. 1944. *Publ. Copenhagen Obs.* No. 138.

Summers, H. P. 1974. *M.N.R.A.S.* **169**, 663.

Svalgaard, L. and J. M. Wilcox 1975. *Solar Phys.* **41**, 46.

Svalgaard, L. and J. M. Wilcox 1976. *Solar Phys.* **49**, 177.

Svestka, Z. 1965. *Astr.Ap.* **3**, 119.

Svestka, Z. 1970. *Solar Phys.* **13**, 471.

Svestka, Z. 1976. *Solar Flares.* Dordrecht: Reidel.

Takakura, T. 1967. *Solar Phys.* **1**, 304.

Takakura, T. 1969. *Solar Phys.* **6**, 133.

Takakura, T. and E. Scalise 1970. *Solar Phys.* **11**, 434.

Takakura, T. *et al.* 1983. *Solar Phys.* **86**, 323.

Tanaka, K. 1974. *IAU Symp.* **56**, 239.

Tanaka, K. 1975. *BBSO preprint #0152.*

Tanaka,K. 1980. *Solar-Terrestrial Predictions Proceedings* ed Donnelly. Vol. 3, C1. NOAA-ERL.

Tanaka, K. 1986. *Publ.Astron.Soc.Japan.* **38**, 225.

Tanaka, K. 1987. *Publ.Astron.Soc.Japan.* **39**, 1.

Tanaka, K. *et al.* 1982. *Hinotori Symp.* p. 43.

Tanaka, K. *et al.* 1983. *Solar Phys.* **86**, 91.

Tanaka, K. and Y. Nakagawa 1973. *Solar Phys.* **33**, 187.

Tanaka, K. and H. Zirin 1985. *Ap.J.* **299**, 1036.

Tandberg-Hanssen, E. 1963. *Ap.J.* **137**, 26.

Tang, F. 1981. *Solar Phys.* **69**, 399.

Tang, F. 1983. *Solar Phys.* **89**, 43.

Tang, F. 1986. *Solar Phys.* **105**, 399.

Tang, F. and R. Moore 1982. *Solar Phys.* **77**, 263.

Tang, F., R. Howard and J. M. Adkins 1984. *Solar Phys.* **91**, 75.

Tarbell, T. D. and A. M. Title 1977. *Solar Phys.* **52**, 13.

ten Bruggencate, P., H. Gollnow, and F. W. Jager 1950. *Zs.Ap.* **27**, 223.

Thomas, R. N. 1957. *Ap.J.* **12**, 260.

Thomas, R. N. and R. G. Athay 1962. *Physics of the Solar Chromosphere.* New York: Interscience.

Timothy, A. F. *et al.* 1975. *Solar Phys.* **42**, 135.

Title, A. and W. Rosenberg 1979. *Research on Spectroscopic Imaging.* Lockheed Palo Alto Res. Lab., LMSC-D674593.

Torres-Peimbert, S. E. Simpson and R. K. Ulrich 1969. *Ap.J.* **15**, 957.

Tousey, R. *et al.* 1973. *Solar Phys.* **33**, 265.

Trottet, G. *et al.* 1982. *Astr.Ap.* **111**, 306.

Tsuneta, S. *et al.* 1983. *Solar Phys.* **86**, 313.

Tucker, W. H. and M. Koren 1971. *Ap.J.* **168**, 283.

Uchida, Y. 1974. *Solar Phys.* **39**, 431.

Ulrich, R. K. 1970. *Ap.J.* **162**, 993.

Ulrich, R. K. 1982. *Ap.J.* **258**, 404.

Ulrich, R. K. and E. J. Rhodes, Jr. 1977. *Ap.J.* **218**, 521.

Ulrich, R. K., E. J. Rhodes, Jr. and G. W. Simon 1977. *Ap.J.* **218**, 901.

Unno, W. *et al.* 1979. *Non-Radial Oscillations of Stars.* Tokyo: Tokyo Univ. Pr.

Vaiana, G. S., *et al.* 1973. *Ap.J.* **185**, L47.

Vainshtein, L. A. *et al.* 1973. *Cross Sections for Excitation of Atoms and Ions by Electrons.* Moscow: Nauka.

Vainstein, L. A. and W. I. Safronova 1978. *At. Data Nucl. Data.* **21**, 49.

van de Hulst, H. C. 1950. *B.A.N.* **11**, 135.

van de Hulst, H. C. 1953. *The Sun*, ed. Kuiper Chicago: Univ. of Chicago.

van Hollebeke, M. A. I. 1979. *Rev. Geophys. & Space Phys.* **17**, 545.

van Hoven, G. and Y. Mok 1984. *Ap.J.* **282**, 267.

Van Regemorter, H. 1962. *Ap.J.* **136**, 906.

Vaughan, A. H. *et al.* 1978. *Pub.A.S.P.* **90**, 267.

Vaughan, A. H. *et al.* 1981. *Ap.J.* **250**, 276.

Veck, N. J. and J. H. Parkinson 1981. *M.N.R.A.S.* **197**, 41.

Vernazza, J., E. H. Avrett and R. Loeser 1976. *Ap.J. Suppl.* **30**, 1.

Vidal-Madjar, A. 1977. *The Solar Output and Its Variations*, ed. O. R. White. Boulder: Colo. Assoc. Univ. Pr. p. 213.

Vrabec, D. 1974. *IAU Symp.* **56**, 201.

Wagner, W. *et al.* 1981. *Ap.J.* **244**, L123.

Wagner, W. 1984. *Ann.Rev.Astr.Ap.* **22**, 267.

Waldmeier, M. 1947. *Publ. Zurich Obs.* **9**, 1.

Waldmeier, M. 1951. *Zs.Ap.* **20**, 323.

Waldmeier, M. 1955. *Ergebnisse und Probleme der Sonnenforschung.* Leipzig: Geest and Portig.

Waldmeier, M. 1962. *Astr.Mitt.Eidg.Stern.* Zurich, No. 248.

Wang, H. T. and R. Ramaty 1974. *Solar Phys.* **107**, 1065.

Wang, J. *et al.* 1985. *Solar Phys.* **98**, 241.

Wang, H. and H. Zirin 1986. *Solar Phys.* submitted.

Wang, H. and H. Zirin 1987. AAS Pasadena meeting.

Wannier. P. G., G. J. Hurford, and G. A. Seielstad 1983. *Ap.J.* **264**, 660.

Weart, S. R. 1970. *Ap.J.* **162**, 887.

Webb, D. F. and H. Zirin 1981. *Solar Phys.* **69**, 99.

Wefer, F. and M. P. Bleiweiss 1980. *Bull.AAS* **8**, 338.

Wesely, M. L. and E. C. Alcaraz 1973. *J.G.R.* **78**, 6224.

White, O. R. and W. C. Livingston, 1981. *Ap.J.* **249**, 798.

Widing, K. 1975. *IAU Symp.* **68**, 153.

Widing K. G. and K. P. Dere 1977. *Solar Phys.* **5**, 431.

Widing, K. G. *et al.* 1986. *Ap.J.* **308**, 982.

Widing, K. G. and J. D. Purcell 1976. *Ap.J.* **204**, L151.

Wilcox, J. M. 1968. *Space Sci. Rev.* **8**, 258.

Wilcox, J. M. and N. F. Ness 1965. *J.G.R.* **70**, 5793.

Wild, J. P. 1950a,b. *Aust.J.Sci.* **A3**, 387, 541.

Wild, J. P. 1969. *Solar Phys.* **9**, 260.

Wild, J. P. and L. L. McCready 1950. *Aust.J.Sci.* **A3**, 387.

Wild, J. P. and H. Zirin 1956. *Aust.J.Phys.* **9**, 315.

Wildt, R. 1939. *Ap.J.* **89**, 295.

Williams, G. E. 1985. *Aust.J.Phys.* **38**, 1027.

Willson, R. C. and H. S. Hudson 1981. *Ap.J.* **244**, L185.

Wilson, O. C. 1963. *Ap.J.* **138**, 832.

Wilson, O. C. 1978. *Ap.J.* **226**, 379.

Wilson, O. C. and M. K. V. Bappu 1957. *Ap.J.* **12**, 661.

Withbroe, G. W. 1975. *Solar Phys.* **45**, 301.

Wlerick, G. and J. Axtell 1957. *Ap.J.* **126**, 253.

Wolf, A. R. 1858. *Astr.Mitt.Eidg.Stern.* **10**, 6.

Wolfenstein, L. 1985. *Phys. Rev. D* **20**, 2634.

Wolff, C. L. 1972. *Ap.J.* **177**, L87.

Wolter, H. 1952. *Ann.d.Phys.* **10**, 94.

Woodard, M. F. 1983. *Nature.* **30**, 189.

Wu, C. Y. R. and H. Ogawa 1985. *J.G.R.* submitted.

Yakovkin, N. A. and M. Yu. Zel'dina 1964. *Sov. Astr.* **7**, 643.

Yeh, S.-H. 1961. *Izv.Krim.Astr.Obs.* **25**, 180.

Zarro, D. M. and H. Zirin 1986. *Ap.J.* **304**, 365-370.

Zheleznyakov, V. V. 1977. *Elektromagnitye Volny v Kosmicheskoi Plasme (Electromagnetic Waves in Cosmic Plasmas.)* Moscow: Nauka, p. 284.

Zirin, H. 1956. *Ap.J.* **123**, 536.

Zirin, H. 1958. *Ap.J.* **127**, 159.

Zirin, H. 1961. *Astr.J.U.S.S.R.* **38**, 5.

Zirin, H. 1964. *Ap.J.* **140**, 1216.

Zirin, H. 1968. *Ap.J* **151**, 383.

Zirin, H. 1969. *Solar Phys.* **7**, 243.

Zirin, H. 1970. *Solar Phys.* **14**, 328.

Zirin, H. 1972. *Solar Phys.* **22**, 34.

Zirin, H. 1974. *Solar Phys.* **38**, 91.

Zirin, H. 1975. *Ap.J.* **199**, L63.

Zirin, H. 1976. *Solar Phys.* **50**, 399.

Zirin, H. 1978a. *Solar Phys.* **58**, 95.

Zirin, H. 1978b. *Ap.J.* **222**, L105.

Zirin, H. 1978c. *IAU Colloq.* **44**, 193.

Zirin, H. 1980. *Ap.J.* **23**, 618.

Zirin, H. 1982. *Ap.J.* **260**, 655.

Zirin, H. 1985. *Ap.J.* **291**, 858.

Zirin, H. 1986. *Aust.J.Phys.* **38**, 961.

Zirin, H. and E. Tandberg-Hanssen 1959. *Ap.J.* **131**, 717.

Zirin, H. and R. D. Dietz 1963. *Ap.J.* **138**, 664.

Zirin, H. and A. Severny 1961. *Observatory.* **81**, 155.

Zirin, H., L. A. Hall, and H. E. Hinteregger 1963. *Space Res.* **3**, 760.

Zirin, H. *et al.* 1971. *Solar Phys.* **19**, 463.

Zirin, H. and A. Stein 1972. *Ap.J.* **178**, L85.

Zirin, H. and K. Tanaka 1973. *Solar Phys.* **32**, 173.

Zirin, H. and B. Lazareff 1975. *Solar Phys.* **41**, 425.

Zirin, H., G. J. Hurford, and K. A. Marsh 1978. *Ap.J.* **224**, 1043.

Zirin, H. and G. Ferland 1979. BBSO. Prepr. #192.

Zirin, H. and D. F. Neidig 1981. *Ap.J.* **248**, L45.

Zirin, H., M. A. Liggett, and A. P. Patterson 1982. *Solar Phys.* **76**, 387.

Zirin, H. and T. Hirayama 1985. *Ap.J.* **299**, 536.

Zirin, H. and Liggett, M. A. 1987. BBSO Prepr #274, submitted to *Solar Phys.*.

Zirin, H. and J. M. Mosher 1987. BBSO Prepr #272, submitted to *Solar Phys.*.

Zirker, J. B. 1977. *Coronal Holes.* Boulder: Colorado Assoc. Univ. Pr. ch. IV.

Zwaan, C. 1966. *Ann.Rev.Astr.Ap.* **6**, 135.

Index